Committee on Health Effects of Waste Incineration

Board on Environmental Studies and Toxicology

Commission on Life Sciences

National Research Council

NATIONAL ACADEMY PRESS
Washington, D.C.

NATIONAL ACADEMY PRESS • 2101 Constitution Avenue, NW • Washington, DC 20418

NOTICE: The project that is the subject of this report was approved by the Governing Board of the National Research Council, whose members are drawn from the councils of the National Academy of Sciences, the National Academy of Engineering, and the Institute of Medicine. The members of the committee responsible for the report were chosen for their special competences and with regard for appropriate balance.

This project was supported by Grant No. EPA-R-822039 between the National Academy of Sciences and the U.S Environmental Protection Agency (EPA), and Grant No. DHHS-U5O/ATU39903 from the Agency for Toxic Substances and Disease Registry, U.S. Department of Health and Human Services, and U.S. Department of Energy. Any opinions, findings, conclusions, or recommendations expressed in this publication are those of the author(s) and do not necessarily reflect the view of the organizations or agencies that provided support for this project.

Library of Congress Cataloging-in-Publication Data

Waste incineration and public health / Committee on Health Effects of
Waste Incineration, Board on Environmental Studies and Toxicology,
Commission on Life Sciences, National Research Council.
    p. cm.
Includes bibliographical references and index.
  . ISBN 0-309-06371-X (casebound)
    1. Hazardous wastes—Incineration—Health aspects. 2. Health risk
assessment. 3. Incineration—Health aspects. 4. Medical
wastes—Incineration. 5. Pollution prevention. I. National Research
Council (U.S.). Committee on Health Effects of Waste Incineration.
    RA578.H38 W37 2000
    363.72'87—dc21

                           00-009914

Waste Incineration and Public Health is available from the National Academy Press, 2101 Constitution Ave., NW, Box 285, Washington, DC 20055 (1-800-624-6242 or 202-334-3313 in the Washington metropolitan area; Internet: http://www.nap.edu).

THE NATIONAL ACADEMIES

National Academy of Sciences
National Academy of Engineering
Institute of Medicine
National Research Council

The **National Academy of Sciences** is a private, nonprofit, self-perpetuating society of distinguished scholars engaged in scientific and engineering research, dedicated to the furtherance of science and technology and to their use for the general welfare. Upon the authority of the charter granted to it by the Congress in 1863, the Academy has a mandate that requires it to advise the federal government on scientific and technical matters. Dr. Bruce M. Alberts is president of the National Academy of Sciences.

The **National Academy of Engineering** was established in 1964, under the charter of the National Academy of Sciences, as a parallel organization of outstanding engineers. It is autonomous in its administration and in the selection of its members, sharing with the National Academy of Sciences the responsibility for advising the federal government. The National Academy of Engineering also sponsors engineering programs aimed at meeting national needs, encourages education and research, and recognizes the superior achievements of engineers. Dr. William A. Wulf is president of the National Academy of Engineering.

The **Institute of Medicine** was established in 1970 by the National Academy of Sciences to secure the services of eminent members of appropriate professions in the examination of policy matters pertaining to the health of the public. The Institute acts under the responsibility given to the National Academy of Sciences by its congressional charter to be an adviser to the federal government and, upon its own initiative, to identify issues of medical care, research, and education. Dr. Kenneth I. Shine is president of the Institute of Medicine.

The **National Research Council** was organized by the National Academy of Sciences in 1916 to associate the broad community of science and technology with the Academy's purposes of furthering knowledge and advising the federal government. Functioning in accordance with general policies determined by the Academy, the Council has become the principal operating agency of both the National Academy of Sciences and the National Academy of Engineering in providing services to the government, the public, and the scientific and engineering communities. The Council is administered jointly by both Academies and the Institute of Medicine. Dr. Bruce M. Alberts and Dr. William A. Wulf are chairman and vice chairman, respectively, of the National Research Council.

COMMISSION ON LIFE SCIENCES

OTHER REPORTS OF THE BOARD ON ENVIRONMENTAL STUDIES AND TOXICOLOGY

Scientific Frontiers in Developmental Toxicology and Risk Assessment (2000)

Modeling Mobile-Source Emissions (2000)

Copper in Drinking Water (2000)

Ecological Indicators for the Nation (2000)

Hormonally Active Agents in the Environment (1999)

Research Priorities for Airborne Particulate Matter: I. Immediate Priorities and a Long-Range Research Portfolio (1998); II. Evaluating Research Progress and Updating the Portfolio (1999)

Ozone-Forming Potential of Reformulated Gasoline (1999)

Risk-Based Waste Classification in California (1999)

Arsenic in Drinking Water (1999)

Brucellosis in the Greater Yellowstone Area (1998)

The National Research Council's Committee on Toxicology: The First 50 Years (1997)

Toxicologic Assessment of the Army's Zinc Cadmium Sulfide Dispersion Tests (1997)

Carcinogens and Anticarcinogens in the Human Diet (1996)

Upstream: Salmon and Society in the Pacific Northwest (1996)

Science and the Endangered Species Act (1995)

Wetlands: Characteristics and Boundaries (1995)

Biologic Markers (5 reports, 1989-1995)

Review of EPA's Environmental Monitoring and Assessment Program (3 reports, 1994-1995)

Science and Judgment in Risk Assessment (1994)

Ranking Hazardous Waste Sites for Remedial Action (1994)

Pesticides in the Diets of Infants and Children (1993)

Issues in Risk Assessment (1993)

Setting Priorities for Land Conservation (1993)

Protecting Visibility in National Parks and Wilderness Areas (1993)

Dolphins and the Tuna Industry (1992)

Hazardous Materials on the Public Lands (1992)

Science and the National Parks (1992)

Animals as Sentinels of Environmental Health Hazards (1991)

Assessment of the U.S. Outer Continental Shelf Environmental Studies Program, Volumes I-IV (1991-1993)

Human Exposure Assessment for Airborne Pollutants (1991)

Monitoring Human Tissues for Toxic Substances (1991)
Rethinking the Ozone Problem in Urban and Regional Air Pollution (1991)
Decline of the Sea Turtles (1990)

Copies of these reports may be ordered from
the National Academy Press
(800) 624-6242
(202) 334-3313
www.nap.edu

Preface

The National Research Council (NRC) established the Committee on Health Effects of Waste Incineration to assess relationships between human health and incineration of hazardous waste, municipal solid waste, and medical waste. In this report, the committee explains its findings and recommendations about waste incineration and public health.

Despite differences in waste composition and incineration processes, the same types of pollutants of concern can be emitted by each kind of incinerator. Therefore, the committee took a generic approach in addressing the dispersion of pollutants from incineration facilities into the environment, pathways of human exposure, possible health effects, social issues, and community interactions. The committee did not compare risks posed by the different types of waste incineration, nor did it assess risks posed by any particular waste-incineration facility. As discussed in this report, even within the same type of waste incineration, there is broad variability in the emission patterns of pollutants, facility-specific emission characteristics (e.g., stack height and local weather conditions that can affect dispersion of released pollutants), the number of people potentially exposed to incineration emissions, and the total contaminant burden of those people resulting from all pollutant sources.

It is also important to keep in mind that the committee was not asked to compare the health risks attributable to waste incineration with those attributable to other waste-management alternatives, such as land disposal. Therefore, the committee took no position on the merits of incineration compared with other waste-management alternatives.

During the course of its deliberations, the committee reviewed scientific literature, government-agency reports, and unpublished data. The committee solicited information from persons representing federal, state, and local governments; academe; technical consulting firms; environmental-advocacy organizations; public-interest groups; and communities with waste incinerators in their environs. Several members toured a facility in Lorton, Virginia that incinerates municipal solid waste. The committee received useful information and perspectives from the following persons, who made presentations to the committee: Germaine Buck, State University of New York at Buffalo; Dorothy Canter, U.S. Environmental Protection Agency; Frank Caponi, County Sanitation Districts of Los Angeles County; Daniel Carey, American Ref-Fuel Company; Fred Chanania, U.S. Environmental Protection Agency; David Doniger, U.S. Environmental Protection Agency; Lawrence Doucet, Doucet & Mainka, Inc.; Heidi Fiedler, University of Bayreuth, Germany; Simon Friedrich, U.S. Department of Energy; Jeffrey Hahn, Ogden Projects, Inc.; Rick Hind, Greenpeace; Wally Jordan, Waste Energy Technologies, Inc.; Steven Kroll-Smith, University of New Orleans; Stephen Mandel, Rosemount Analytical, Inc.; Melanie Marty, California Environmental Protection Agency; Peter Park, Center for Community Education and Action; Mark Pollins, U.S. Environmental Protection Agency; Jerome Nriagu, University of Michigan; Juan Reyes, Agency for Toxic Substance and Disease Registry, U.S. Public Health Service; Philip C. Sears, Allee, King, Rosen & Fleming; Terri Swearingen, Tri-State Environmental Council; and Stormy Williams, Desert Citizens Against Pollution.

This report has been reviewed in draft form by individuals chosen for their diverse perspectives and technical expertise, in accordance with procedures approved by the NRC's Report Review Committee. The purpose of this independent review is to provide candid and critical comments that assist the NRC in making the published report as sound as possible and to ensure that the report meets institutional standards for objectivity, evidence, and responsiveness to the study charge. The review comments and draft manuscript remain confidential to protect the integrity of the deliberative process. The committee wishes to thank the following individuals for their participation in the review of this report: John C. Bailar III, University of Chicago; A. John Bailer, Miami University; Gaylon Campbell, Washington State University; A.J. Chandler, A.J. Chandler & Associates, Ltd.; Caron Chess, Rutgers University Center for Environmental Communication; Walter Dabberdt, National Center for Atmospheric Research; Donald Hornig, Harvard University; Kathryn Kelly, Delta Toxicology Inc.; Richard Magee, New Jersey Institute of Technology; Jonathan Samet, Johns Hopkins University; and Kenneth Sexton, University of Minnesota.

The individuals listed above have provided many constructive comments and suggestions. It must be emphasized, however, that responsibility for the final content of this report rests entirely with the authoring committee and the NRC.

The committee is thankful for the useful input of Kun-Chieh Lee and Sanford S. Penner into its deliberations early in the study. We also wish to express our appreciation to the following National Research Council staff members for their effective support of our work: Raymond Wassel, Carol Maczka, James Reisa, Bonnie Scarborough, Ruth Crossgrove, Ruth Danoff, Tracie Holby, Katherine Iverson, Catherine Kubik, Eric Kuchner, and others.

Donald R. Mattison, *Chair*
Committee on Health Effects of Waste Incineration

Contents

xv

WASTE INCINERATION
&
PUBLIC HEALTH

Executive Summary

Incineration is widely used to reduce the volume of municipal solid waste, to reduce the potential infectious properties and volume of medical waste, and to reduce the potential toxicity and volume of hazardous chemical and biological waste. In the United States, more than 100 facilities incinerate municipal solid waste, and more than 1,600 facilities incinerate medical waste. Also, almost 200 incinerators and industrial kiln facilities, and many industrial boilers and furnaces combust hazardous and nonhazardous waste.

Whether incineration is an appropriate means of managing waste has been the subject of much debate in this country. A major aspect of the debate is the potential risk to human health that might result from the emission of pollutants generated by the incineration process; some of those pollutants have been found to cause various adverse health effects. Although such effects have generally been observed at much higher ambient concentrations than those usually produced by emissions from an incineration facility, questions persist about the possible effects of smaller amounts of pollutants from incineration facilities, especially when combined with the mix of pollutants emitted from other sources. The possible social, economic, and psychologic effects associated with living or working near an incineration facility also have been topics of concern.

This report was prepared by the National Research Council's Committee on Health Effects of Waste Incineration. The committee was formed to assess relationships between waste incineration and human health and to consider specific issues related to the incineration of hazardous waste, municipal solid waste, and medical waste. The committee was asked to consider various design, siting, and operating conditions at waste-incineration facilities with respect to releases of potentially harmful pollutants to the environment. It was also asked to consider appropriate health-based approaches for demonstrating that an incineration facility meets and maintains established levels of health protection. Issues related to communication of information on waste incineration were also within the study charge. The committee was asked to consider types of information that should be provided to government officials, industry managers, and the general public to help them in future efforts to understand and weigh the risks associated with waste incineration and its alternatives. Finally, the committee was asked to consider factors that might affect public perceptions of waste incineration.

The committee was not charged to assess risks posed by any particular waste-incineration facility or to compare the risks of incineration with risks posed by various waste-management alternatives, such as landfilling. The committee focused its attention on wastes that have reached an incineration facility—it was not asked to address the collection or storage of wastes at, or their transportation to, any incineration facility; nor was it asked to consider treatment of residual ash away from a facility.

WASTE-INCINERATION PROCESSES AND EMISSIONS

The principal gaseous products of waste incineration, like other combustion processes, are carbon dioxide and water vapor. And, like many combustion processes, incineration also produces byproducts such as soot particles and other contaminants released in exhaust gases, and leaves a residue (bottom ash) of incombustible and partially combusted waste that must be emptied from incinerator chambers and properly disposed. The composition of the gas and ash byproducts is determined, at least in part, by the composition of the wastes fed into an incineration facility. This feedstream composition can be altered by other waste-management activities, such as reducing the amount of waste generated, reusing materials, and recycling waste materials for use as feedstocks for various manufacturing processes.

The exhaust gases from waste incineration facilities may contain many potentially harmful substances, including particulate matter; oxides of nitrogen; oxides of sulfur; carbon monoxide; dioxins and furans; metals, such as lead and mercury; acid gases; volatile chlorinated organic compounds; and polycyclic aromatic compounds. Some pollutant emissions are formed, in part, by incomplete combustion that may in turn lead to the formation of pollutants such as dioxins and furans. The formation of products of incomplete combustion is governed by the duration of the combustion process, the extent of gas mixing in the combustion chamber, and the temperature of combustion. Good combustion efficiency depends upon maintaining the appropriate temperature, residence time, and turbulence in the incineration process. Optimal conditions in a combustion chamber must be maintained so that the gases rising from the chamber mix thoroughly and continuously with injected air; maintaining the optimal temperature range involves burning of fuel in an auxiliary burner during startup, shutdown, and process upsets. The combustion chamber is designed to provide adequate turbulence and residence time of the combustion gases.

Operation of the incinerator also affects the emission of heavy metals, chlorine, sulfur, and nitrogen that may be present in the waste fed into the incinerator. Such chemicals are not destroyed during combustion, but are distributed among the bottom ash, fly ash, and released gases in proportions that depend on the characteristics of the metal and the combustion conditions. Mercury and its compounds, for example, are volatile, so most of the mercury in the waste feed is vaporized in the combustion chamber. In the cases of lead and cadmium, the distributions between the bottom ash and fly ash depend on operating conditions. At higher combustion-chamber temperatures, more of the metals can appear in the fly ash or gaseous emissions. Therefore, combustion conditions need to maximize the destruction of products of incomplete combustion and to minimize the vaporization and entrainment of heavy metals, especially when adequate control of emissions is lacking. Formation of oxides of nitrogen is promoted by high temperatures and the presence of nitrogen-containing wastes.

In addition, air-pollution control devices can greatly influence emissions from waste-incineration facilities. For example, airborne particles can be controlled with electrostatic precipitators, fabric filters, or wet scrubbers. Hydrochloric acid, sulfur dioxide, dioxins, and heavy metals can be controlled with wet scrubbers, spray-dryer absorbers, or dry-sorbent injection and fabric filters. Oxides of nitrogen can be controlled, in part, by combustion-process modification and ammonia or urea injection through selective catalytic or noncatalytic reduction. Concentrations of dioxins and mercury can be reduced substantially by passing the cooled flue gas through a carbon sorbent bed or by injecting activated carbon into the flue gas.

With current technology, waste incinerators can be designed and operated to produce nearly complete combustion of the combustible portion of waste and to emit low amounts of the pollutants of concern under normal operating conditions. In addition, using well-trained employees can help ensure that an incinerator is operated to its maximal combustion efficiency and that the emission-control devices are operated optimally for pollutant capture or neutralization. However, for all types of incinerators, there is a need to be alert to off-normal (upset) conditions that might result in short-term emissions greater than those usually represented by typical operating conditions or by annual national averages. Such upset conditions usually occur during incinerator startup or shutdown or when the composition of the waste being burned changes sharply. Upset conditions can also be caused by malfunctioning equipment, operator error, poor management of the incineration process, or inadequate maintenance.

Typically, emissions data have been collected from incineration facilities during only a small fraction of the total number of incinerator operating hours and generally do not include data during startup, shutdown, and upset conditions. Furthermore, such data are typically based on a few stack samples for each pollutant. The adequacy of such emissions data to characterize fully the contribution of incineration to ambient pollutant concentrations for health-effects assessments is uncertain. More emissions information is needed, especially for dioxins and furans, heavy metals, and particulate matter.

Recommendations

Government agencies should continue to improve—or in some cases should begin—the process of collecting, and making readily available to the public, substantially more information on the following:

- The effects of design and operating conditions on emissions and ash. Such information should show how specific emissions and ash characteristics are affected by modifying the operating conditions of an incinerator to maximize its combustion efficiency. It should also indicate the types and combinations of operating conditions that optimize the effectiveness of emission-control devices.
- New combustor designs; continuous emission monitors; emissions-control technologies; operating practices; and techniques for source reduction, fuel cleaning, and fuel preparation, including records of demonstrated environmental performance and effects on emissions and ash.
- Emission and process conditions during startup, shutdown, and upset conditions. Emissions testing has usually been performed under relatively steady-state conditions. However, the greatest emissions are expected to occur during startup, shutdown, and malfunctions. Such emissions need to be better characterized with respect to possible health effects. Therefore, data are needed on the level of emissions, the frequency of accidents and other off-normal performance, and the reasons for such occurrences.

ENVIRONMENTAL PATHWAYS OF HUMAN EXPOSURE

After pollutants from an incineration facility disperse into the air, some people close to the facility may be exposed directly through inhalation or indirectly through consumption of food or water contaminated by deposition of the pollutants from air to soil, vegetation, and water. For metals and other pollutants that are very persistent in the environment, the potential effects may extend well beyond the area close to the incinerator. Persistent pollutants can be carried long distances from their emission sources, go through various chemical and physical transformations, and pass numerous times through soil, water, or food.

Dioxins, furans, and mercury are examples of persistent pollutants for which incinerators have contributed a substantial portion of the total national emissions. Whereas one incinerator might contribute only a small fraction of the total environmental concentrations of these chemicals, the sum of the emissions of all the incineration facilities in a region can be considerable. Many older incinerators have been closed down and replaced by modern low-emitting units, so the relative contribution of incineration to the current concentrations of chemicals in the environment is uncertain.

Results of environmental monitoring studies around incineration facilities have indicated that the specific facilities studied were not likely to be major contributors to local ambient concentrations of the substances of concern, although there are exceptions. However, methodological limitations of those studies do not permit general conclusions to be drawn about the overall contributions of waste incineration to environmental concentrations of those contaminants.

Although emissions from incineration facilities can be smaller than emissions from other types of sources, it is important to assess incinerator emissions in the context of the total ambient concentration of pollutants in an area. In areas where the ambient concentrations are already close to or above environmental guidelines or standards, even relatively small increments can be important.

Computational models for the environmental transport and fate of contaminants through air, soil, water, and food can provide useful information for assessing major exposure pathways for humans, but, in general, they are not accurate enough to provide estimates of overall environmental contributions from an individual facility within a factor of 10. The models suggest that fish consumption is the major pathway of human exposure to mercury, and that meats, dairy products, and fish are potentially the major exposure pathways for dioxins and furans. For assessment of persistent pollutants, there is usually a poor correlation between total ambient concentrations and local emissions from an incinerator.

Recommendations

- Environmental assessment and management strategies for emissions from individual incineration facilities should include a regional-scale framework for assessing dispersion, persistence, and potential long-term impacts on human health.
- Better material balance information—including measurements of source emissions to air and deposition rates to soil, water, and vegetation—are needed to determine the contribution of waste-incineration facilities to environmental concentrations of persistent chemicals. The variation of these emissions over time needs to be taken into account: for the short term to determine if any important emission increases occur at an incineration facility, and for the long term to measure changes due to the replacement of less-efficient incinerators with modern, lower-emitting units.

- To facilitate evaluation of the overall contributions of incinerators to pollutants in the environment, estimates of dispersion of incinerator emissions into the environment should be gathered. The additional information would allow conversion of emissions estimates into environmental concentration estimates.
- Government agencies should link emissions and facility-specific data from all incineration facilities to characterize better the contributions of incinerators to environmental concentrations. Existing databases should be linked to provide easy access to specific operating conditions of an incinerator, height and diameter of the emission stack, flow rate and temperature of the gases leaving the stack, local meteorological conditions, air-dispersion coefficients as a function of distance from a facility, and precise geographic location of the emission point. Data should be standardized for uniform reporting.

HEALTH EFFECTS

Few epidemiologic studies have attempted to assess whether adverse health effects have actually occurred near individual incinerators, and most of them have been unable to detect any effects. The studies of which the committee is aware that did report finding health effects had shortcomings and failed to provide convincing evidence. That result is not surprising given the small populations typically available for study and the fact that such effects, if any, might occur only infrequently or take many years to appear. Also, factors such as emissions from other pollution sources and variations in human activity patterns often decrease the likelihood of determining a relationship between small contributions of pollutants from incinerators and observed health effects. Lack of evidence of such relationships might mean that adverse health effects did not occur, but it could also mean that such relationships might not be detectable using available methods and data sources.

Pollutants emitted by incinerators that appear to have the potential to cause the largest health effects are particulate matter, lead, mercury, and dioxins and furans. However, there is wide variation in the contributions that incinerators can make to environmental concentrations of those contaminants. Although emissions from newer, well-run facilities are expected to contribute little to environmental concentrations and to health risks, the same might not be true for some older or poorly run facilities.

Studies of workers at municipal solid-waste incinerators show that workers are at much higher risk for adverse health effects than individual residents in the surrounding area. In the past, incinerator workers have been exposed to high concentrations of dioxins and toxic metals, particularly lead, cadmium, and mercury.

Recommendations

- To increase the power of epidemiologic studies to assess the health effects of incinerators, future multi-site studies should be designed to evaluate combined data from all facilities in a local area as well as multiple localities that contain similar incinerators and incinerator workers, rather than examining health issues separately site by site.
- In addition to using other exposure-assessment techniques, worker exposures should be evaluated comprehensively through biological monitoring, particularly in combination with efforts to reduce exposures of workers during maintenance operations.
- Assessments of health risks attributable to waste incineration should pay special attention to the risks that might be posed by particulate matter, lead, mercury, dioxins and furans.
- Health risks attributable to emissions resulting from incinerator upset conditions need to be evaluated. Data are needed on the levels of emissions during process upsets as well as the frequency, severity, and causes of accidents and other off-specification performance to enable adequate risk assessments related to these factors. Such information is needed to address whether off-normal emissions are important with respect to possible health effects.
- The Environmental Protection Agency (EPA) and the Occupational Safety and Health Administration (OSHA) should continue striving to improve coordination of enforcement activities between the two agencies to protect the health of incineration workers.

REGULATION OF WASTE-INCINERATION FACILITIES

Waste-incineration facilities are required to comply with a combination of federal, state, and local regulations that vary from place to place and time to time. EPA has proposed or has promulgated separate regulations for incineration of medical, hazardous, and municipal solid wastes to reduce emissions to values achieved by the best-controlled 12% of incinerators. This standard is known as "maximum achievable control technology," or MACT. Facilities that meet the MACT requirements are generally expected to have substantially lower emissions. The intended reduction in emissions would lower exposures and possible risks to populations surrounding incinerators, especially for particulate matter, lead, mercury, and other metals. However, the effects of such regulations are less apparent when emissions of the most-important pollutants from all incineration sources are considered on a regional scale. For example, the collective contribution of dioxins from multiple incineration sources might remain problematic despite MACT regulations. Because the collective effects of incinerators on metropolitan or regional scales are largely unknown, it is uncertain whether implementation of MACT standards for incinerators will substantially reduce the actual risks posed by persistent environmental pollutants at those scales.

Based on estimates of incinerator emissions, environmental transport and fate, potential total exposure, and relative toxicity of the individual substances inferred from studies not involving incineration, the committee concludes

- Compliance with MACT regulations is expected to reduce substantially local population exposures, especially for particulate matter, lead, mercury and other metals, acidic gases, and acidic aerosols.
- Substantial concerns about regional dioxin and furan exposures and moderate concerns about regional exposures to metals are not expected to be relieved by MACT regulations, because the regulations may not adequately reduce risks attributable to cumulative emissions on a regional basis.
- Substantial concerns about workers' exposures to particulate matter, lead, mercury, and dioxins and furans are not expected to be relieved by MACT compliance, because those regulations were not designed to affect workers' exposures.

Recommendations

- Technologies used in other countries for combustion, emission control, continuous emission monitoring, and public dissemination of information, as well as optimum operating practices, should be actively studied and considered for adoption in the United States.
- All regulated medical-waste incinerators and municipal solid-waste combustors should have uniform limits for each pollutant, irrespective of plant size, design, age, or feedstock, as is the case for hazardous-waste combustors. The same technology for air-pollution control is applicable to small and large facilities. Allowing less-stringent limitations for some designs or sizes is inconsistent with the principle of minimizing risks of health effects.
- Government agencies should encourage research, development, and demonstration of continuous emission monitors (CEMs), dissemination technologies, and computer programs that automatically analyze, summarize, and report CEM data. In addition to the CEMs already required in municipal solid waste incinerator rules, requirement of CEMs for hydrochloric acid and particulate-matter should be considered on such incinerators. Also, as soon as a mercury monitor that measures ionic and metallic forms of mercury emissions has been proven reliable, EPA should consider its use for domestic incinerators. The same approach should be used for

other monitors, including those for other heavy metals and dioxins and furans. EPA should also explore the utility of technologies such as direct electronic transmission and display to disseminate CEM data to regulatory authorities and the public. Providing such data and data summaries on the Internet should be considered.

- In future regulatory decision-making, greater consideration should be given to emission levels achieved in actual performance of incinerators, including process upset conditions (described earlier). In monitoring for compliance or other purposes, data generated during the intervals in which a facility is in startup, shutdown, and upset conditions should be included in the hourly emission data recorded and published. It is during those times that the highest emissions may occur, and omitting them systematically from monitoring data records does not allow for a full characterization of the actual emissions from an incineration facility.

SOCIAL ISSUES AND COMMUNITY INTERACTIONS

In addition to possible physical-health effects, a waste-incineration facility may have other effects on individuals, groups, or the entire population in the surrounding area. The effects might be economic (such as job creation or decrease in property values), psychological (such as stress or stigma), or social (such as community factionalization or unity). However, there is little rigorous information on those impacts of waste-incineration facilities.

Citizen concerns need to be heard and understood. Conflicts can increase the time and expense of conducting waste incinerators and other facilities that might be potentially beneficial to society. Opposition to the facilities also can indicate that important concerns are not being addressed adequately.

Much public opposition to waste incineration might be due to a lack of understanding of the relative health risks posed by incineration in comparison with other waste-management methods. But health is not the only issue, and the differences between expert and public perceptions are not due merely to differences in information and understanding; they can also be due to differences in social values. People's perceptions are often extraordinarily resistant to change, in part because they reflect underlying values. Efforts that ignore or try to change these perceptions radically are likely to fail. Risk communication should accept as legitimate the perceptions and concerns of various members of the public and involve them in consultative, participatory processes. Not only do members of the public have a right and responsibility to be involved in the assessment and management of hazards in their communities, but such involvement might result in improved assessments and management strategies.

Developing effective participatory programs is very difficult, but some general principles are beginning to emerge. The process of public involvement should be open, inclusive, and substantive, and members of the public in an affected area should be involved early and often. Major concerns are likely to include issues of safety, compensation, and local oversight and control. Satisfying the public's need for information on incinerator safety requires continual assessment and demonstration of regulatory compliance with existing standards.

Recommendations

- The social, psychological, and economic effects of proposed and existing waste incineration facilities should be assessed, and mitigation of or compensation for such effects should be considered where appropriate.
- The boundaries of an area potentially affected by a waste incinerator should not be defined at the outset by a particular community's political boundaries or jurisdiction. Instead, the assessment area should be based on the geographic extent over which various effects could reasonably occur. Such an approach permits a more-accurate analysis of the impacts. It also permits a better understanding of problems that might arise in connection with information exchange among all persons involved with, or affected by, decision-making concerning the facility.
- Proponents of an incineration facility should assume, in dealing with local communities, that they (the proponents) should make the case for a new or expanded facility, especially if a waste combustor is not used solely within a manufacturing facility to incinerate waste on site.
- If a new or expanded facility is contemplated, local citizens might consider conducting their own assessments of the proposed facility and its effects through various approaches, including, for example, hiring independent consultants that members of the community trust, seeking technical-assistance grants from the government, or finding technical advisers who are acceptable to both sides.
- Particular attention should be paid to equity issues when a facility is to be placed in a community that is already experiencing disproportionate health, environmental, or socioeconomic burdens.
- Government agencies should improve—or in some cases begin—to collect and make readily available, information on site-specific and large-scale empirical research on possible socioeconomic impacts of waste-incineration facilities on their host areas. To the extent practicable, efforts should be expanded to gather such information routinely before and during operation of incineration facilities that have the potential to have more than a minor socioeconomic impact.

UNCERTAINTY AND VARIABILITY

Incineration facilities vary with regard to types of waste incinerated, operating practices, allowable magnitudes of emission, emission-control technologies, types of substances emitted, environmental conditions, proximity to other sources of contaminants, and frequency of process upsets. The people who might be exposed to the contaminants are likely to differ in their susceptibilities and activity patterns. Some uncertainties are specific to waste incineration, and some are inherent in any activity that releases contaminants into the environment.

Some of the uncertainties and variability can be reduced or better accounted for; others will remain intractable. The most-effective decisions concerning the siting, design, operation, and regulation of incineration facilities are the ones that take uncertainty and variability fully into account.

Recommendations

- Incinerator risk assessments should include the following components of uncertainty and variability analyses
  — An estimate of the variability and uncertainty distributions of all input values and their effects on final estimates.
  — A sensitivity analysis to assess how model predictions are related to variations in input data.
  — Variance-propagation models that show how the variability and uncertainty of final results are tied to the uncertainties and variabilities associated with the various models, their inputs, and assumptions used throughout the risk assessment.

The committee's evaluation of waste incineration and public health was substantially impaired by the lack of available compilations of the ambient concentrations of pollutants resulting from incinerator emissions. In addition, large variabilities and uncertainties associated with risk-assessment predictions often limit the ability to define risks posed by incinerators.

1

Scope of the Committee's Effort

In this century, incineration has been used widely to reduce the volume of municipal-solid waste and produce electric energy or steam, to reduce the volume and potential infectious nature of contaminated medical waste, and to reduce the potential toxicity and volume of hazardous chemical and biological waste. Although various forms of incineration are widely used for waste management, pollution control, or energy recovery, there has been increased public debate in the last several decades over the expected benefits mentioned above and the potential risk to human health that might result from the emission of pollutants generated by the incineration process.

Unfortunately, there have been only a few studies of human populations that investigated the attribution of certain adverse health effects to particular incinerators. Most studies were unable to detect any effects. Those studies, of which the committee is aware that did report finding health effects, had shortcomings and failed to provide convincing evidence. Therefore, those reported effects are still open to many other possible explanations (see Chapter 5).

Debate over the expected benefits and the potential health risk of waste incineration has led to substantial polarization of opinions with respect to regulatory decisions about incineration facilities. This report, by the Committee on Health Effects of Waste Incineration, of the National Research Council's Board on Environmental Studies and Toxicology, addresses scientific and technical aspects of the design and operation of facilities that burn waste, releases of pollutants from such facilities and transport through the environment, possible human health effects of exposure to those pollutants in the environment, and relevant regulatory and sociological considerations. For this report, "incineration" is a general term that refers to the process of burning waste through the use of incinerators, industrial boilers, or furnaces, kilns, or other facilities.

CHARGE TO THE COMMITTEE

The committee was specifically asked to assess relationships between various aspects of waste combustion and estimates of human health risk. The committee was asked to consider, to the extent practicable, the following issues for the combustion of hazardous, nonhazardous, and hospital wastes:

- Relationships between human health risk estimates and various design, siting, and operating conditions at waste-combustion facilities, including incinerators, cement kilns, industrial furnaces, and industrial boilers.
- Operating practices at combustion facilities and expectations regarding technology and the release of hazardous substances.
- Appropriate methods for assessing the siting, design, and operation of combustion facilities.
- Appropriate health-based performance criteria for demonstrating that a combustion facility meets and maintains agreed upon health-risk tolerance levels.
- Types of scientific, technical, and other information that should be provided to government officials, industry managers, and the general public to help them understand and weigh the risks associated with waste combustion and its alternatives including innovative ways of oxidizing waste.
- Public perceptions of waste combustion and their bases.

It is important to note that the committee was not asked to assess the magnitude of health risks associated with individual waste-incineration facilities. Also, the committee was not asked to develop its own health-risk tolerance levels for incineration facilities.

The committee formed to address this charge was composed of persons with expertise in incineration technology, emission characterization, transformations and fate of environmental pollutants, exposure and dose characterization, public health, health risk assessment, sociology, risk perception and risk communication, and law. Biographical information on the members is provided in Appendix A.

COMMITTEE'S APPROACH TO ITS CHARGE

In developing an approach to its task, the committee received oral and written testimony from interested or affected citizens; community activists; industry representatives; environmental advocates; professional scientists and engineers; and local, state, and federal government officials. The committee also gathered and considered relevant available information.

The committee focused on three types of waste streams for which incineration has been used as an option for waste management: municipal solid wastes, hazardous wastes, and medical wastes. The committee structured its efforts by using an emission-to-receptor framework that is a modification of the risk assessment frameworks presented by NRC (1983, 1994). The committee's framework included consideration of the following components:

- Emission characteristics and various factors that could affect emissions resulting from waste-incineration facilities.
- Transformation and fate of certain emitted contaminants in various environmental media (i.e., air, water, and soil).
- Contributions of incineration to environmental concentrations of contaminants.
- Human populations that might be exposed to contaminants of concern, and the pathways through which exposure can occur.
- Health responses that might be expected from exposures to contaminants of concern.
- Characterization of relationships between waste incineration and health risks.

Other aspects of the committee's study that are not explicitly included in the emission-receptor framework are public perceptions of waste combustion and their bases, sociological considerations of incineration facility siting, and communication of information for understanding and weighing the risks associated with waste combustion.

The committee did not consider the potential health effects on organisms other than humans. In addition, the committee did not consider incineration of waste streams consisting solely of sewage sludge, wood wastes, radioactive waste, or industrial waste that is considered nonhazardous.[1] It is possible, however, that some of those types of waste are fed to a facility that incinerates municipal solid waste.

The committee also did not address, to any great degree, the effects of waste-management activities, such as waste recycling or reuse, except for how such practices might affect the characteristics of waste streams fed to an incineration facility and the resulting emissions to the environment. The committee focused its attention on wastes that have reached an incineration facility—not on the collection, storage, or transportation of wastes to a facility and not on transportation of residual ash away from a facility. However, if one were to perform

---

[1] Various aspects related to the use of incineration for destroying the U.S. stockpile of extremely hazardous chemical agents and munitions have been addressed by another NRC committee (see NRC 1999a and related reports cited therein).

a comparative assessment of the total environmental impacts of waste incineration as a management option, the above-mentioned considerations should be included.

Despite past efforts to characterize the potential health risks at numerous individual existing and proposed incineration sites, the committee has carried out this study with rather sparse information on the relationship between human exposure to pollutants released to the environment through waste incineration and the occurrence of health effects because such information is generally unknown. One reason, for example, for the lack of information is that few epidemiologic studies have been conducted to investigate exposures to incineration emissions and their human health consequences. Although the committee was not asked to and did not attempt to perform its own epidemiologic studies or risk assessments at individual waste-incineration facilities, the committee used available data on human and animal exposures to specific substances to examine the implications for incinerator sites of dose-response projections (see Chapter 5).

The committee conducted this study with the understanding that society faces the challenge of choosing among various waste-management alternatives. Although a comparative health risk assessment of alternatives for managing waste streams was not part of its charge, the committee conducted this study with the intent of informing the public debate over appropriate uses of waste incineration.

ORGANIZATION OF THE REPORT

The emission-receptor framework, described earlier, was used to develop and organize the chapters of this report. The six components of the charge to the committee are addressed within these chapters. Chapter 2 presents an overview of waste generation, waste stream composition, and waste-management activities that affect the characteristics of waste fed into incineration facilities. Chapter 3 discusses various incineration processes used to burn waste, characteristics of emissions of certain contaminants, operating practices and design options most likely to affect emissions, and expectations regarding the technology and release of hazardous contaminants from incineration facilities. Chapter 4 discusses the environmental transport and fate of pollutants once they have left an incineration facility, and considers the contribution of incineration to ambient concentrations of such pollutants.

Chapter 5 examines the techniques used to evaluate the potential for health effects from incineration, and discusses some of the results obtained with those techniques. It also relates human health risk estimates to issues discussed in Chapters 3 and 4 with respect to various design and operating conditions at waste-combustion facilities. In addition, Chapter 5 identifies important considerations for developing health-based performance criteria to demonstrate that a combustion facility meets and maintains agreed-upon health-risk tolerance levels.

Chapter 6 discusses the regulations that affect waste incineration in the United States. Chapter 7 discusses important social issues, including public perceptions of waste combustion and their bases. The chapter also discusses risk communication about incineration. Chapter 8 addresses important uncertainties involved in assessing possible relationships between exposure to environmental concentrations of pollutants released from waste-incineration facilities and occurrences of human health effects.

Chapters 3 through 8 provide conclusions and recommendations regarding appropriate methods for assessing the siting, design, and operation of combustion facilities. They also identify various types of scientific, technical, and other information that should be provided to decisionmakers and other interested or affected parties to help them understand and weigh the risks associated with waste combustion.

2

Waste Incineration Overview

This chapter provides an overview of waste generation, waste stream composition, and incineration in the context of waste management. Communities are faced with the challenge of developing waste-management approaches from options that include reduction of waste generated, incineration, landfilling, recycling, reuse,[1] and composting. Waste-management options other than incineration are discussed here to illustrate that a combination of options, such as recycling, with incineration can alter the characteristics of waste streams fed to incineration facilities. However, the committee was not charged to undertake a comparative assessment of waste-management options.

In general, any incineration facility will incorporate the following processes: waste storage and handling, processing to prepare waste, combustion, air-pollution control, and residue (ash) handling. The types of waste-incineration facilities discussed in this report include incinerators, industrial boilers, furnaces, and kilns (see Chapter 3). There is a large variety of technology, varying from stationary facilities designed to combust millions of tons of waste per year collected from a broad geographical area, down to mobile incinerators used to remediate wastes from specific sites that are contaminated by hazardous waste.

TYPES OF WASTE INCINERATED

Three types of waste to which incineration is applied extensively are municipal solid waste, hazardous waste, and medical waste. Incineration of those three

---

[1] Reuse refers to using a material more than once in its original manufactured form (e.g., refilling a returned glass bottle).

TABLE 2-1   Waste Generation in the United States, Numbers of
Incineration Facilities, and Amounts of Waste Combusted

Type of Waste	Amount Generated (million tons/yr)	Number of Incineration Facilities	Amount of Waste Combusted (million tons/yr)
Municipal solid waste	209[a]	122[b]	36[a]
Hazardous waste	276[c]		3[c]
On-site Incinerators		129	
Commercial Incinerators		20	
Industrial Boilers and Furnaces		950[d]	
Cement Kilns		18	
Light Weight Aggregate Kilns		5	
Medical waste	—[e]	1,655[f]	0.8[g]

[a] Estimate is for 1996 as presented in "Characterization of Municipal Solid Waste in the United States: 1997 Update" Franklin Associates 1998.

[b] The Integrated Waste Services Association reports that there are 103 waste-to-energy facilities operating in the United States http://www.wte.org. In addition, Franklin Associates (1998) reported that 19 facilities incinerated municipal solid waste without energy recovery.

[c] Estimate for 1991, presented by OECD 1996. Amount generated is largely in aqueous form. Does not include soil contaminated by hazardous waste. EPA estimates that hazardous waste incinerators burn 1.5×10^6 tons per year (Fed. Regist. 61(April 19):17358-17536).

[d] EPA (1997a) estimates that there were around 900 boilers in the United States in 1993. There were less than 50 hazardous waste-burning industrial furnaces operating in the United States during that time.

[e] OTA (1990) reports that estimate for medical waste, exclusive of that generated from home health-care, range from 0.3 to 2% of the total municipal solid-waste stream.

[f] Brian Strong and Katie Hanks, MRI, Feb. 22, 1999, memorandum "Emissions Inventory for Hospital, Medical, Infectious Waste Incinerators Covered by the Proposed Section 11(d)/129 Federal Plan." EPA Docket number A-98-24, item II-B-1.

[g] Estimate of 845,500 is based upon 2,373 incinerators. Brian Hardee and Katie Hanks, MRI, July 16, 1997, memorandum "Revised Impacts of the Regulatory Options for New and Existing Medical Waste Incinerators (MWIs)." EPA Docket number A-91-61, item IV-B-072.

types is the focus of this discussion. Table 2-1 presents estimates of the amounts of those wastes generated, numbers of incineration facilities, and amounts combusted in the United States.

Municipal Solid Waste

Municipal solid waste is defined as the solid portion of the waste (not classified as hazardous or toxic) generated by households, commercial establishments, public and private institutions, government agencies, and other sources. This waste stream includes food and yard wastes, and a multitude of durable and nondurable products and packaging. Figure 2-1 illustrates the composition of municipal waste in the United States in 1997. Almost 40% of the municipal waste stream is composed of paper and paperboard, about 10% plastics, about 13% metals and glass, and about 13% yard trimmings. The remainder consists of miscellaneous materials (wood, rubber, textiles, and so on). Municipal solid waste does not include segregated medical waste, but does include some medical waste that is mixed in.

The quantity of municipal solid waste in the United States has been increasing (see Table 2-2) despite government attention to the practices of waste reduction at the source and to recycling. Factors that contribute to the rate increase include the following: the U.S. population is growing (from 180 million in 1960 to 249 million in 1990 to a projected 276 million in 2000); per capita generation of waste has increased because of increasing consumption of nondurable, disposable items, and durable items, as well as extensive use of packaging.

As shown in Table 2-2, the per capita generation of municipal solid waste in the United States increased from 1960 to 1990, but decreased from 1994 to 1996. The decrease in per capita generation is attributable to increased on-site composting of organic materials from 1990 to 1996. Despite the results of recycling and composting, the nation faces the challenge of increased total waste generation as long as population continues to increase (Figure 2-2). The amount of discards after recovery was higher in 1996 than in 1970, which indicates that the

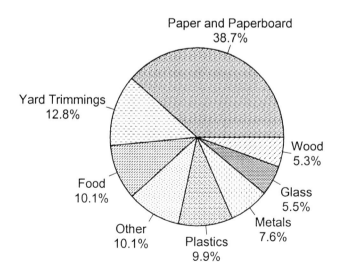

FIGURE 2-1   Municipal solid-waste composition by weight, 1997. (Total weight = 217 million tons.) Source: Franklin Associates 1998.

TABLE 2-2 Per Capita Generation, Materials Recovery, Incineration, and Discards of Municipal Solid Waste, 1960 to1996 (in pounds per person per day; population in thousands)[a]

	1960	1970	1980	1990	1992	1994	1995	1996
Generation	2.68	3.25	3.66	4.51	4.49	4.51	4.41	4.33
Recovery for recycling and composting	0.17	0.22	0.35	0.74	0.87	1.07	1.15	1.18
Discards after recovery	2.51	3.04	3.31	3.77	3.62	3.44	3.26	3.15
Incineration	0.82	0.67	0.33	0.70	0.70	0.68	0.70	0.70
Discards to landfill, other disposal	1.69	2.36	2.98	3.07	2.92	2.75	2.56	2.44
Resident population	179,979	203,984	227,255	249,398	255,011	260,372	262,890	265,284

[a] Details may not add to totals due to rounding. Adapted from: Franklin Associates 1998.

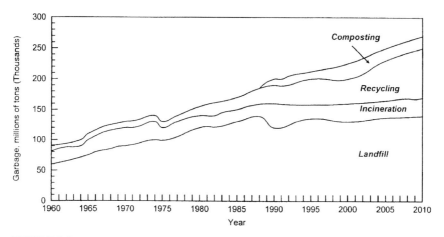

FIGURE 2-2   Trends in municipal solid-waste generation and management in the United States, 1960-2010. Source: Franklin Associates 1997.

increased generation since 1970 has more than compensated for reductions due to recycling and composting.

Although it is still the predominant method of solid-waste management in the United States, the fraction landfilled was smaller in 1996 at 56% than it was in 1985 at 83% (Franklin Associates 1998). Incineration rates have varied over the last few decades (Figure 2-2 and Table 2-2). In 1960, combustion in low-efficiency combustors without energy recovery or advanced pollution-control technology burned 31% of the municipal solid waste generated. In 1980, incineration was down to 9%. However, because of increased emphasis on waste-to-energy conversion, by 1990, incineration had increased to 16% of total waste generation. By 2000, incineration is projected by EPA to decrease slightly to 15.6%.

A decrease in the total capacity of municipal-waste incinerators is thought to have occurred for several reasons: the continued availability of lower-cost disposal alternatives (such as landfilling); opposition from local advocacy groups, which has resulted in municipal planners' rejection of waste-incinerator construction at many locations; mandatory recycling programs and increasing confidence in reduction and reuse as options; and the loss of flow control of municipal wastes.[2]

Uncontrolled combustion of municipal solid waste has been practiced for many years by individual homeowners burning trash, and by managers of hotels

---

[2] "Flow control" refers to legal provisions that allow state and local governments to designate the places where municipal solid waste is taken for processing, treatment, or disposal.

and community housing units burning their waste in small incinerators. But large-scale incineration in specially designed furnaces, with or without energy recovery, is of more-recent origin in the United States. The first U.S. waste-to-energy plant was operating in 1905 in New York City. Redesign of these systems for specialty use on municipal wastes led to today's generation of furnaces, most of which represent proprietary technologies from European manufacturers (IAWG 1995). According to the Integrated Waste Services Association (IWSA) in 1999, 103 waste-to-energy facilities in the United States combust about 15% of the nation's municipal solid waste (http://www.wte.org).

Hazardous Waste

Hazardous waste is defined by EPA under the Resource Conservation and Recovery Act (RCRA) as a waste material that can be classified as potentially dangerous to human health or the environment on the basis of any of the following criteria:

- It might ignite easily, posing a fire hazard.
- It might be corrosive, capable of damaging materials or injuring people.
- It might be reactive—likely to explode, catch fire, or give off dangerous gases when in contact with water or other materials.
- It might be toxic, capable of causing illness or other health problems if handled incorrectly.
- It might be on a list of specific wastes or discarded compounds that EPA has classified as hazardous.

Hazardous wastes are generated by entities such as manufacturers, service and wholesale-trade companies, universities, hospitals, government facilities, and households. They are generated both by the chemical manufacturing industry and by users. However, pollution-prevention programs are proliferating. Such programs are used in industry and are encouraged by federal agencies. They can be designed to study each manufacturing process with an eye to reducing hazardous materials used or generated, and thereby, reducing the amount of hazardous materials that could be released as air pollution, water pollution, or hazardous solid wastes. Such programs are used in industry and are encouraged by federal agencies. Not only can they reduce environmental effects at lower expense than do on-site emission control devices and water-treatment facilities, but they can save manufacturers money.

When a hazardous waste is generated, the generator can either manage the waste on site or move it off site for treatment, disposal, or recycling. Before the establishment of EPA and the enactment of stricter environmental laws and regulations in the 1970s, dumping of chemical wastes into inadequately designed landfills or simply onto the land or into rivers or oceans was common. Before there was awareness of potential health hazards associated with soil contamination, land dumping was often seen as the most practical way to dispose of chemical wastes.

Concern over contamination of air, surface water, and groundwater from uncontrolled land-disposal sites provoked the emergence of tougher regulations for land disposal. Those provided incentives for industry to use a wide variety of traditional and advanced technologies for managing hazardous wastes. The regulations also require that many wastes be treated by incineration or other methods to reduce organic content to specified levels before the wastes can be disposed of in a secured landfill.

EPA estimates that regulations forbidding land disposal of any hazardous waste that contains liquid will substantially increase the quantity of hazardous waste directed to incinerators, boilers, and furnaces. Although industrial growth is also likely to increase hazardous-waste generation, increasing emphasis on waste minimization and recycling is likely to exert pressure to reduce such generation.

Many kinds of hazardous waste are fed to incinerators, boilers, and industrial furnaces essentially as received. These wastes are often difficult to handle because of their consistency or hazardous nature, so minimal handling is preferred. Where feasible, however, pretreatment operations are desirable to facilitate homogenization of the waste and continuous feeding to the combustor. Common pretreatment operations for liquid wastes are blending and solids filtration; for solids, screening and size reduction (crushing or shredding); and for wastes in containers, liquid-phase decanting and shredding to allow continuous auger feeding.

Several types of industrial furnace systems are used to incinerate hazardous waste to recover energy or material. The major ones are cement kilns, lightweight-aggregate kilns, halogen-acid furnaces, and metal-recovery and smelting furnaces.

Cement is produced by feeding raw materials into a rotary kiln and burning them with fuel under controlled-temperature conditions. Suitable hazardous waste is used as an auxiliary or replacement fuel. Lightweight aggregate is produced much like cement, in a kiln configured and fueled much like a cement kiln using feed stocks that include special clays, pumice, scoria, shale, and slate. It is used to make insulation and monostructural and lightweight concrete. Halogen-acid furnaces are typically modified firetube boilers that process secondary waste streams containing 20-70% chlorine or bromine. The combustion gases are "scrubbed" with water to produce a halogen-acid product.

Cement kilns have been used to burn hazardous waste since 1972, when PCBs were combusted in Ontario, Canada. Since then, the use of waste-fueled kilns has become widespread in the United States, Belgium, and Switzerland. An intended benefit of combustion of waste as fuel in kilns is the recovery of energy from the waste and the consequent conservation of nonrenewable fossil fuels. Moreover, there is a strong economic incentive in that the kiln operators are paid to take the waste, rather than having to pay for fuel. A substantial portion of the energy supplied by coal in a cement kiln can be replaced with waste-derived fuel.

Other common combustors used for hazardous wastes are mobile incinerators and industrial boilers. Mobile incinerators are most commonly used for contaminated-soil remediation projects. A typical mobile incinerator that is used at a site to treat contaminated soil consists of an incineration module, an air pollution control system, and other site-specific ancillary systems. As of March 1992, a survey showed that mobile incinerators were in various stages of remediating 2,139,700 tons of contaminated soil at 56 sites (Dempsey and Oppelt 1993). The sites included CERCLA-Superfund sites, RCRA sites, and spill cleanup sites, although the survey did not include underground storage tank sites that were contaminated by leaked material.

The boilers used to burn hazardous waste are standard industrial boilers widely used for steam generation in the process industries. They include fuel and combustion air delivery systems, waterwall furnace sections, and convective heat transfer sections. The waste feeding system is usually the only nonstandard equipment added to burn hazardous waste. Many natural gas-fired and oil-fired and a smaller number of coal-fired boilers also burn hazardous waste. Most gas-fired and oil-fired boilers burn only wastes that contain essentially no ash or chlorine, because they do not have air-pollution control devices.

Medical Wastes

Medical (biomedical) wastes can have infectious or toxic characteristics that, with improper disposal, pose public-health concerns. Medical waste is created by a wide variety of activities. Essentially every aspect of the health-care delivery system contributes, but hospitals are the largest medical-waste producers, generating up to about 26 pounds of waste per bed per day (Lawrence Doucet, Doucet & Mainka, Inc., pers. commun., September 1995). Large quantities of waste are also generated in analytic laboratories, medical and dental offices, and other primary and secondary health-care facilities. However, because of the potential handling dangers (for example, of blood-borne pathogens), few studies have directly analyzed the physical and chemical composition of medical waste. It is extremely heterogeneous, and its chemical composition and combustion characteristics, are determined largely by where it is generated. Nevertheless, Figure 2-3 indicates an average composition in 1995. By law (RCRA, subtitle J), infectious waste (e.g., microbiological cultures) should be incinerated or disposed. Noninfectious medical waste may be disposed together with regular municipal wastes. Public concerns with management of medical waste have increased by the possibility of spreading Acquired Immune Deficiency Syndrome (AIDS), by poorly operated hospital-waste incinerators, and improperly disposed medical debris washing up on public beaches in the 1980s.

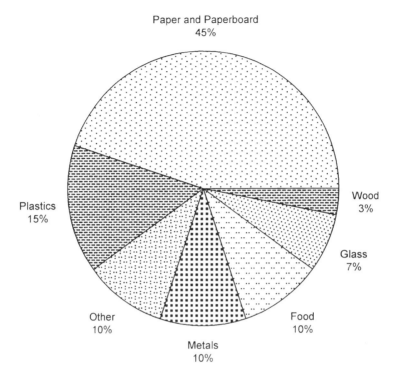

Paper and Paperboard 45%

Plastics 15%

Wood 3%

Glass 7%

Other 10%

Metals 10%

Food 10%

FIGURE 2-3   Medical waste composition. Data from AHA 1993.

Typically, in hospitals today, about 15% of the total hospital waste is "red-bag" waste—waste that is considered infectious, and so must be incinerated or otherwise sterilized to prevent the spread of disease (AWMA 1994). However, some hospitals arbitrarily treat as much as 90% of their waste as red bag, apparently due to a lack of standard practices in health-care facilities for separating wastes that are truly red-bag from other waste. A 1992 study of New York City medical waste determined that much of the waste put into red bags did not qualify as red-bag waste and that much waste placed in brown bags should have been put into red bags. It was determined that the amount of red-bag waste can fall to less than 5% of the total when proper procedures are followed (Waste-Tech 1991). Such procedures include segregating recyclable material and compostable material (such as cafeteria food waste), implementing purchasing procedures that reduce the use of disposable material (for example, by installing electric hand dryers to replace or augment paper towels), and educating personnel to correctly dispose of waste as infectious or noninfectious. The study indicated that one key to reducing red-bag waste is a thorough waste characterization for a facility, followed by the implementation of a plan for waste minimization and recycling.

A study by the Waste Energy Technologies, Inc. and the New York City Department of Sanitation indicates the degree to which red-bag waste is recyclable and preventable if changes in procurement and good maintenance practices are implemented (Wally Jordan, Waste Energy Technologies, Inc., pers. commun., July 1999). If such recycling becomes widespread, the need for medical-waste incinerators could be reduced to the quantity corresponding to the amount of red-bag waste.

Today, the two most-common medical-waste management practices are autoclaving (steam sterilization) and incineration followed by landfilling. In autoclaving, bags of infectious waste are placed in a sealed chamber, which is sometimes pressurized, and heated by direct contact with steam to sterilize them.

As a waste-management strategy, properly conducted and controlled incineration of medical waste will reduce both its volume and its infectious character. Most facilities that incinerate medical waste burn a diverse mixture that might include pathologic waste. Most of the materials that make up the general medical waste stream burn readily and, given the proper conditions, will continue to burn once ignited. Metal and glass sharps do not burn but do not greatly impede combustion of other materials. Pathologic waste has a very high moisture content and will not support self-sustained combustion, but it will burn if adequate heat is applied to drive off excess moisture. However, given the quantity of chlorinated or metals-laden (e.g., cadmium) plastics placed in medical-waste incinerators, antiquated and inefficient designs of some incinerators, and suboptimal operation and emission monitoring at many on-site facilities, the potential exists for relatively high emissions of various pollutants including dioxins and furans.

WASTE MANAGEMENT

The committee was not charged to evaluate integrated waste-management strategies. However, discussion of incineration in a community requires some consideration of the many alternative options and technologies available. The use of different management strategies can substantially affect the size and composition of the waste stream that is fed to an incineration facility.

Historically, waste has been disposed in the cheapest available manner. In areas with ready access to cheap land and transport, such as most of the United States, wastes were simply landfilled. In areas with little available land and high transport costs, incineration of wastes was adopted as a way to reduce the volume to be transported and landfilled. The addition of a waste-to-energy component to such incinerators was adopted as a way to reduce the overall costs of the process. Public concern about environmental deterioration as global and local phenomena, engendered concern about reliance on incineration and on landfill practices, have led to calls for increased efforts to prevent, recycle, and compost waste.

Such calls have led to more attention to solid-waste management options as well as a transition from old-style incinerators (those which do not recover energy and do not have advanced combustors or emission-control systems) to modern waste-to-energy (WTE) plants.

In 1988, the federal government and several states set goals or enacted laws requiring the development and implementation of integrated waste-management plans. EPA devised a hierarchy of waste-prevention and waste-management methods that could be implemented in an integrated system and that would reflect the relative desirability of each method from the point of view of environmental protection. These methods include volume source reduction, reuse, recycling, composting, waste-to-energy conversion, and landfilling.

Some methods—such as recycling, composting, reuse, and reduction—decrease the amount of waste for disposal, and so have been considered to be environmentally preferable. Simple waste-prevention practices, such as reducing paper use by copying on both sides, or taking reusable bags to the supermarket, are examples of waste-reduction actions that also conserve natural resources. Similarly, for products made directly from raw materials, use of smaller amounts of the raw materials is expected to lessen environmental effects of the manufacture, transportation, treatment, and disposal of packaging. Waste reduction, reuse, recycling, and composting are all designed to reduce the quantity of material that must be incinerated and ultimately landfilled, and they are also likely to change the characteristics of the wastes that are incinerated and landfilled. The object is to reduce the quantities of toxic materials that may be released to the environment, from incinerator or landfill emissions.

Thus, various factors help determine whether a waste-incineration facility can be sited and operated successfully, including the amount of incineration capacity relative to the amount of waste that will not be committed to reduction, reuse, recycling, and composting; the extent to which an incinerator is designed with the most-effective technologies available and designed to reduce emissions, as much as technologically feasible; and the operation of the facility so as to properly use the advanced designs to maximize combustion and emission-control efficiency.

Volume Source Reduction

Volume source reduction methods are used to reduce waste generation. Such reduction is accomplished by changing the design of products and packaging (for example, to eliminate or reduce packaging, stimulate reuse, increase the durability of products, and eliminate disposables) and by modifying consumers' purchasing habits (for example, leading them to purchase fewer disposables, more durables, and products in less packaging).

Durables (consumer products that are designed and manufactured to have a useful life of at least 3 years, and that are easily maintained, repaired, and re-used) made up about 15% of the U.S. waste stream in 1996, nondurables and throwaways 27%, containers and packaging about 33%, and yard waste and food waste, both of which are compostable, about 24% (Franklin Associates 1998). Yard waste has decreased as a percentage of the waste stream, from 23% in 1960 to 13% in 1996, but has increased in tonnage from 20 to 28 million tons/year in that time. But yard waste composting is expected to account for a 0.1% decrease by 2000 in overall per capita generation of waste that would be landfilled or incinerated, according to EPA. That will be the result of programs that encourage homeowners to leave grass clippings on the lawn and compost yard waste on-site and of limitations on landfilling of yard wastes. However, this effect could be offset by increased generation of waste from packaging and nondurable and durable products.

Nondurables, containers, and packaging make up roughly 55% of the waste stream and are considered the main targets for volume source reduction. Since 1960, smaller numbers of packages made of heavy materials, such as glass and metal, have been replaced with larger numbers of packages made of lighter but more voluminous plastic and paper. But nondurables have been increasing as a percentage of the waste stream since 1960, from 18.5% to 26.7% in 1993, and account for much of the overall increase in total and per capita generation. Much of the overall increase is doubtless due to the replacement of durables with single-use, disposable products.

Reuse

The durables component of municipal solid waste is the prime target of reuse programs. Such programs include encouragement of remanufacturing of items (e.g., electronics, furniture, and automobiles), and particularly encouragement of design of new items specifically with remanufacturing in mind, as well as encouragement of repair, cleaning, and recharging. In 1993, this component of durables constituted 17.2% of the waste stream, and they have been increasing steadily since 1960, when durables contributed 11.1% to the waste stream. Reuse methods may involve changing consumer habits with respect to the repair, cleaning, and other means of extending the lifespan of products already purchased; or perhaps a change in consumers' disposition of products that would otherwise have been thrown away (for example, resale, renting, donation, swapping, lending and borrowing, or return to the manufacturer for re-manufacture).

Recycling and Composting

Recycling and composting both involve the recovery of some substantive value from materials discarded in municipal solid waste. Once waste-prevention methods, such as source and toxic reduction and reuse, have been applied where appropriate, the remainder of the waste stream is a candidate for recycling and composting strategies. Recycling entails recovery, processing, and refining of materials in the waste stream to create new products. Recyclable materials generated in 1993 include much of the glass (6.6%), several metals (8.3%), several resins of plastics (9.3%), most grades of paper (37.6%), and wood (6.6%). New materials (textiles, for example) are entering the recyclables stream with the development of new technologies.

Composting involves conversion of the organic materials in municipal solid waste to compost, a soil conditioner, through the exploitation of bacterial and fungal decomposition. The quality, and therefore use, of compost is limited by the types of materials used in the composting process. Any organic waste is compostable, the principal ones in the waste stream being soiled paper and cardboard (38%), yard waste (13%), and food waste (10%) (Franklin Associates 1998). Because much of the paper is also recyclable, and paper makes up a considerable fraction of the waste stream, there is considerable overlap between the potentially recyclable and compostable fractions. (In fact, paper waste, can be prevented, recycled, composted, incinerated, or landfilled.)

Some methods for increasing recycling and composting rates include the following:

- Economic incentives, such as paying people and businesses for recyclables and compostables or, not charging for collection of recyclables and compostables.
- Changes in packaging design to increase the availability to consumers of recyclable packaging and recyclable products.
- Consumer education designed to motivate people and businesses to separate products and packaging for recycling and organic materials for composting. Also, to purchase products and packaging made from recycled materials and purchase products and packaging that are themselves recyclable.
- Mandatory recycling requirements (for residential, institutional, and commercial sectors) with penalties for noncompliance.
- Economic incentives for manufacturing of products and packaging to use secondary (recycled content) materials as feedstock.

Sizing An Incineration Facility

Deciding whether to use incineration as a municipal solid waste-management option within a community involves the difficult process of weighing economic, social, and public-health considerations. One problem is deciding the amount of incineration capacity that is desired, given the lifetime of typical incinerators (20-30 years); the need to recover construction and operating costs;

and the concern that too large an incinerator might distort a community's options for source reduction, reuse, recycling, and composting (for example, through contractual requirements with the incinerator builder or operator).

Sizing an incineration facility in this era of changing waste stream characteristics and changing waste-management programs is not straightforward. Municipal solid-waste incinerators designed in the 1980s to burn 100% of the waste stream would be expected to see at least some decline in demand (about 30% under current conditions) as reuse, recycling, and composting took place during the 20-30 years of their design life. From a national perspective, recycling and composting alone accounted for about 27% of the waste stream in 1998 and it continues to grow. But that percentage is more in the many areas that have local or state requirements for higher recycling, composting, and reduction rates. Some states (such as California, New Jersey, and New York) have set goals of 50-60%, and a number of localities (such as Minneapolis and Newark) have reduced their disposal rates by 50% through waste stream reduction, recycling, reuse, and composting (e.g., Sudol 1994). The ultimate potential for all four methods could be well over 80% for some places, assuming intensive programs designed to address as many categories of materials, products, and packaging as possible and to educate the general public and businesses to the greatest extent feasible. That potential was determined through an extensive 46-material waste-composition study done for New York City in 1989, which suggested that 80% of that city's waste stream was theoretically recyclable or compostable and some additional quantity was reusable or repairable. Roughly half, or 40% of the waste categories, were recyclable in the city's early curbside-collection program (DOS 1992); other considered categories fell under its intensive recycling program. New York City now recycles newspapers, magazines, catalogs, phone books, corrugated cardboard, milk and juice cartons, household metallic items, plastic containers, glass containers, metal cans, and aluminum foil.

THE DEVELOPMENT OF POLLUTION PREVENTION, COMBUSTION CONTROLS, AND EMISSION CONTROLS

Waste-incineration technology, in general, and emission control, in particular, have improved substantially over the past years. The technologies and practices for controlling and processing the waste stream, for incinerating waste, and for controlling and managing the emissions and ash output have changed substantially in the last two decades. Today, in many cases, incineration takes place in the context of waste prevention (to reduce both the volume and the toxicity of the waste generated), recycling, and composting. Such activities affect the types and quantity of wastes incinerated and emissions generated.

Before the era of recycling, waste prevention, and emission controls, a typical large incinerator might be fed a heterogeneous mix of unprocessed municipal solid waste and lower-volume waste streams from different sources (such as medical waste). The combustion process was largely unregulated with respect to temperature and oxygen control; consequently, waste was often not completely burned. There were considerable emissions of air pollutants and uncontrolled handling of the ash residue.

The principal products of waste incineration are carbon dioxide (CO_2) and water vapor, as for almost all other combustion processes, because the major process occurring is oxidation of the carbon and hydrogen in the waste. Also in common with other combustion processes, incineration produces byproducts, such as residual (bottom) ash and fly ash (from incombustible materials), and trace organic and inorganic compounds in the exhaust gases. The composition of these residues is determined by the composition of the incinerated waste stream, by the combustion process, and by reactions occurring in the waste gases after combustion. Because the input wastes may have higher concentrations of nonfuel components (including metals, chlorine, sulfur, and nitrogen compounds[3]) than fuels or biomass, there is more solid residue (bottom ash and fly ash), and the concentrations of trace compounds in the waste gases tends to be higher than for combustion with an equal heat output from fuel or biomass.

Pollution Prevention

A first step in controlling emissions is to minimize their creation in the incinerator. Measures for pollution prevention include reductions of pollutant precursors in the waste stream (for example, metals, chlorine, sulfur, and nitrogen). Such reductions can be brought about by means of product and packaging redesign, the reuse of products and packaging that contain precursors or catalysts for production of trace toxics, and recycling products and packaging, especially those containing such precursors. With smaller amounts of pollutant precursors entering an incinerator, their availability to produce air pollutants and ash in the incinerator is reduced; with larger amounts of such precursors entering an incinerator, greater and costlier effort is needed to prevent their escape to the air from the control devices.

Reduction of the quantity of toxic elements in the waste stream or reduction of elements that are transformed into, or catalyze production of, pollutants of concern upon incineration are often-overlooked components of source reduction. Heavy metals are found in batteries, pigments, leather, solder, and cans; chlorine is contained in PVC plastics and some bleached paper; polystyrenes might contain chlorofluorocarbons; sulfur is in tires and gypsum wallboard; and nitrogen is

---

[3] Metals may themselves be toxic, or may catalyze the production of toxic inorganic or organic trace compounds in flue gases, for example, toxic chlorinated compounds like the dioxins and furans. Sulfur in the input stream will produce sulfur oxides in the flue gases, and nitrogen compounds will produce nitrogen oxides.

in food and yard waste. Consumer products and packaging are also responsible for heavy-metals in incinerator ash and in leachate from landfills.

It is expected that most of the metals in the waste stream are contained in metallic items, such as cans. However, the heavy-metal and chlorine content in plastics and paper is especially relevant in a discussion of source reduction. The use of plastic and paper is among the fastest-increasing in the production of nondurables and packaging. Development of products and packaging that do not require any metals and other pollutant precursors, or that use fewer or less-toxic precursors (such as less-toxic substitutes for metals as pigments and stabilizers in plastics) would have the greatest effect on reducing the toxic precursors of the waste stream. But volume source reduction of packaging and nondurables in the waste stream would also result in reduction of some pollutant precursors.

Research by Franklin Associates for EPA (1994a) showed that after 80% recycling, lead-acid car batteries still contribute 66% of the lead in the waste stream, with contributions also from electronic items (from the solder that is used), leaded glass (particularly from TV and computer monitor tubes), leaded ceramics, and leaded plastics. Batteries are a similarly large contributor for cadmium: even after nickel-cadmium battery recycling, such batteries still contribute 54% of the cadmium in the waste stream. The second leading contributor of cadmium in the waste stream is plastics, with further contributions from electronic items, appliances, and pigments. Similar results were found for mercury. Batteries are the leading contributor of lead, and fluorescent tubes, thermometers, thermostats, and a few other categories contribute most of the rest.

However, the committee emphasizes that such data by themselves can not be used to predict the metals composition in the emissions of an incinerator combusting the waste. Minimizing the use of precursors in consumer products via design changes (and maximizing the recycling of materials that contain precursors) would reduce, but not eliminate, the environmental effects of solid-waste disposal by incineration and landfilling.

CONCLUSIONS

- The use of different waste management strategies can substantially affect the size and composition of the waste stream that is fed to an incineration facility. Waste reduction, reuse, recycling, and composting are all designed to reduce the quantity of material that must be incinerated and ultimately landfilled, and they are also likely to change the characteristics of the wastes that are incinerated and landfilled.
- A first step in controlling emissions is to minimize their creation in the incinerator. Measures for pollution prevention include reductions of pollutant precursors in the waste stream (for example, metals, chlorine, sulfur, and nitrogen) by means of product and packaging redesign, the reuse of products and packaging that contain precursors or catalysts for pro-

duction of trace toxics, and recycling products and packaging, especially those containing such precursors. Reduction of the quantity of toxic elements in the waste stream or reduction of elements that are transformed into, or catalyze production of, pollutants of concern upon incineration are often-overlooked components of source reduction.

3

Incineration Processes and
Environmental Releases

Waste incineration is one of many societal applications of combustion. As illustrated in Figure 3-1, the typical waste-incineration facility includes the following operations:

- Waste storage and feed preparation.
- Combustion in a furnace, producing hot gases and a bottom ash residue for disposal.
- Gas temperature reduction, frequently involving heat recovery via steam generation.
- Treatment of the cooled gas to remove air pollutants, and disposal of residuals from this treatment process.
- Dispersion of the treated gas to the atmosphere through an induced-draft fan and stack.

There are many variations to the incineration process, but these unit operations are common to most facilities.

This chapter addresses the combustion and air-pollution control operations commonly used in municipal solid-waste, hazardous-waste, and medical-waste incineration facilities. The intent is to identify, and briefly discuss, the design features and operating parameters that have the greatest influence on emissions. Waste storage, feed preparation, and gas temperature reduction (which may involve heat-recovery operations) are addressed to a lesser extent.

This chapter also addresses the air pollutants emitted from incineration processes that are of primary concern from a health effects standpoint (see Chapter

5). Formation mechanisms and emission-reduction techniques are discussed. Information is provided on stack emission rates during normal operation vs. off-normal operating scenarios such as startup, shutdown, and process upset conditions. Fugitive emissions, residual ash, and scrubber water handling are briefly discussed.

WASTE STORAGE, FEED PREPARATION, AND FEEDING

Table 3-1 lists the common waste storage, waste staging, feed preparation and feeding practices for municipal solid-waste, hazardous-waste, and medical-waste incinerators. These practices are highly waste- and facility-specific.

Proper design and operation of these "front-end" plant operations are important for several reasons:

- While the plant is operating, the potential for worker exposure to hazardous materials is the greatest in this part of the facility. Without appropriate engineered and administrative controls, including personnel protective equipment, operators can be exposed to hazardous dust and vapors.
- This part of the plant is the highest potential source of fugitive dust and vapor emissions to the environment, and the greatest potential fire hazard.
- Without proper waste preparation and feeding, the furnace combustion performance may be impaired.

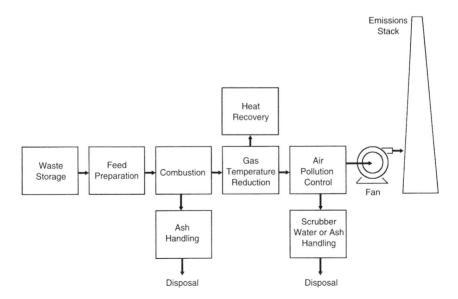

FIGURE 3-1   Typical waste-incineration facility schematic.

TABLE 3-1 Common Waste Storage, Feed Preparation, and Feeding Practices in Municipal Solid-Waste, Hazardous-Waste, and Medical-Waste Incineration Facilities

Waste Type	Storage/Staging	Feed Preparation	Feeding to Incinerator[a]
Municipal Solid Waste	• Pit • Tipping floor in piles	• Removal of oversize noncombustibles and special wastes (e.g., auto batteries) • In refuse-derived fuel plants, shredding, air classification, metals recovery	• Crane or bucket to vertical hopper/ feedchute • Front-end loader to hopper/ram feeder • In refuse-derived fuel plants, by pneumatic conveyance
Hazardous Waste—Solids	• Tarped rolloff bin • Drums	• Screening for debris removal • Shredding • Decanting liquids from drums	• Drop chute with double-gate airlock • Ram feeder • Auger feeder
Hazardous Waste—Liquids	• Tanks, segregated by chemical compatibility • Drums	• Aqueous/organic phase separation • Blending of compatibles • Preheating to reduce viscosity • Solids filtration, usually in-line	• Pump to burner or atomizing nozzle
Medical Waste	• Red bags or puncture-resistant boxes • Refrigerated room (Preferably)	• None	• Manual (especially for smaller units) • Ram feeder

[a] There are many variations in the methods of waste feeding in small or old municipal solid-waste incineration facilities.

There are many regulations and guidelines for the design and operation of waste storage, handling, and feeding systems. Organizations that develop such regulations and guidelines include the U.S. Occupational Safety and Health Administration (OSHA), U.S. Environmental Protection Agency (EPA), and National Fire Protection Association (NFPA).

COMBUSTION PROCESSES

General Considerations

Combustion is a rapid, exothermic reaction between a fuel and oxygen (O_2). In incineration applications, the fuel is predominately waste (although fossil fuels may be co-fired) and the oxygen source is air. Combustion produces many of the same stable end products, whether the material burned is natural gas, coal, wood, gasoline, municipal solid waste, hazardous waste, or medical waste. The flame zone of a well-designed incinerator is sufficiently hot to break down all organic and many inorganic molecules, allowing reactions between most volatile components of the waste and the oxygen and nitrogen (N_2) in air. The predominant reactions are between carbon (C) and oxygen, producing carbon dioxide (CO_2), and between hydrogen (H) and oxygen, producing water vapor (H_2O). Incomplete combustion of organic compounds in the waste feedstream produces some carbon monoxide (CO) and carbon-containing particles. Hydrogen also reacts with organically-bound chlorine to produce hydrogen chloride (HCl). In addition, many other reactions occur, producing sulfur oxides (SO_x) from sulfur compounds, nitrogen oxides (NO_x) from nitrogen compounds (and, a little, from the nitrogen in the air), metal oxides from compounds of some metals, and metal vapors from compounds of others.

The furnace is designed to produce good mixing of the combustion air and the gases and vapors coming from the burning waste. Nevertheless, in parts of the furnace where combustion is not complete (for example, near the walls of the furnace), combustible components of organic compounds are burned off, leaving the incombustible particulate matter known as fly ash entrained in the flue gas. The incombustible portion of the waste (known as bottom ash) is left behind.

Incineration facilities incorporate a number of general methods for ensuring proper combustion and reducing emissions. A steady situation with no major fluctuations in the waste-feed supply rate, combustion-air flows, or other incineration conditions promotes efficient combustion. Inefficient combustion can result in higher levels of products of incomplete combustion. Similarly, the more often a facility is started up and shut down (for maintenance or because of inadequate or varying waste stream volume), the more uneven the combustion and the greater the potential for increased emissions.

Optimal design and operation of a furnace requires attention to incineration temperature, turbulence of the gas mixture being combusted, and gas-residence time at the incineration temperature. To achieve efficient combustion, every part of the gas stream must reach an adequately high temperature for a sufficient period of time, and there must be adequate mixture of fuel and oxygen.

The temperature achieved is the result of heat released by the oxidation process, and has to be maintained high enough to ensure that combustion goes to completion, but not so high as to damage equipment or generate excessive nitrogen oxides. Typically, temperatures are controlled by limiting the amount of material charged to the furnace to ensure that the heat-release rate is in the desired range, and then tempering the resulting conditions by varying the amount of excess air.

Turbulence is needed to provide adequate contact between the combustible gases and oxygen across the combustion chamber (macroscale mixing) and at the molecular level (microscale mixing). Proper operation is indicated when there is sufficient oxygen present in the furnace, and the gases are highly mixed. Cool spots can occur next to the furnace's walls; where heat is first extracted from the combustion process. Such cool spots on walls are more substantial in waterwall furnaces than in refractory-lined furnaces.

A number of new design features and operating techniques have been adopted to increase temperature, extend residence time, and increase turbulence in waste incinerators in order to improve combustion efficiency and provide other benefits like improved ash quality. They include high-efficiency burner systems, waste-pretreatment practices such as shredding and blending, and oxygen enrichment in addition to the features and methods discussed below. Considerable attention has also been given to measurement and control of key process operating conditions to allow better control of the whole combustion process.

Furnace Types

Table 3-2 lists the types of furnaces used for municipal solid-waste, hazardous-waste, and medical-waste incineration. Municipal solid-waste furnace designs have evolved over the years from simple batch-fed, stationary refractory hearth designs to continuous feed, reciprocating (or other moving, air-cooled) grate designs with waterwall furnaces for energy recovery. The newer municipal solid-waste incinerators are waste-to-energy plants that produce steam for electric power generation.

The predominant hazardous-waste incinerator designs are liquid-injection furnaces and rotary kilns. Hazardous wastes are also burned in cement kilns, light-weight aggregate kilns, industrial boilers, halogen-acid recovery furnaces, and sulfuric-acid regeneration furnaces.

Medical wastes are burned in fixed-hearth incinerators, with the primary chamber operated in the starved-air mode (newer "controlled air" designs) or excess air mode (older Incinerator Institute of America (IIA) design). Both designs incorporate secondary, afterburner chambers. The smallest medical-waste incinerators are single-chamber, batch-operated devices.

TABLE 3-2   Furnace Designs for Municipal Solid-Waste, Hazardous-Waste, and Medical-Waste Incineration

Waste Type	Furnace Design Type	Application

Municipal Solid Waste	Mass burn
	Waterwall furnace
	Reciprocating or other
	   continuous moving grate
		Most newer municipal-scale
		   facilities

	Mass burn Refactory furnace lining Various grate or stationary hearth designs	Old or small facilities
	Refuse-derived fuel Spreader-stoker/cyclone furnaces	Few facilities in United States
	Fluidized bed	Foreign applications
Hazardous Waste	Liquid injection Rotary kiln with secondary combustion chamber	Common Common

	Fluidized bed	Few in United States. More
		   common for biosludge
		   incinerators

	Fixed hearth with secondary chamber	Mostly with plant trash co-feed
Medical Waste	Multiple chamber	Old IIA design for older facilities

	Controlled-air primary
	   chamber with afterburner
		Predominant design in United
		   States since 1970s

	Rotary kiln with afterburner	Few in United States

For illustrative purposes, the following discussion focuses on the basic design and operating considerations for one type of furnace.

Furnace Design Considerations For Municipal-Waste Incinerators

The design of the furnace is critical to optimal combustion. Furnace configurations depend on what they were designed to burn. Older designs, many of which are still used, do not generally permit as efficient combustion as newer designs.

Sizing

Poor combustor design can prevent stable, optimal combustion conditions. Sizing a furnace to match the quantity of waste fed to the incinerator is important with respect to temperature, turbulence, and time. If the heat input from the waste is too low for the furnace size, the temperature in the furnace may drop to such an extent that complete combustion is not achieved, particularly in water-wall furnaces. If the furnace is too small for the quantity of waste fed, the temperature will be high and there may be difficulty in supplying sufficient oxygen for complete combustion, and the quantities of unburnt residues might be increased.

Grates

In older incinerator systems, traveling grates simply transported refuse into the combustion zone. Newer grate systems are designed to agitate the waste in various ways, causing it to be broken into smaller pieces as combustion proceeds. This process permits exposure of a larger surface area of waste to air and high temperatures, assisting complete combustion by preventing unburnt material from simply being transported through on the grate.

Air-Injection Systems

For complete combustion to occur, air must be injected into the furnace in at least two locations: under the grate that carries burning waste (primary or under-fire air) and above the grate to mix additional oxygen with the combustion gases (secondary or over-fire air). Additional controls have been provided in modern municipal solid-waste incinerators to better regulate both the under-fire air at various points on the grate, depending upon burning conditions, and the over-fire air in response to temperature and heat transfer taking place in the furnace. In such advanced systems, primary air is injected into the drying, burning, and burnout zones of the grate, with a separate system for secondary air. Control may be effected by manual or automatic adjustments to dampers. The latter method is preferred, because it allows for automatic control loops with continuous monitoring devices. The temperature and oxygen needs of the furnace can be controlled by adjusting the quantity of primary and secondary air entering the furnace. In plants built before the middle 1980s, particularly those with holes in the furnace walls, the entry of primary and secondary air is not as well controlled, and the excess-air rates required for adequate combustion can be several times the amount that would be required with a more modern design. This can result in larger volumes of flue gas to be treated for contaminant removal, and reduced efficiency of utilization of the exhaust heat.

Arches and Bull Noses

To achieve complete combustion, gases produced must remain in the high-temperature zone of the furnace for a minimal residence time, usually 1-2 seconds. Achieving that residence time is usually accomplished by designing the furnace to retard the upward flow of gases, for example, by installing irregularities into the furnace walls. Modern facilities are configured to achieve improved combustion efficiency by using arches and bull noses. Arches, which are structures above the burning and burnout zones, are used to prolong the stay of combustion gases above the grate area. Bull noses are protrusions that are built into the furnace walls, usually near the point of injection of over-fire air, to upset the normal upward flow of the heated gases volatilizing from the burning waste. The induced gas redirection retards the movement of the combustion gases out of the furnace and promotes mixing with air.

Flue-Gas Recirculation

Flue-gas recirculation systems are used to recycle into the furnace relatively cool flue gas (extracted after the heat exchangers have reduced its temperature) that contains combustion products and an oxygen concentration lower than air. The process is used to lower nitrogen oxide formation by limiting the flame temperature and by slightly diluting the flame oxygen concentration. Care must be taken to ensure that not too much flue gas is recirculated, lest the combustion process be adversely affected.

Auxiliary Burners

Waste feedstock, particularly municipal solid waste, is heterogeneous, and its components, or even the whole waste stream, may vary in combustibility. That can make it difficult to maintain the minimal temperature necessary throughout a furnace. In modern combustors, maintenance of temperature can be aided by auxiliary burners that are typically set to come on automatically when the furnace temperature falls below a predetermined point; the threshold is usually set between 1,500 and 1,800°F at the location of the auxiliary burner, which is close to the chamber exit. The auxiliary burners are fed fossil fuels and are particularly intended to be used during system startup, shutdown, and upsets.

GAS-TEMPERATURE REDUCTION TECHNIQUES

The most common combustion-gas cooling techniques for incinerators are waste-heat boilers, and direct-contact water-spray quenches. Waste-heat boilers are employed on all new municipal solid waste-to-energy plants, many hazardous-waste incinerators, and some of the larger medical-waste incinerators.

Waste-to-energy plants have radiant waterwall furnaces as well as convective boiler sections. Hazardous-waste and medical-waste incinerators usually have just convective boiler sections, typically of fire-tube rather than water-tube design.

Most hazardous-waste and medical-waste incinerators, particularly the smaller units, do not have heat-recovery boilers. Combustion gases are quenched by water sprays atomized into the hot gas flow. Other, less common, gas-temperature reduction methods include air-to-gas heat exchangers and direct gas tempering with air.

Gas cooling techniques are integral to incineration system design, and can be important with respect to emissions of certain pollutants. As discussed later in this chapter, emissions of mercury and dioxins and furans can be affected by the rate of gas cooling and the air pollution control device (APCD) operating temperature. Dry APCDs, including scrubbers and particulate control devices, achieve the highest degree of reduction of mercury, dioxins and furans, and acid gases when flue-gas temperatures are lowered to about 300°F or less at the APCD inlet.

AIR-POLLUTION CONTROL TECHNIQUES

Historically, incinerator APCDs were designed to remove two classes of pollutants which are particulate matter and acid gases. More recently, some method for improving the removal of dioxins and/or mercury is considered necessary. Also, as discussed in Chapter 6, NO_x emission limits have been established for some incinerators. In several instances in European plants, increasingly stringent regulations have resulted in use of more than one particulate-control device or more than one type of scrubber in a given incineration facility, and emissions have typically been reduced more than would be expected with the single device alone.

Modern municipal solid-waste incinerators in the United States are equipped for particulate, acid gas, and, in many cases, dioxin and mercury removal. These municipal solid-waste incinerators typically employ fabric filters or dry electrostatic precipitators (esp) for particulate removal. ESPs became common in the 1970s. In the 1980s, fabric filters, also known as baghouses, started to replace, or be used in tandem with, ESPs as the preferred design for particulate removal because of their improved capacity for filtering finer particles. Spray dryer absorbers and dry-lime injection systems are used for acid gas—HCl and sulfur dioxide (SO_2)—removal. Dry powdered activated carbon injection systems provide dioxin and furan and mercury removal.

Many small old municipal-waste incinerators do not have effective air-pollution control systems. Some have only a particulate-control device, often a relatively ineffective one designed to meet old standards for emissions of particulate. Newer ones have both particle and acid-gas-control devices, such as wet scrubbers.

Hazardous-waste incinerators in the United States have traditionally used wet air-pollution control systems. Recently, however, there has been a trend toward fabric-filter systems (particularly in larger incineration facilities) because of their superior fine-particle-emission and metal-emission control efficiencies and their ability to produce a dry residue rather than a scrubber wastewater stream. Wet ESP devices may be favored in the future for existing wet APCDs to meet emission-control regulations.

Cement kilns and coal-fired boilers that burn waste as fuel have traditionally used either fabric filters or dry electrostatic precipitators as active control techniques. Passive controls include the neutralization of acid gases by cement materials and the recycling of cement kiln dust back into the process.

Particulate Collectors

Fine-particle control devices fall into three general categories, which are filtration collectors, including fabric filters (baghouses); electrostatic collectors, including dry and wet electrostatic precipitators (ESPs) and ionizing wet scrubbers; and wet inertial-impaction collectors, including venturi scrubbers and advanced designs that use flux-force condensation-enhancement techniques. When properly designed and operated, all of them are capable of effective fine-particle control, but they are not all equally effective.

Fabric filters are used at relatively low flue-gas temperatures (about 280-400°F). Flue gas containing particles passes through suspended filter bags. The particles suspended in the gas streams are collected on the filters and periodically removed and fed to a collection hopper.

Fabric filters are widely used today in municipal solid-waste incineration facilities, cement kilns, and lightweight-aggregate kilns because of their highly efficient collection of fine particles. They are used in a smaller number of hazardous-waste incinerators and medical-waste applications. The performance of fabric filters is relatively insensitive to particle loading, or to the size distribution and physical and chemical characteristics of the particles. They are limited to an operating temperature range between the gas dew point on the lower end and the bag-material thermal-stability limit on the upper end. A typical and practical operating-temperature for this technology in municipal solid-waste applications is about 300°F, but the best environmental performance is achieved at lower temperatures (to minimize dioxin and furan production within the APCD itself).

The primary factors affecting the performance of fabric filters are fabric type and weave, air-to-cloth ratio (gas flow rate to total bag surface area), cleaning method and frequency, bag cake formation and maintenance, and bag integrity with respect to mechanical, thermal, and chemical breakdown. The fabric type must be matched to the temperature range of the application and the chemical composition of the gas for good performance and bag longevity. Maximal air-to-cloth ratio for good performance is also a function of fabric type and weave. The method, intensity, duration, and frequency of the bag-cleaning cycles are important to maintain mechanical integrity of the bags and good cake formation. Good cake formation (as measured by baghouse pressure differential) is required for good performance of woven and felted bags; it is less critical for laminated membrane bags, which can function using surface filtration alone.

In properly designed and operated fabric-filter systems, maintaining bag integrity is the critical determinant of day-to-day performance. Bag integrity can be monitored via pressure drop, visual stack-opacity inspections, continuous on-line stack-opacity monitors, or other continuous monitoring techniques that use optical sensing or triboelectric sensing.

During shutdowns, bag integrity can be checked by visual examination of the clean-gas plenum for localized dust buildup. More-sensitive techniques involve the use of fluorescent submicrometer powder and black-light examination of the plenum.

Dry ESPs are widely used today in municipal solid-waste incineration facilities and on cement kilns and coal-fired boilers that burn hazardous waste. Wet ESPs are less widely used and are primarily in hazardous-waste incineration applications. Dry ESPs operate above the dewpoint of the gas. Wet ESPs are constructed from materials that resist acid corrosion and operate under saturated-gas conditions.

Dry ESPs are less effective than fabric filters for collection of submicrometer particulate matter (0.1-1.0 μm) but are nevertheless very effective collection devices. Their performance is influenced by a variety of design characteristics and operating conditions, including the number of electric fields used, charged electrode wire (or rod) and grounded collection plate (or cylinder) geometry, specific collection area (ratio of collection surface area to gas flow rate), electrode design, operating voltage and current, spark rate, collector cleaning method (to limit buildup or re-entrainment of dust), fluctuations in gas flow rate and temperature, particulate-loading fluctuations, particle-size distribution, and particle resistivity (less important for wet ESPs). Wet ESPs have superior submicron particle collection capabilities because they do not suffer rapping re-entrainment and dust layer back-corona problems associated with dry ESPs.

In a properly designed unit, the important monitoring and process-control measures are inlet gas temperature (dry ESPs only), gas flow rate, electrical conditions (voltage, current, and spark rate), cleaning intensity and frequency, and hopper-ash level (dry ESPs only).

Wet inertial-impaction scrubbers, primarily venturi scrubbers, have historically been the particulate matter control technology of choice for most hazardous-waste and medical-waste incinerators. They are inherently less efficient for submicrometer particulate matter than fabric filters or ESPs, but nonetheless can meet regulatory requirements in many applications.

The primary performance criterion for most wet inertial-impaction scrubbers is the gas-pressure drop, a measure of the energy applied to atomize scrubbing liquid and create fine droplets for particle impaction. For injector venturi scrubbers, the corresponding criterion is liquid-nozzle pressure drop. Other important design and operating characteristics are the liquid-to-gas ratio, inlet gas temperature (to avoid scrubber-liquid evaporation), solid content of recirculated scrubber liquid, mist eliminator efficiency, materials of construction to avoid corrosion and erosion, particulate loading, and particulate-size distribution. In a properly designed unit, the most-important monitoring and process control measures are pressure drops, liquid and gas flow rates, and liquid blowdown rate (blowdown is used to control solids buildup).

A few designs use steam injection or scrubber-liquid subcooling to enhance flux force and condensation. For those designs, steam-nozzle pressure and scrubber-liquid temperature are additional useful monitoring measures.

Acid Gas Scrubbers

A commonly used APCD for removal of acid gases is a packed-bed absorber. A scrubbing liquid is trickled through a matrix of random or structured packings through which the gas is simultaneously passed, resulting in gas-liquid contact over a relatively large surface area. The scrubbing liquid can be water or an alkaline solution, which reacts with the acid-gas constituents to form neutral salts. The wastewater discharge from the packed-bed absorber is a salt-water brine that must be managed properly. This effluent may contain unreacted acids, trace organics, metals, and other solids removed from the gas stream.

Packed bed absorbers have been used for decades in the United States, primarily in hazardous-waste and medical-waste incineration applications. They have been used in Europe for municipal solid-waste applications. The European installations include duel-stage wet absorbers, in which the first stage is operated with an acidic scrubber liquid and the second stage is operated with an alkaline scrubber liquid. Acid gases, such as HCl, that are highly water soluble are largely collected in the first stage. Acid gases, such as SO_2, that are not very water soluble are effectively collected in the second, alkaline stage.

The important design and operating criteria for wet acid-gas absorbers are gas velocity, liquid-to-gas ratio, packing mass transfer characteristics, pH of the scrubbing liquid, and materials of construction (to prevent corrosion).

In recent years, municipal solid-waste and a few larger hazardous-waste and medical-waste incineration facilities have used spray-dryer scrubbers for acid-gas control. The spray dryers use slurries of lime, sodium carbonate, or sodium bicarbonate as the alkaline reagent. The water in the atomized slurry droplets evaporates, cooling the gas, and the alkali particles react with the acid-gas constituents to form dry salts. The salts and unreacted alkali must be captured in a downstream fabric filter or electrostatic precipitator. Dry-injection scrubbers, which use an alkaline reagent without water, have also been used in recent years, although only rarely in United States municipal-waste, hazardous-waste, and medical-waste incinerators. They are typically not as efficient as spray-dryer absorbers at removing emissions. The important design and operating criteria for spray-dryer absorbers and dry-alkali scrubbers include gas temperature in the reagent contacting zone, reagent-to-acid gas stoichiometry, reagent distribution in the gas, and reagent type.

NO_x Controls

NO_x emissions can be reduced by combustion-furnace designs, combustion-process modifications, or add-on controls. Combustion-furnace designs that reduce thermal NO_x include a variety of grate and furnace designs, bubbling and circulating fluidized-bed boilers, and boiler designs, especially those with automatic controls, that permit flue-gas recirculation. Combustion-process modifications that reduce NO_x formation include controlling the amount of oxygen available during the combustion process, and operating within a specific temperature range. For minimizing NO_x production in the combustion process, it is recommended that there be a lower-oxygen condition just above the grates (or in the primary chamber of a dual-chamber facility) coupled with a higher excess-oxygen condition at the location of overfire air injection (or in the secondary chamber of a dual-chamber facility). Municipal solid-waste incineration facilities tend to create the most NO_x when furnace temperatures are higher than is necessary (higher than 2,000°F) to destroy products of incomplete combustion (PICs). To minimize NO_x formation, and the formation of PICs, the furnace should be operated within fairly narrow ranges of temperature and excess oxygen (9-12%) with turbulent (well-mixed) conditions.

Some NO_x formation is inevitable from nitrogen present in the fuel and from atmospheric nitrogen, and it may be necessary to use flue-gas controls to achieve further reduction of these emissions. Add-on NO_x flue-gas control systems include selective noncatalytic reduction (SNCR), selective catalytic reduction (SCR), and wet flue-gas denitrification.

SNCR reduces NO_x by injecting ammonia or urea into the furnace via jets positioned at the location where temperatures are about 1600-1800°F. In the proper temperature range, the injected ammonia or urea combines with nitrogen oxide to form water vapor and elemental nitrogen.

SCR operates at a lower flue gas temperature than SNCR, and in addition uses a catalyst. Ammonia is injected into the flue gases when they are at about 600°F, and the mixture is passed through a catalyst bed. The catalyst bed may be shaped in a variety of forms (honeycomb plates, parallel ridged plates, rings, tubes, and pellets), while the catalyst can be one of a variety of base metals (such as copper, iron, chromium, nickel, molybdenum, cobalt, or vanadium). Each combination has advantages and disadvantages with respect to catalyst-to-NO_x contact, fouling of the catalyst, and pressure drop through the catalyst. The biggest disadvantage of SCR for incineration applications is that the combustion gas must always be reheated to the required 600°F temperature range after cooling below this level to remove particulate matter. The catalyst beds required for SCR must be installed downstream of highly effective particulate removal devices to avoid fouling.

Wet scrubbers for NO_x removal are comparable to wet acid gas absorbers in configuration. They use strong oxidizers in aqueous solution to convert NO to NO_2 (which is water soluble in caustic solution) or NO_3-(nitrate), which is water-soluble. The exact chemistries of these systems are considered proprietary by the vendors.

Carbon Adsorption and Other Dioxin and Mercury Removal Techniques

Carbon injection refers to the injection of finely divided activated carbon particles into the flue gas stream ahead of the particulate APCD. The carbon particles adsorb pollutants on their surface, and then the carbon particles are themselves captured in the particulate APCD. Activated carbon has a large surface-area-to-volume ratio, and is extremely effective at adsorbing a wide range of vapor-phase organic-carbon compounds, and also some other vapors (like mercury) that are otherwise hard to control. Maximum effective use of the technique requires optimization of the rate of injection of activated carbon (Brown and Felsvang 1991). Studies in Europe and practical experience in the United States and elsewhere indicate that this technique can substantially reduce emission of dioxins and furans and of mercury. Also, Lerner (1993) reported that cadmium chloride is effectively removed from a flue gas stream by using activated carbon.

Dioxins and furans are removed along with mercury by injection of powdered activated carbon in a number of municipal-waste incinerators and a few hazardous-waste incinerators. This is a widely used form of emissions control in the United States and is quite effective for PCDD/PCDF removal. Removal efficiency is a complex function of carbon type, dosage, gas temperature, and gas-to-solid contact efficiency. Other add-on control technologies used outside the United States are adsorption in granular activated carbon or coke beds, catalytic oxidation in SCR units (which are also the most efficient NO_x controls demonstrated commercially), and injection of an inhibitor of dioxin-formation catalyst.

For high efficiency mercury removal, many municipal solid-waste incinerators and a smaller number of hazardous-waste and medical-waste incinerators have adopted powdered activated-carbon injection upstream of dry particle collection devices, usually fabric filters. As for dioxin removal, the effectiveness of powdered activated-carbon injection is determined by the carbon type, dosage, gas temperature, and gas-contact efficiency.

Other processes for mercury removal are granular activated-carbon filtration in fixed-bed reactors, selenium porous-media filter, gold-amalgamation filter beds, sodium sulfide injection, and wet scrubbing with mercury-reactive solutions. None of those techniques is used commercially in the United States, but fixed-bed carbon adsorbers used in Europe often produce mercury and dioxin removal efficiencies that are higher than conventional technologies used alone (e.g., scrubber/fabric filter with injection of activated carbon).

SYSTEM OPERATION

Many variables that affect incinerator operation are controlled by operators, so the combustion conditions that control emission rates may be substantially affected by operator decisions. Poor operator control either of the furnace (by permitting temperature or oxygen concentration to decrease) or of the stoking operation can cause reduced combustion efficiency. In most incinerators, mixing and charging of waste into the incinerator, grate speed, over-fire and under-fire air-injection rates, and selection of the temperature setpoint for the auxiliary burner are entirely or partially controlled by plant personnel.

In addition, the extent of emission control achieved by post-combustion APCDs depends on how the devices are operated. Suboptimal operation can be caused by poorly trained or inattentive operators, faulty procedures, and equipment failure. Operators must be attentive to the flow rate of waste into the incinerator and furnace operation so as to allow for effective function of APCDs.

Although some of the most-modern incineration equipment has been automated, there will always be a need for operators to deal with unexpected situations. In addition, automated equipment requires calibration and maintenance, and combustor parts can wear out or malfunction. Examples of what can go wrong include clogged air injection into the incineration chamber, fouled boiler tubes, a hole in the fabric filters, and a clogged scrubber nozzle.

Worker Training

Before the 1980s, there were no uniform national standards for hazardous-waste combustor maintenance or worker-training. The extent and adequacy of maintenance and worker training programs were company-specific and site-specific.

In 1986, the Occupational Safety and Health Administration 1910 regulations were promulgated, requiring worker training in hazardous-material management. Classroom training courses are now required for hazardous-waste workers at remediation sites and plant facilities. Annual refresher courses are required, as is supplemental training for supervisory personnel.

Resource Conservation and Recovery Act (RCRA) regulations impose federal requirements for inspection plans and worker-training plans for all facilities that manage hazardous waste, including combustion facilities. The inspection plans address facility maintenance, leak inspections, and calibration schedules for monitoring equipment. The training plans are intended to address hazardous-material safety and facility operations.

The American Society of Mechanical Engineers has developed a certification guideline for hazardous waste-incinerator operators.

Monitoring and Data Collection

For the most recently completed waste incinerators, particularly hazardous-waste incinerators, environmental regulations have led to extensive monitoring of key incineration process conditions, including waste feed rates; feed rates of ash, chlorine, and toxic metals (determined by sampling and analysis of the waste stream); combustion temperatures; gas velocity (or gas residence time); facility-specific air-pollution control-system operating measures; and stack-gas concentrations of O_2, CO, total hydrocarbons, HCl, NO_x, and SO_x, and opacity (see Chapter 6). Computerized systems collect and record process data, automatically control such process conditions as combustion temperature (by varying fuel feed and air flow rates), and automatically cut off waste feeds if operating conditions stray outside limits set by permits. For example, a low combustion temperature or high stack-gas CO concentration might initiate an automatic waste-feed cutoff.

RCRA regulations for hazardous-waste incinerators require continuous monitoring of important air-pollution control-system operating conditions, including pressure drops across venturi scrubbers, pH of acid-gas absorber scrubbing solutions, voltage or power supplied to electrostatic collectors, and fabric-filter pressure drops or triboelectric sensor readings.[1] Stack-gas monitors are often used to monitor the performance of the air-pollution control system directly for such measures as HCl, SO_2, NO_x, and opacity.

With electronic transmission of such sensor outputs, the performance of the control and monitoring systems could be more-readily displayed and monitored. Reliable continuous emission monitors (CEMs) for dioxins and furans or for metals would be desirable, because automatic devices electronically linked to such devices (for example, to optimize the injection of alkaline and carbon reagents and water in the emissions control devices) could directly control those emissions of greatest potential health consequence. Such arrangements have been in use for continuous automatic control of acid gases for some time. CEMs for mercury have undergone in-use testing in Europe, for example see Felsvang and Helvind (1991).

---

[1] Triboelectricity refers to an electric charge that is generated by friction.

PROCESS EMISSIONS

The principal products of combustion are CO_2, water vapor, and ash, which are respectively oxidation-reaction products of carbon, and hydrogen, and non-combustible materials in the fuel. However, when the combustion reactions do not proceed to their fullest extent, other substances, some of which are potentially harmful, can be produced. The types and concentrations of contaminants in the waste stream (flue gas) flowing from any incineration process depend on the process type, the waste being burned, and combustion conditions. Such pollutants derive from three sources: they or their precursors are present in the waste feed, they are formed in the combustion process because of incomplete oxidation, or they are created by reformation reactions in the gas cooling or APCD.

As discussed in Chapter 5, the products of primary concern, owing to their potential effects on human health and the environment, are compounds that contain sulfur, nitrogen, halogens (such as chlorine), and toxic metals. Specific compounds of concern include CO, NO_x, SO_x, HCl, cadmium, lead, mercury, chromium, arsenic, beryllium, dioxins and furans, PCBs, and polycyclic aromatic hydrocarbons. In addition, the total quantities of particulate matter and acid particles (which may largely be liquids condensed after emission) that escape the APCD are also considered independently. The following discussion focuses on the source and control of the following pollutants: particulate matter, acid gases, mercury (Hg), lead (Pb), and products of incomplete combustion. They are used to represent the pollutants from incineration that are of concern for possible health effects.

Particulate Matter

Particulate matter consists primarily of entrained noncombustible matter in the flue gas, and the products of incomplete combustion that exist in solid or aerosol form (and discussed separately later). Particle concentrations in the flue gas in the absence of control devices have been found to range from 180 to more than 46,000 mg per dry standard cubic meter (0.08 to more than 20 grains per dry standard cubic foot).

Particulate matter from waste combustors includes inorganic ash present in the waste and carbonaceous soot formed in the combustion process. The inorganic-ash fraction of the particulate matter consists of mineral matter and metallic species. These materials are conserved in the combustion process and leave the combustion chamber as bottom ash or fly ash. Soot is a product of incomplete combustion that consists of unburned carbon in the form of fine particles or as deposits on inorganic particles. High-molecular-weight organic compounds condense on the surface of the particles, particularly on the carbon, downstream of the combustor.

There are four general methods for limiting particulate emissions from waste combustors

- Limiting the ash content of the waste feed via source control or selection.
- Designing and operating the primary combustion chamber to minimize fly-ash carryover.
- Designing and operating the combustion chamber(s) in accordance with good combustion practice to minimize soot formation.
- Using well-designed and well-operated fine-particle APCDs.

Source control of ash-producing waste constituents is an obvious method to reduce particulate emission, but it is impractical for most waste combustors. However, some incinerators and boilers burning liquid hazardous waste are able to meet particulate matter emission limits by stringent source selection alone.

The first three methods listed above are effective in reducing particle loadings in the combustion gas but are generally not sufficient by themselves to meet current and proposed maximum-available-control-technology (MACT) emission standards for particulate matter. Add-on particulate control is expected to be needed to meet the proposed MACT standards for waste incinerators.

Fine-particle control devices are in three general categories: filtration collectors, including primary fabric filters (baghouses); electrostatic collectors, including dry and wet electrostatic precipitators (ESPs) and ionizing wet scrubbers; and wet inertial-impaction collectors, including venturi scrubbers and advanced designs that use flux-force condensation-enhancement techniques.

Acid Gases

Acid gases are flue-gas constituents that form acids when they combine with water vapor, condense, or dissolve in water. Acid gases include NO_x, SO_x, HCl, hydrogen bromide, hydrogen fluoride, and hydrogen iodide. HCl and SO_2 are often present in uncontrolled flue-gas streams in concentrations ranging from several hundred to several thousand parts-per-million-by-volume. The concentrations of NO_x, hydrogen fluoride, and sulfur trioxide are typically below several hundred parts-per-million-by-volume. Free halogens such as chlorine, bromine, and iodine can also be produced at low concentrations from combustion of wastes that contain compounds of those elements.

Emissions of SO_2, HCl, and the other halogen acids can only be controlled through the use of add-on APCDs, which have been previously described in this chapter.

There are two sources of NO_x from incineration (and other combustion) processes, commonly referred to as thermal NO_x and fuel NO_x. Thermal NO_x is formed by the reaction of nitrogen and oxygen in the combustion air. Its formation is favored by high temperature (i.e., flame zone temperature), relatively large residence time at this temperature, and higher oxygen concentration.

Fuel NO_x is formed by the oxidation of chemically-bound nitrogen in the waste (or fuel). Conversion of bound nitrogen to NO_x is strongly influenced by the localized oxygen concentration; it is less sensitive to temperature than thermal NO_x formation. Fuel NO_x formation can exceed thermal NO_x formation by an order of magnitude in incinerators burning wastes containing bound nitrogen.

NO_x formation can be reduced, to a degree, by furnace design and combustion process changes as described earlier in the chapter. Add-on controls are required for more effective removal.

Mercury

Heavy metals in waste are not destroyed by incineration. Metallic elements with high vapor pressures, or with compounds that have high vapor pressures, can be converted to the vapor phase in the combustion chambers and tend to condense as the flue gas is cooled. They can adsorb onto fine (generally submicrometer) particles. It is likely that mercury remains in the vapor phase in the air-pollution control section of the incineration process, depending on temperature, and the same may be true for some of the more-volatile metal compounds.

Mercury emission from waste combustors is determined largely by the mercury feed rate and by whether mercury-specific APCDs are used. Virtually all mercury species found in wastes are volatile at combustion temperatures, so there is a high degree of partitioning to the gas phase, regardless of the chemical form of mercury or the combustion-system operating conditions. There is evidence that mercury is present primarily as elemental mercury vapor at incinerator combustion temperatures. The rate of cooling in the air pollution system and the HCl/Cl_2 concentrations in the gas affect the conversion of elemental mercury to water-soluble mercuric chloride (Gaspar et al. 1997; Chambers et al. 1998; Gaspar 1998).

Mercury emission has been limited through operator control of waste feed rates. Conventional APCDs—such as fabric filters, ESPs, inertial-impaction scrubbers, and other wet scrubbers are at best only partially effective for mercury removal at normal operating temperatures. Traditional wet-scrubber APCDs have provided moderate (20-90%) mercury control efficiencies. The most-modern facilities use powdered activated-carbon injection into the gas stream for mercury removal. The best performances of conventional APCDs are typically those of wet scrubbers operating at saturation temperature or lower. Lower scrubber-water temperatures lead to vapor condensation, and reduced mercury vapor pressure. Soluble forms of mercury, such as $HgCl_2$, are preferentially removed in wet scrubbing systems.

For high efficiency (>90%) mercury removal, many municipal solid-waste combustors and a smaller number of hazardous-waste and medical-waste incinerators have adopted powdered activated-carbon injection in tandem with alkaline reagents upstream of dry particle collection devices, usually fabric filters.

Lead

Lead (Pb) emissions from waste incinerators are influenced by the concentration of Pb in the waste feed, the chemical form of Pb, the physical matrix of the waste, the degree of ash carryover from the primary combustion chamber, thermal conditions in the primary and secondary combustion chambers that affect Pb volatilization, and the air-pollution control system efficiency for fine-particle removal from the gas. The method of feeding waste to the combustor chamber (in batches vs. continuous feeding) can have an indirect effect on Pb emissions.

The concentration of Pb in the waste is important because Pb is conserved in the combustion process; all the Pb fed to the combustor exists with the bottom ash, is collected as fly ash, or is emitted as fine particles in the stack gas.

The chemical form of Pb, the feed location and physical waste matrix, and local temperature in the combustion system are important because they affect the extent to which Pb is vaporized in the combustion process. Volatile forms of Pb, such as $PbCl_2$, might vaporize completely in the combustion process, whereas nonvolatile species, such as PbO, tend to partition to the bottom ash in the primary combustion chamber. Pb in liquid wastes fed through burners is exposed to flame temperatures and is, thus, more likely to vaporize than Pb in solid wastes. Pb in combustible solid wastes (e.g., paper or plastics) will vaporize to a greater extent than Pb in mostly noncombustible items, such as glass. The combustion-chamber temperature profile also affects the vapor pressure and degree of volatilization of the Pb species.

The extent of Pb vaporization in the combustion process is important because it affects the distribution of Pb among the fly-ash particle-size fractions. Pb that does not vaporize during combustion either partitions to the bottom ash or carries over as fly ash with a particle-size distribution characteristic of the incoming waste material. Pb that does vaporize, however, recondenses in the cooler downstream air-pollution control environment and adsorbs to the finer particles. The finer particles are more difficult to remove from the gas. Thus, Pb-removal efficiency tends to be lower than the overall particle-removal efficiency. The behavior of Pb and other metals in the combustion environment has been extensively studied by EPA and others (Campbell et al. 1985; Barton et al. 1987, 1990, 1996; Fournier et al. 1988; Fournier and Waterland 1989; Carroll et al. 1995).

The design and operation of the primary combustion chamber as they affect ash carryover and the design and operation of the APCD also influence Pb emissions. The principles are the same as those described earlier for particulate-matter emission control.

In summary, there are four general methods for limiting Pb emissions from waste combustors:

- Limiting the Pb content of the waste feed via source control.
- Designing and operating the combustion process to minimize Pb vaporization.
- Designing and operating the primary combustion chamber to minimize fly-ash carryover.
- Using well-designed and properly operated APCDs.

From a practical standpoint, the second method is likely to be the most difficult to implement because the objective of the combustion process is to burn all the waste completely. The most-reliable methods of limiting Pb emissions are source control and good particulate APCD performance.

Products of Incomplete Combustion

Organic and inorganic substances that are broken down into free-radicals (molecular species possessing an unpaired electron) in the combustion unit sometimes do not combine with oxygen or hydroxyl radicals and instead combine among themselves to form many organic compounds. Most of these compounds can be destroyed in the postflame zone of a well-designed incineration system. Such compounds that are not combusted and released into the exhaust gas are called products of incomplete combustion (PICs). PIC emissions heavily depend on combustion conditions, which, in turn, depend on the design and operation of the combustion device. Depending on the temperature, some of the heavy organic constituents can condense onto fine particles. Examples of PICs are CO and trace organic chemicals. (The latter can also be remnants of the original feed stream.) PICs include simple compounds (e.g., methane, ethane, acetylene, and benzene), dioxins and furans, partially oxidized organic compounds (e.g., acids and aldehydes), and polycyclic aromatic hydrocarbons.

Dioxins and Furans

As discussed in Chapter 5, dioxins and furans are the most-hazardous organic PICs that have been found in the flue gas of any combustion device. ("Dioxins and furans" refers collectively to polychlorinated dibenzodioxins (PCDDs) and polychlorinated dibenzofurans (PCDFs)). For poorly designed and poorly operated incineration facilities, the flue-gas dioxin and furan concentrations can be much higher than those generated by typical combustion devices. The polybrominated analogues have also been found in incineration emissions (see for example, Sovocool et al. 1989).

Modern incinerators produce dioxins and furans from three points in the process: stack-gas emissions, bottom ash, and fly ash. Often, bottom ash and fly ash are mixed for waste management purposes, but they may contain different amounts of dioxins and furans. With the exception of a few older wet-scrubber units, most municipal solid-waste incineration facilities are able to achieve zero discharge with respect to aqueous waste, so there are no major contaminated waste water streams.

Three possible sources of dioxin and furan emissions are the following: (1) uncombusted components of the original fuel (dioxins and furans are present in the materials that are thermally treated, and some quantity of this material survives thermal treatment); (2) formation from precursor compounds (dioxins and furans are formed from the thermal breakdown and molecular rearrangement of particular precursor compounds); and (3) de novo synthesis (dioxins and furans are synthesized from a basic chlorine donor, a molecule that takes chlorine to the predioxin molecule, and the formation and chlorination of a precursor) (EPA 1994b).

All types of organic chemicals, including polychlorinated dioxin and furans, can be destroyed under high-temperature oxidizing conditions. Destruction can occur at around 1800°F or higher if oxygen and organic molecules are well mixed as in practical combustion devices. Destruction of polychlorinated dioxins and furans present in the waste feed stream can take place at temperatures as low as 1350°F if oxygen and organic molecules are perfectly mixed (Duvall and Rubey 1977; Dellinger et al. 1984). However, dioxins and furans are also produced within the incineration process from precursors that are not destroyed below 1,800°F. Lahl et al. (1990) suggest that, although dioxins and furans may be present in the incoming mixture, most of the dioxins and furans in the exhaust gases are the products of formation within the incinerator and not persistence of the compounds present in the waste stream.

It is known that the presence of catalytic metals (e.g., copper, nickel, zinc, iron, and aluminum and their salts) and the temperature range of 450-750°F can promote dioxin and furan formation (e.g., Stieglitz and Vogg 1987; Vogg et al. 1992). Traditionally, many APCDs are operated within that temperature range to avoid acid-corrosion problems. Other requirements for dioxin and furan formation include prolonged gas-residence time in the stated temperature range, the presence of carbon as gaseous PICs or particles, and the presence of chlorine as HCl, Cl_2, or metal salt. Some types of organic compounds, such as chlorophenols and chlorobenzenes, tend to act as precursors for this type of secondary dioxin and furan formation. There is evidence that sulfur and ammonia can inhibit dioxin and furan formation.

As noted above, three sources have been proposed for the dioxins and furans found in the products of combustion. In addition, a substantial amount of research has been performed on effects of combustion conditions, facility configuration, waste stream composition, and pollution-control equipment. Siebert et al. (1988) investigated various factors associated with the operation of municipal solid-waste combustors and found APCD outlet temperature, presence of acid-gas controls, and the startup year of the facility to be the most-important determinants of dioxin and furan formation. Fangmark et al. (1993) studied the effect of bed temperature, oxygen concentration, variations in HCl and water, and temperature and residence time in the postcombustion zone on dioxin and furan formation and concluded that postcombustion temperature was the most important. A study conducted for the American Society of Mechanical Engineers, ASME (1995), indicated that there was no statistically significant cross-incinerator correlation between chlorine content of the waste stream fed to incinerators, and the dioxin and furan concentration in the emissions of those incinerators. Numerous factors have been associated with dioxin and furan formation, including the presence of particulate carbon, metal catalysis, combustion efficiency, temperature, and presence of precursors. The only consensus at this point seems to be that good combustion efficiencies and low postcombustion temperatures reduce the secondary dioxin formation.

Dioxin and furan emissions can be controlled through good combustion practice and rapid cooling of the combustion gas to air-pollution control system temperatures (generally ranging from 285°F to 300°F). Rapid combustion-gas cooling is inherent in many wet-scrubbing system designs, except for units equipped with waste-heat boilers. A number of hazardous-waste incinerators equipped with wet scrubbers might meet regulatory standards without other add-on control.

For cement kilns, analysis of the characteristics traditionally used to measure combustion efficiency (CO and total hydrocarbons) indicates that there is no substantial relationship between good combustion practice and dioxin emissions (CKRC 1995). There are two possible reasons for that. It is likely that the total hydrocarbons and CO are associated with the raw-mineral feedstock, rather than the fuel, and CO can result from nonequilibrium conditions in the kiln due to high combustion temperatures.

Dioxins and furans, as well as mercury, are removed by injection of powdered activated carbon in a number of municipal-waste incinerators and a few hazardous-waste incinerators.

STACK EMISSION RATE INFORMATION

Normal Operation

For several types of incinerators, measured emissions have been compiled for the purposes of selecting allowable emissions under regulatory standards. EPA compiled a database (dated February 1996) containing the results of hazardous-waste combustor trial burns and facility operating and design characteristics as part of the development of the April 1996 proposed "Maximum Achievable Control Technology" (MACT) standards for hazardous-waste combustors

(Fed. Regist. 61(April 19):17358). The database contained information from hazardous-waste facilities in three source categories: incinerators, cement kilns, and lightweight-aggregate kilns. The database also contained data on boilers, although the last were not subject to the proposed rule. However, test data are not available for all pollutants from all of these sources. The database was updated in December 1996 to correct entries and add new test data (Fed. Regist. 62(Jan. 7, 1997):960), so that it contains information on 122 incinerators, 43 cement kilns, and 17 lightweight-aggregate kilns. Not all these facilities burn hazardous waste, so not all were used in setting MACT standards (http://www.epa.gov/epaoswer/hazwaste/combust/v1 4tsds.htm gives access to a Portable Document File version of the database; the committee believes that a version in Paradox database format is also available). This test database remains the most extensive published source of emissions data for hazardous-waste combustors in the United States. However, there are certain limitations to these data that should be noted.

All these data are the result of discrete stack-sampling events, not continuous emissions monitoring that would reflect day-to-day operation. There is no reliable representative data base of continuous emissions measurements for any of the pollutants examined here.

Many of the emissions data are from trial burns, which do not reflect typical day-to-day operation. Trial burns of hazardous-waste incinerators are intended to establish operating permit limits as well as to measure emissions performance. To meet this purpose, trial burns are usually conducted at extreme combinations of operating conditions, such as minimum combustion temperature for organics emission testing; maximum combustion temperature for metals emission testing; minimum combustion residence time and maximum gas flow rate; maximum feedrates of ash-bearing waste, halogens, and metals; and worst-case air pollution control system operating conditions. As a result, the emissions data in the database may overstate normal operating emissions. Conversely, trial burns are likely to be better controlled and more highly supervised than the day-to-day operation. As a result, upset conditions may be less prevalent during the stack-sampling events, and such events are not characterized by this EPA data base.

The database was primarily compiled to evaluate the range of stack-gas concentrations found at hazardous-waste incinerators. Although there is sufficient information to estimate total emission rates, there is no information recorded on the subsequent efficiency of dispersion of those emissions (which is facility-specific, and not usually recorded in typical emission test reports), so that it is not possible to reliably estimate resultant population-exposure concentrations.

For medical-waste incinerators, EPA has estimated emission factors based on a limited number of emission tests as reported in a memorandum (EPA 1996a). This document cites the reports for the emissions tests used, but does not list the test results In addition, the stack-gas concentration information was given only in summary form in the report, although stack flow-rates are given for some facilities. As a result, it is impossible to estimate facility-specific total emission rates or resultant population-exposure concentrations, although an attempt was made to estimate total national dioxin emissions using these (and other) data (National Dioxin Emissions from Medical-Waste Incinerators, Item IV-A-7 in docket A-91-61 at http://www.epa.gov/ttn/uatw/129/hmiwi/rihmiwi. html).

For municipal-waste incinerators, EPA has summarized stack-concentration test data for U.S. incinerators from a total of 104 reports dated 1987-1991 in a 1993 document "Emission Factor Documentation for AP-42, Section 2.1, Refuse Combustion." (available at http://www.epa.gov/ttn/chief/ap42c2.html) Five incinerator designs (waterwall, refuse-derived fuel, modular starved-air, mass burn-refractory wall, and mass burn-rotary waterwall) are represented, and various control technologies are separately evaluated. In connection with proposed MACT rules, EPA compiled data on U.S. municipal incinerators in Dockets A-89-08, A-90-45, and A-97-45 (see http://www.epa.gov/ttn/uatw/129/medical wastec/rimedical wastec.html). The EPA presented to the committee an update to January 1995 on the stack-gas concentration information for dioxin (Compilation of MWC Dioxin Data, Office of Air Quality Planning and Standards, July 27, 1995). That update included information on the latest test reports for 122 units at 71 facilities (there were approximately 160 facilities operating at that time); although the data were obtained by telephone and so may suffer from some quality control problems; and it appears that information for some facilities was averaged across multiple units; and some units had been modified specifically to reduce dioxin formation after the date of the last available test.

Stack-gas concentrations for dioxins (total tetra through octa CDDs and CDFs—toxic equivalent (TEQ) values can be obtained approximately by dividing these values by 50) spanned approximately a 20,000-fold range in 1995. Figure 3-2 shows the distribution of stack-gas concentrations (ng/dscm at 7% oxygen) for the 122 units mentioned in the EPA update to the committee (and also shows where the proposed MACT standards would fall).

This large range in stack-gas concentrations is apparently due to dioxin formation within APCDs if the temperatures range from 450 to 750°F. The range of stack-gas concentrations would be even larger than shown were it not for some corrective actions already taken by 1995 and reflected in the test information shown in the figure, and further actions were already agreed at that time for the highest emitters. For example, at the Pulaski, MD facility, 1993 tests showed concentrations of 3,313 to 9,045 ng/dscm in the five units. Interim measures (principally modification of water sprays to reduce the gas temperature into the ESPs, together with modification of combustion conditions) reduced the concentrations to 37 to 1,500 ng/dscm (reductions of approximately 4-fold for 4 units, and 240-fold for the fifth), and then-current regulations required a reduction to less than 60 ng/dscm by 1996.

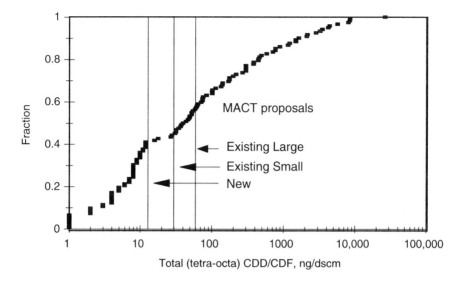

FIGURE 3–2   Distribution of stack-gas concentrations (ng/dscm at 7% oxygen) for 122 municipal solid-waste combustor units.   The figure shows where the MACT standards would fall.

Similarly, initial tests at the Norfolk, VA facility showed dioxin concentrations of 21,129 to 42,995 ng/dscm before water sprays were installed to reduce temperatures to the ESPs, lowering concentrations to the range of 1,640 to 4,210 ng/dscm.   One unit had already been retrofitted with a spray dryer/fabric filter APCD by March 1995, with all others to be retrofitted by 1996.

The reduction in stack-concentrations occurring after retrofit can be illustrated by the Detroit, MI facility.   Initially all three units were equipped with ESPs, and a retrofit was initiated to spray dryer/fabric filters in 1987.   In 1993 and 1994, the two retrofitted units showed stack concentrations in the range 2 to 10 ng/dscm.   The third, nonretrofitted but otherwise similar unit, showed a stack concentration of 2,851 ng/dscm in 1993—the unit was to be retrofitted by 1996.

Webster and Connett (1998) evaluated the emissions from 81 (of about 160) municipal-waste incineration facilities over the period 1985 to 1995 using the same EPA memorandum augmented with some additional individual test reports.   Their calculations confirmed the large range of total emissions from different facilities, the importance for national emission estimates of the largest emitters, and the large effect on such estimates of reducing emission rates for the large emitters.   Retrofit or closing of several incinerators indicated a substantial decrease in total atmospheric emissions of dioxins from municipal-waste incinerators at the end of that period.

Off-Normal Operation

Stack-gas testing is usually performed under relatively steady-state, relatively normal conditions. For hazardous-waste incinerators, stack tests required in the permitting process are designed to be at the outer limits of normal operations, an approach that might result in higher-than-average emissions. However, there is always the option for stack testing under normal operating conditions (i.e., during normal plant operations), and this option has been used to develop emission estimates for use in evaluating average emissions for risk assessment purposes. Both types of testing are likely to miss periods of off-normal operation, including upsets, malfunctions, startups, and shutdowns. The last three terms have regulatory definitions (40 CFR 60), malfunctions specifically referring to sudden and unavoidable failures (not caused in whole or part by poor maintenance, careless operation, or other preventable upset conditions or preventable equipment breakdown).

Emissions during startup and shutdown are likely to be different in nature from those during regular burning of waste. For hazardous-waste and medical-waste incinerators, at least, startup and shutdown periods (without malfunctions) are defined in regulations to include only periods when the waste is not being burned (using auxiliary fuels to bring the facility to operational temperature, for example). However, emissions might also differ for the periods just after the beginning of feeding of waste into the incinerators, because this will induce some variations in operating conditions. Upsets may include any variation from normal operational conditions, and may or may not affect emission rates. Various attempts have been made to evaluate the effect of upset conditions on emission rates.

The effect of transient combustion upsets was tested in a Dow Chemical Company hazardous-waste incinerator in Louisiana (Trenholm and Thurnau 1987) that was burning solids (ram-fed drums every 4 minutes into the kiln, alternating types of solids), organic liquids (continuous feed to kiln and secondary chamber), and aqueous liquids (continuous feed to the kiln). It was found difficult to induce upset conditions (CO levels did not change on spiking the drums with 10 gallons of volatile hydrocarbons, or suddenly increasing the liquid waste feed). The final method was to triple the feed rate (2 gals/min to 6 gals/min) of liquid organic waste to the secondary combustion chamber for 7 seconds. The transients did not change average process conditions, but CO spikes to 700 ppm were obtained, increasing the average CO from around 0-3 ppm to 10-15 ppm, with highly variable total hydrocarbons (barely increased from a baseline 0-8 ppm in one run, increased to 60-150 ppm in two other runs). Particulate matter concentrations increased on average approximately 2-fold, while average concentrations of individual volatile organic hydrocarbons varied both up and down in a compound-specific manner.

In a series of tests on the Marion County WTE, the EPA evaluated the effect of running at various operating conditions including low and high total air, low and high overfire air, and combinations of low load and high or low total or overfire air. Excess air varied approximately 2-fold from the baseline of about 72%, and CO stack concentration varied approximately 5-fold down and 1.15-fold up from the baseline of about 11 ppmv. Measured particulate stack concentrations were reduced around 25% under the off-normal conditions tested. Emissions testing was also performed on this facility during startup (beginning measurement when the waste ignited, and continuing for 4 hours) and shutdown (beginning 5 minutes before cessation of waste feed, and continuing for 3 hours, just after the forced draft fan was shut off). Baseline stack-gas total concentration of CDD/CDF was 2.2 ng/dscm (at 12% oxygen), with total concentrations of 3.47 ng/dscm during shutdown and 11.7 ng/dscm during startup. There was a considerable shift in congener distribution during these periods, with the corresponding 2,3,7,8-TEQs (I-TEF/87) being 0.063, 0.008, and 0.013, respectively.

Various tests have been performed on incinerators in addition to the empirical results found during the interim corrective measures described above. Most of these have been to evaluate the effect of process variations, rather than process upsets, but the results have implications for upsets. Six examples of such tests are summarized here.

During testing of the Prince Edward Island facility (Environment Canada 1985), low combustion temperatures were associated with increased dioxin emissions (see Appendix B, Box B-1). The Pittsfield, MA facility (NYSERDA 1987) was tested under variable conditions (see Appendix B, Box B-2). The results showed dioxin increasing with both too much and too little excess oxygen, that low primary combustion temperature substantially increased dioxin emission rates, and that high CO concentrations usefully indicated combustion conditions that also correlated with high dioxin concentrations. During testing at the Peekskill incinerator in Westchester County (NYSERDA 1989), two approximately hour long tests were performed during cold-start conditions (see Appendix B, Box B-3). Total CDD/CDF concentrations were 18-51 times baseline (normal operation) values at the ESP inlet, and 40-96 times normal values at the ESP outlet. CO was also elevated (5-57 ppmv normal, cold starts 114 and 180 ppmv at the superheater exit). A comparative report of these three early tests (Visalli 1987) stated that "test results indicate that levels of dioxin and furan in the flue gas entering a pollution control device are affected by different plant operating conditions if the conditions deviate sufficiently from normal operations," that furnace temperature can be used as a gross indicator of total dioxin and furan emissions, and that operating an incinerator at excess oxygen levels below about 5% may cause an increase in dioxin and furan emissions.

The Quebec City mass burn incinerator (Finkelstein et al. 1987) was tested under various operating conditions, some characterized as very poor (primary/secondary air ratio 90/10, high excess air of 115%), and poor (furnace temperature 850°C, 130% excess air, and primary/secondary air ratio 60/40) (see Appendix B, Box B-4). Three other operating condition combinations, under low, design, and high load, were designated as good. Dioxin and furan emissions "become exponential" and were correlated with over 125% excess air, and also with departure from full operating load.

The Oswego mass-burn facility was tested in groups of three runs (NYSER-DA 1990) to evaluate the effect of a clean combustion chamber (right after startup) versus end-of-campaign (just prior to maintenance shutdown), and (two groups of runs) the effect of secondary chamber temperature (see Appendix B, Box B-5). Low furnace temperatures were correlated with high dioxin and furan concentrations (5- to 6-fold increase) at the secondary chamber outlet and ESP inlet. The dioxin and furan concentrations were also highly correlated with high CO levels, particularly with upper percentiles of distributions of CO levels from a continuous emission monitor.

The Hartford refuse-derived fuel facility was tested to determine generation of trace organics and metals in the furnace under different process operating conditions (EPA 1994c) (see Appendix B, Box B-6). Steam flow rate (an indicator of load) and combustion air flow rates/distributions were the primary independent variables defining operating conditions as "good," "poor," and "very poor." Dioxins, furans, CO, total hydrocarbons, PCBs, chlorobenzenes, chlorophenols, and PAHs were measured. Multiple-regression models were developed to evaluate the effect of various continuously monitored emission and process parameters on dioxin emissions (prediction models) and the effect of various combustion control measures on dioxin emissions (control models). The best prediction model showed that CO, NO_x, moisture in the flue gas at the spray-dryer inlet, and furnace temperature explained 93% of the variation in uncontrolled dioxin emissions, with CO explaining 79% by itself. The best control model showed that refuse-derived fuel moisture, rear wall overfire air, underfire air flow, and total air explained 67% of the variation in uncontrolled dioxin emissions. Since CO was found to be such a strong predictor of dioxin emissions, the relationship was explored further. It was found that the fraction of time the CO level was over 400 ppm was quite strongly correlated to the amount of uncontrolled dioxins generated, particularly when examining only those runs where there was poor combustion.

In summary, these test results and empirical demonstrations, together with other lines of evidence (including other tests and laboratory demonstrations), show that dioxin and furan concentrations exiting the furnace are controlled by combustion conditions. Subsequently, dioxins and furans may be produced by reactions on surfaces in the flue-gas duct or in APCDs, with production rates increasing substantially above a certain temperature. Production of dioxins and furans during upset conditions are thus expected to rapidly increase outside a window of good-combustion conditions. Various monitors of these conditions (including CO emissions and temperatures throughout the flue-gas train) should thus correlate with dioxin and furan emissions, even during upset conditions.

FUGITIVE EMISSIONS

In most state-of-the-art municipal-waste incinerators, fugitive emissions, consisting of vapors or particles from waste tipping, waste feeding, incineration, and ash handling are mitigated by designing buildings to be under negative pressure. Air is drawn from the waste-handling areas into the combustion chamber, where it is mixed with the combustion gases. Potential fugitive emissions collected in this manner and drawn through the combustion chamber and emission-control devices leave the plant with odors virtually destroyed and dust removed by the particle-control devices.

Fugitive dusts can also be created in the bottom-ash pits and the fly-ash hoppers. Enclosed ash-handling areas are part of state-of-the-art municipal-waste incinerator designs, but older incinerators may not have such advanced enclosed ash handling. In the modern systems, emissions created in the ash-handling areas (bottom ash and fly ash) are drawn through the emission control devices so that workers are not unnecessarily exposed to dust from the ash. Such dusts, particularly fly-ash dusts from particulate APCDs, may be enriched in toxic metals and contain condensed organic matter.

At hazardous-waste incineration facilities, the most common fugitive emissions are (from liquid wastes) vapors from tank vents, pump seals, and valves; and (from solid wastes) dust from solid-material handling, together with possible fugitives from particulate APCDs. The magnitude of those emissions and their control mechanisms are similar to those in other process industries that handle hazardous materials and are therefore regulated under RCRA subpart BB. However, the high-temperature seals on rotary-kiln incinerators are a potential source of vapor and dust emissions peculiar to such incineration facilities; these emissions are controlled by maintaining a negative pressure in the kiln.

ASH AND OTHER RESIDUES

Types of Ash and Other Residues

Residues generated by incinerators include bottom ash, fly ash, scrubber water, and various miscellaneous waste streams.

Bottom ash is the remains of the solid waste that is not burned on the grate during the combustion process and consists of unburned organic material (char), large pieces of metal, glass, ceramics, and inorganic fine particles. Bottom ash is collected in a quench pit beneath the burnout section of the grate.

Fly ash is the solid and condensable vapor-phase matter that leaves the furnace chamber suspended in combustion gases and is later collected in APCDs. The APCDs in use since the middle 1980s capture a high percentage of the contaminants in the flue-gas stream. Fly ash is a mixture of fine particles with volatile metals and metal compounds, organic chemicals, and acids condensed onto particle surfaces. It can also contain residues from reagents, such as lime and activated carbon, themselves with condensed or absorbed contaminants. Fly ash is collected in hoppers beneath the APCDs.

Scrubber water is a slurry that results from the operation of wet scrubbers and contains salts, excess caustic or lime, and contaminants (particles and condensed organic vapors) scrubbed from the flue gas.

In addition, there are various other waste streams that may be generated by the incinerator. For example, waste-to-energy plants produce blow-down water from the heat recovery boilers; some municipal solid-waste incinerators recover small quantities of condensed metals (e.g., lead alloys) from parts of their flue gas system. The initial sorting of municipal-solid waste produces a stream of large items unsuitable for burning (such as whole refrigerators, gas stoves, and auto batteries).

In 1995, the International Ash Working Group reviewed the available scientific data and developed a treatise on municipal solid-waste incinerator-residue characterization, disposal, treatment, and use (IAWG 1995). It found that the different temperature regimes in a municipal solid-waste incineration facility impart different characteristics to the residues collected from the various operational steps in a facility. Its report concluded that the development of management strategies for municipal solid-waste incinerator residues requires knowledge of the intrinsic properties of the material, including the physical, chemical, and leaching properties.

Cement kilns burning hazardous waste are in a class by themselves. All cement kilns are major sources of particulate emissions and are regulated as such by EPA and the states. Kiln-exhaust gases contain large amounts of entrained particulate matter known as cement-kiln dust, a large fraction of which is collected in APCDs. The kiln dust so collected is generally recycled to the kiln feed. Under the current BIF regulation, residue generated primarily by the combustion of fossil fuels may be exempted as RCRA hazardous waste provided that the facility operator can demonstrate that such wastes are no different from normal process residues or that any change caused by the combustion of hazardous waste as supplemental material in the fuel will not cause harm to human health or the environment. Cement-kiln dust is in that category.

Ash Handling

Two concerns of on-site ash management at incineration facilities are the safety of workers and the possibility that fugitive ash will escape into the environment during handling or removal of the ash for disposal. Both concerns require that the ash be contained at all times both inside and outside the facility, as described above. In the facility, water is used to quench the ash, simultaneously reducing dust generation and minimizing the possibility of ash-dust inhalation or ingestion by workers. In modern systems, a closed system of conveyors to transport the ash from the furnace to trucks helps to minimize worker exposure. Although some facilities have partially closed ash-removal systems, few have completely enclosed ash-handling systems throughout the plant.

Ash and Scrubber-Waste Disposal

Fly ash from municipal-waste incineration is characteristically more likely than bottom ash to exhibit the toxicity characteristic as defined by the RCRA leaching test as a result of high concentrations of lead or cadmium. Since 1994, it has been required that municipal-incinerator ash be tested to determine whether it is hazardous. If it is hazardous according to RCRA definitions, it must be disposed of as hazardous waste.

All residues generated by hazardous-waste incineration, except waste burned for metal recovery, are considered hazardous waste. That stems from the "derived-from" rule, which states that residues generated by the treatment of hazardous waste remain hazardous until delisted. Ash from hazardous-waste combustion must be handled and disposed in a secure hazardous-waste landfill that is designed to ensure that there will be no groundwater pollution. Under some circumstances, the ash can be classified as nonhazardous after a comprehensive test procedure, as provided under RCRA regulations.

The most common management method for ash generated by municipal solid-waste incineration is landfill disposal, either commingled with municipal solid waste or alone in an ash monofill, although some ash is used in production on construction materials, roadbeds, or experimental reefs.

Dry and spray-dry scrubber waste is incorporated in the fly ash, because the APCD is where the injected material is collected. Wet-scrubber wastewater is discharged to on-site wastewater-treatment systems, or discharged to the municipal sewer, after whatever pretreatment is required by local regulations.

SUMMARY

The pollutants of concern including dioxins and furans, heavy metals (in particular, cadmium, mercury, and lead), acid gases, and particulate matter, either are formed during waste incineration or are present in the waste stream fed to the incineration facility.

Emissions of dioxins and furans result, in part, by the processes in the combustion chamber that lead to the escape of products of incomplete combustion (PICs) that react in the flue gas to form the dioxins. PICs are formed when combustion reactions are quenched or incompletely mixed. The combustion chamber for incineration must therefore be designed to provide complete mixing of the gases evolved from burning of wastes in the presence of air and to provide adequate residence time of the gases at high temperatures to ensure complete reactions.

The operation of the combustion chamber also affects the emission of pollutants, such as heavy metals, that are present in the waste feed stream. Such compounds are conserved during combustion and are partitioned among the bottom ash, fly ash, and gases in proportions that depend on the compounds' volatility and the combustion conditions. Mercury and its salts, for example, are volatile, so most of the mercury in the waste feed is vaporized in the combustion chamber. In the cases of lead and cadmium, the partitioning between the bottom ash and fly ash will depend on operating conditions. More of the metals appear in the fly ash as the combustion-chamber temperature is increased. In general, there is a need for the combustion conditions to maximize the destruction of PICs and to minimize the vaporization of heavy metals. It is also important to minimize the formation of NO_x (which is favored by high temperatures or the presence of nitrogen-containing fuels).

In addition to the composition of the waste feed stream and the design and operation of the combustion chamber, a major influence on the emissions from waste-incineration facilities is their air-pollution control devices. Particulate matter can be controlled with electrostatic precipitators, fabric filters, or wet inertial scrubbers. Hydrochloric acid (HCl) and sulfur dioxide (SO_2) can be controlled with wet scrubbers, spray dryer absorbers, or (to a lesser extent) dry-sorbent injection and downstream bag filters. NO_x can be controlled, in part, with combustion-process modification and with ammonia or urea injection through selective or nonselective catalytic reduction. Concentrations of dioxins and mercury can be reduced substantially by injecting activated carbon into the flue gas, or by passing the flue gas through a carbon sorbent bed, which adsorbs the trace gaseous constituents and mercury.

The application of improved combustor designs, operating practices, and air-pollution control equipment and changes in waste feed stream composition have resulted in a dramatic decrease in the emissions that used to characterize uncontrolled incineration facilities. For example, emission of dioxins from uncontrolled incinerators exceeded 200 nanograms/TEQ per dry standard cubic meter (200 ng/TEQ-dscm) in a number of commercial units. It has been reduced to below 0.1 ng/TEQ-dscm in many modern units. Rates of emission of mercury have decreased, at least in part, as a consequence of changes in the waste feed streams resulting from the elimination of mercury in some waste stream components, such as alkaline batteries.

To maximize combustion efficiency, it is necessary to maintain the appropriate temperature, residence time, and turbulence in the incineration process. Optimal combustion conditions in a furnace ideally are maintained in such a manner that the gases rising from the grate mix thoroughly and continuously with injected air; the optimal temperature range is maintained by burning of auxiliary fuel in an auxiliary burner during startup, shutdown, and upsets; and the furnace is designed for adequate turbulence and residence time for the combustion gases at these conditions. The combustion efficiency of an incinerator

---

BOX 3-1
Best Practices for Reducing Incineration Emissions[a]

- Screen incoming wastes at the plant to reduce incineration of wastes (such as batteries) that are noncombustable and are likely to produce pollutants when burned.
- Maintain a continuous, consistent thermal input rate to the incinerator to the extent possible. In municipal solid-waste facilities, optimize mixing of waste in pit or on tipping floor (to homogenize moisture and BTU content).
- Optimize furnace operation, including temperature, oxygen concentration, and carbon monoxide concentration. In municipal solid-waste incinerators, this can be done by optimizing grate speeds; underfire and overfire air-injection rates, locations, and directions; and operating auxiliary burners.
- Survey furnace emission-control devices and related equipment regularly to ensure that they continue to be operative and properly sealed and insulated.
- Select correct type of nitrogen-reducing reagent (either ammonia or urea) and optimize the injection rate and location, if add-on of NO_x control is required.
- In dry air pollution control systems, optimize flue-gas temperature in control devices (to minimize dioxin formation and to maximize condensation and capture of pollutants while avoiding gas dewpoint problems.
- Select correct alkaline reagent (e.g., lime slurry, dry lime, Na_2CO_3 or $NaHCO_3$) to maximize absorptive capacity and optimize injection rate and location.
- Optimize type of sorbent (such as carbon) used (to maximize adsorptive capacity) and optimize injection rate and location for removal of mercury and dioxins and furans.
- Optimize voltage and other electric conditions of an ESP (to maximize capture of particles).
- Optimize baghouse pressure drop, bag-break detection, wet-scrubber pressure drop, pH, and liquid-to-gas ratio.
- Maintain a maximum gas flow-rate limit to ensure adequate residence time in the combustion chamber and proper operation of the air pollution control equipment.
- Implement a training and certification program for plant operators.
- Inspect and calibrate continuous emission monitors and other process instrumentation.

---

[a] Optimization with respect to cost was not considered because it was not within the committee's scope of work.

---

thus depends both on the design of the furnace and on operating practices. Furthermore, adequate operator training and certification is needed with monitoring of performance conditions to ensure that emission targets are met.

The committee has identified specific best practices for reducing incineration emissions primarily from municipal solid-waste incineration; see Box 3-1.

CONCLUSIONS

- Waste-incineration technology and practice can be implemented under conditions that meet currently applicable and proposed emissions limits and other environmental regulatory constraints.
- Emission data needed to fully characterize environmental concentrations for health-effects assessments are not readily available for most incineration facilities. Such information is lacking especially for dioxins and furans, heavy metals (such as lead, mercury, and cadmium), and particulate matter. The variation of these emissions over short and long time frames needs to be taken into account to characterize environmental concentrations fully, but the data are not available. Variations over short periods can result from process upset conditions; variations over long periods can result from replacement of less-efficient incineration facilities with modern low-polluting units.
- The characteristics of incineration emissions and residual ash are affected by the wastes fed to an incineration facility, its combustion efficiency, and the degree of emission control of that facility.
- Improving the combustion efficiency of an incineration process by optimizing combustor operations will reduce the quantity of soot produced and will lessen the formation of PICs, such as dioxins and furans. However, one must take into account the potential to increase the heavy-metal content in the emissions due to volatilization resulting from the higher combustion temperatures needed to improve combustion efficiency.
- Emissions from incineration facilities are reduced by modifying operating characteristics—such as furnace temperature, air-injection rate, flue-gas temperature, reagent type, and injection rate, and by selecting optimal combustor designs and emission-control technologies.
- Use and continued calibration and maintenance of continuous monitors of emissions and process characteristics provide real-time feedback and facilitate maintenance of optimal operating conditions at all times by incineration operators.
- Continuous emission monitors (CEMs) for CO, O_2, SO_x, NO_x, and HCl are available and have been certified by jurisdictions in this country and in other countries. CEMs for particulate matter and total mercury are under development and are in the process of being certified.
- Emissions from incineration facilities can be reduced by choosing advanced combustion designs and emission-control technologies for the pollutant of concern and by having well-trained and certified employees who can help to ensure that the combustor is operated to maximize combustion efficiency and that the emission control devices are operated to optimize conditions for pollutant capture or neutralization.
- If emission rates are desired that are lower than current and proposed regulations require, incineration and emission-control technologies and

operating practices exist to reduce emissions further. Some modifications involve purchase of equipment, and others require greater use of reagent or changes in other process conditions. Others simply require vigilant monitoring and adjustment of process conditions.

RECOMMENDATIONS

- Government agencies should conduct studies of incineration facilities to characterize emissions and process conditions during startup, shutdown, and other upset conditions. Studies should consider variations in the waste's heating value over normal operating range, and variability over winter and summer conditions.
- Emissions from small, as well as larger, incineration facilities need to be evaluated. In conducting post-MACT assessments, EPA should evaluate all types of facilities, even those exempt from the regulations.
- Government agencies should encourage the development and adoption of continuous-emission-monitoring technology. These data should be made easily available to the public routinely. Experiences of other countries should be considered. Continuous emission monitoring of particulate matter and other pollutants of concern from incineration processes, such as mercury, should be implemented when practical and reliable techniques are available.
- Consideration should be given to establishing a certification procedure for municipal solid-waste incinerator control-room operators. Certification standards have been developed as part of the American Society of Mechanical Engineers, standards for qualification of resource recovery operators, medical-waste incinerator operators, and hazardous-waste incinerator operators. Renewal of certification should include retesting on new techniques, practices, and regulations.
- Government agencies should gather and disseminate information on the effects on emissions and ash as a result of various operating conditions, such as furnace and downstream flue-gas temperatures, reagent types and injection rates, and air-injection adjustments. Such guidance should show how specific emissions and ash characteristics are affected by modifying these process conditions. Government agencies should also compile and disseminate to the public information regarding new combustor designs, continuous emission monitors, emission-control technologies, operating practices, and source reduction/fuel cleaning/preparation techniques, including records of environmental performance and effect on emissions and ash as shown in pilot and full-scale tests.
- Emissions and facility-specific data should be linked to better characterize the contributions to environmental concentrations for health-effects assessments. Existing databases should be linked to provide easy access

to specific operating conditions of an incinerator, including temperature at inlet of air-pollution control device, exit-gas temperature, furnace temperature, stack height, stack diameter, flow rate, and temperature of the gases leaving the stack, injection rate of all reagents, local meteorological conditions, air-dispersion coefficients as a function of distance from the facility, and precise geographical location of the emission point (to within 10 meters). When it is appropriate, data should be standardized to 7% oxygen and measurements to units of dry standard cubic meters. All data collected should be easily accessible by the public.

4

Environmental Transport and Exposure Pathways of Substances Emitted from Incineration Facilities

The main pathway for pollutants to get into the environment from a waste-incineration facility is, as for many other sources, through emission to the atmosphere. A large number of substances have been detected—most of them at very low concentrations—in the gaseous and particulate emissions from waste incineration. Among the emitted pollutants are metals and other noncombustible matter; acid gases; and products of incomplete combustion that include a large number of organic compounds as well as oxides of nitrogen, sulfur, and carbon. These pollutants are partitioned among the gas and particulate phases of the stack emissions from an incineration facility. As the pollutants disperse into the air, facility workers and people close to a facility might be exposed directly through inhalation or indirectly through consumption of food or water contaminated by deposition of the pollutants to soil and vegetation. Other people can be exposed through a different mix of environmental pathways after the pollutants travel some distance in the atmosphere; go through various chemical and physical transformations; or pass through soil, water, or food. As part of estimating the amount of incineration-released contaminants that people are exposed to and the patterns of such exposure, investigators seek to track the concentration and movement of, and changes that occur in, the contaminants as they move through the environment from the incineration facility to a point of contact with people. Such information is also helpful in determining the contribution of incineration to the mix of environmental contaminants from all sources.

This chapter provides a review of the environmental dynamics of substances emitted from waste-incineration facilities and the pathways that could result in human exposure to such contaminants. The chapter is not intended to provide a

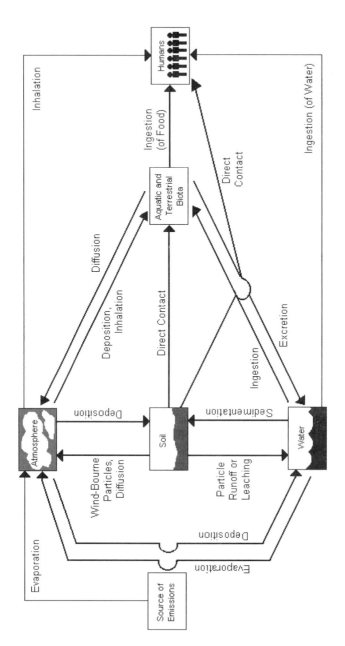

FIGURE 4-1 Possible pathways from emission of substances to human exposure. Source: Adapted from NRC 1991a.

comprehensive examination of the many aspects considered because such an examination is beyond the committee's task. To illustrate some of the important considerations with respect to environmental dynamics and exposure, particular attention is given to the main substances of concern that are discussed in Chapter 5 from a health-effects perspective. The chapter also examines approaches for estimating environmental concentrations that are used to estimate human exposures. As an illustration of how incineration facilities contribute to environmental concentrations at different geographical scales and for different agents, information is provided on particulate matter, various metals (cadmium, arsenic, mercury, and lead), dioxin-like compounds, carbon monoxide, and hydrogen chloride.

TRANSPORT PATHWAYS IN THE ENVIRONMENT

Substances released from combustion sources are ultimately dispersed among, and can at times accumulate in, various environmental compartments (e.g., soils, vegetation, indoor dusts, animals, and humans). Some contaminants that are released from incineration facilities are likely to contribute primarily to environmental compartments on a local scale (within 10 km). However, others that are more persistent in the environment, can be distributed over much greater distances—even up to a regional scale over hundreds of kilometers. Most of the substances released from incineration facilities to air do not remain in air but are deposited to soil, vegetation, or surface water and can come into contact with humans through a series of complex environmental pathways that include transport through several environmental media (see Figure 4-1).

As discussed in the great detail in Chapter 5, understanding the potential health impacts of waste incineration requires an understanding of the relative contribution of indoor, local, and regional sources of many pollutants. Therefore, an investigation must account for transport of pollutants through environmental compartments, and should examine large space- and time-scales, in addition to a combination of local environmental media over the short term. The required characterization of concentrations of contaminants in an environmental medium, such as air, involves accounting for the gains (or inputs to) and losses from that medium, and transport through it. For example, Table 4-1 lists the types of gains and losses that are considered in estimating the concentrations of contaminants in air.

In order to take account of the multimedia nature of pollutant transport, assessments usually examine multiple pathways that define the movement of a pollutant from a source, through a linear sequence of environmental compartments, to a receptor. An example of a particular pathway is a source emitting a pollutant to air (an air compartment), transport of the pollutant to the air above a field (another air compartment), deposition on vegetation (a vegetation compartment), eating of the vegetation by cows (an animal compartment), and drinking

TABLE 4-1   Modes of Gains and Losses of Contaminants in the Lower Atmosphere

Gains	Losses
Emission sources resulting from human activity, including incineration	Washout by rainfall
	Convection to higher levels in the atmosphere
Diffusion from soil	
	Deposition on soil
Diffusion from plants	
	Resuspension of deposited soil particles
Diffusion from surface water	
	Deposition on plants

of the cows' milk by humans (a receptor, exposed by the ingestion route). The pathway may be elaborated to almost any arbitrary degree, depending on how it is to be evaluated. For example, the deposition of a pollutant from air to vegetation may incorporate additional air compartments like a boundary layer of air around the vegetation, and a laminar flow layer of air above that, and so forth.

Multiple pathways may intersect one another in various environmental compartments, although each pathway individually usually does not self-intersect. Evaluation of each pathway individually is generally simpler than attempting to evaluate all simultaneously. The complexity of multiple connected compartments is reduced by examining pair-wise interactions between them.

The correctness of an approach for assessing the environmental transport of a specific substance depends on the linearity of the physical and chemical processes involved in pollutant transport with respect to pollutant concentrations in each compartment—fluxes between compartments usually depend linearly on the concentrations in connected compartments. If this linearity holds, a transport network through multiple compartments may be represented by the linear superposition of non-self-intersecting pathways. Where nonlinearities occur, the approach becomes less useful, and all compartments may have to be examined simultaneously, although it sometimes may be possible to contain all the nonlinearities within more-complex components of single pathways.

Persistence and Spatial Scale

Persistent air pollutants, such as dioxins, furans, and mercury, can be dispersed over large regions—well beyond the local areas and even the countries from which the sources first emanate. Many such pollutants are semi-volatile organic compounds (SVCs), with vapor pressures typically between about 10^{-1}-10^{-5} Pascal at ambient temperature (Wania and Mackay 1993), but they also include high-vapor-pressure metallic compounds (e.g., of mercury) and very low-vapor-pressure materials (e.g., most metals) incorporated in fine particles. The organic persistent SVCs tend to be lipophilic so that they readily partition into carbon and lipid tissues of plants and animals, and will often largely partition to the fine particles in ambient air. If they are resistant to physical, chemical, and biological degradation processes, they can persist for many years—such compounds have been labeled "persistent organic pollutants" (POPs) (Wania and Mackay 1993). Their small, but still significant, vapor pressure allows them to continually be re-emitted from the environmental sinks into which they partition. It has been postulated that, when released into the atmosphere, POPs tend to undergo a repeated deposition to and re-emission from soils, vegetation, and water, with transport effected in the vapor phase or adsorbed to ambient fine particles (Wania and Mackay 1993). ATSDR (1998a) reported a surprising extent of large-scale distribution and mixing for one class of POPs, the dioxins, which have been attributed to waste incineration. Similarly, efforts to assess pesticide POP use in various regions of the globe by sampling tree bark revealed a significant transport of such persistent compounds over a very large distances (Simonich and Hites 1995).

The recognition of POPs has created a need for environmental assessment and management strategies that provide an appropriate regional-scale framework for assessing the dispersion, persistence, and potential long-term impacts on human health and ecosystems. What is also needed is a process by which field data can be used to calibrate and validate models so that they can be used to inform control-strategy decisionmakers. For example, Scheringer (1996) has shown that the spatial scale needed to characterize the multimedia dispersion of organic chemicals is chemical dependent and should address the competition among reaction, atmospheric dispersion, and deposition. It should also address the impact of chemical partitioning into soil, vegetation, and surface water on the effective dispersion velocity in the air. According to Scheringer (1996), the effective dispersion velocity of a chemical is no greater than the average velocity of a parcel of air moving along the land surfaces. It is essentially equal to the velocity of the associated air parcel for high vapor-pressure compounds. However, the effective velocity is slower than the air-parcel velocity to the extent that a chemical partitions to particles, vegetation, surface water, and surface soil.

Efforts to move from an existing qualitative characterization of the large-scale disposition of POPs to a more-quantitative characterization are hindered by a number of scientific obstacles. One problem is the lack of a modeling framework that includes coupled mass exchange at boundaries of various environmental compartments and appropriately links the space- and time-scales involved in long-range transport. The low quality of many measurements of the large-scale partitioning of these chemicals between air and airborne particulate matter, between air and soils, and between air and vegetation, is another problem. For example, the measurements of vapor-particle partitioning that have been made are known to have large artifactual biases due to the sampling methods used (Gundel et al. 1995).

Quantification of Local Air Dispersion

Substances in outdoor (or ambient) air are dispersed by atmospheric advection and diffusion. Meteorological conditions, local terrain, and facility designs have an overwhelming influence on the behavior of contaminants in the lower atmosphere. Wind (direction, speed, and turbulence) and atmospheric stability are the most important. The standard models for estimating the local time and spatial distribution of contamination in the atmosphere from point sources are the Gaussian statistical solutions of the atmospheric diffusion equation. These models are obtained from solution of the classical differential equation for time-dependent diffusion in three dimensions. Pasquill (1961) has discussed the physical basis, analytical solutions, and the use of these equations. Turner (1970) and Hanna et al. (1982) have compiled workbooks on applications of these solutions to air pollution problems, including the application of the Gaussian models to area and line sources. There are numerous computer programs available and many papers describing algorithms for assessing the dispersion of point (e.g., stack), line (e.g., roadway) and area (e.g., shopping mall) air pollution sources. The output of a standard Gaussian plume model can be expressed as the ratio of the atmospheric concentration to the source strength release rate. Typical units are $\mu g/m^3$ per $\mu g/sec$, or sec/m^3. This ratio is typically estimated using screening-level models such as SCREEN3 (EPA 1995a,b), or more complex, site-specific, models such as the Industrial Source Complex (ISC) models (EPA 1995a). (Such models are easily obtained from the U.S. Environmental Protection Agency at the following website address: http://www.epa.gov/scram001/.) For example, SCREEN3 provides a high-end estimate for the worst-case 1-hour average of this ratio as large as 0.05 sec/m^3 for ground-level releases in urban areas, but the ratio typically decreases with the height of release. The annual average concentration-to-source ratio is likely to be about 0.08 (± 0.02) times the maximum 1-hour average (EPA 1995b). The ISC models can provide specific estimates for any given location, and can also take account of simple, intermediate, and complex terrain; dry deposition; wet deposition; and plume depletion.

Simpler approaches to estimating the dispersion of substances in the atmosphere may be based on the application of a mass balance to a volume element, parcel, or box of air. This gives rise to the "box" models. In this approach, the region to be studied is divided into cells or boxes. The concentration in each box is assumed to be uniform and is a function of the box volume, the rate at which material is being imported, emission rates within the box, and the rate at which material is exported from the box. Such simplified approaches may be more appropriate than the Gaussian plume models in circumstances where dispersion is not describable by Gaussian plumes. Also, such approaches may be sufficient to demonstrate that it is not necessary to go to the expense of employing more-complex models.

Deposition on and Accumulation in Soil

Soil is formed from the weathering action of climate on rocks and minerals and from the actions of living organisms. It is a mixture of minerals, water, air, and organic substances. The proportion of these components and the characteristics of the contaminants of concern determine, to a large extent, how such a contaminant is transported or transformed in soil. A contaminant can enter soil water, soil solids (mineral and organic phases), and soil air. Soils are characteristically heterogeneous in the vertical direction, so that a trench dug into soil typically reveals several horizontal layers that have different colors and textures.

Studies of radioactive fallout in agricultural land-management units have revealed that, in the absence of tilling, particles deposited from the atmosphere initially accumulate in and are resuspended from a surface-soil layer that is 0.1-1 cm thick (Whicker and Kirchner 1987). Over the long term, there is mechanical transport deeper into the soil (e.g., by earthworms, ants, rabbits, anything else that burrows, and by frost heave and wetting/drying cycles). Particles in the surface layer can be transported mechanically in the horizontal direction by run-off to nearby surface waters or be blown by wind. Surface-soil contaminants can be transported (on particles) by wind erosion, by volatilization to the atmosphere, by diffusion, leaching, and mechanical movement deeper into the soil, by erosion (attached to particles) or dissolution in runoff, and may be transferred to plant surfaces by rain splash or via resuspension and deposition. They can also be transformed through photolysis by sunlight, through chemical degradation, and through degradation by microorganisms (biodegradation).

The roots of most plants are typically confined within the top 3 ft (about 90 cm) of soil. Contaminants in this root-zone soil, below the surface layers, are transported upward by vapor-and liquid-phase diffusion, root uptake, and by capillary motion of water; they are transported downward by vapor- and liquid-phase diffusion and leaching; and chemically transformed primarily by biodegradation, hydrolysis, and other liquid and solid phase chemical reactions.

Deposition on and Uptake by Plants

By mass, the dominant component of the terrestrial biota is land plants. Plants generally have contact with two environmental media—air and soil. Uptake of contaminants by plants can occur directly from air via particle deposition or by foliar uptake of contaminant vapors. Particle deposition and foliar vapor uptake can also take place from contaminated soil (itself contaminated through various pathways from air contamination), through evaporation or suspension, or through rain splash. Uptake from soil through roots is a relatively minor pathway for many pollutants emitted from incineration facilities. Whereas many inorganic chemicals enter plants via root uptake from soil, the translocation of many organic chemicals from soil through roots appears to be a relatively minor pathway for their accumulation in plants (Fiedler et al. 1991; Trapp and Matthies 1997). For modeling purposes, there has been a reliance on simple bioconcentration factors (BCFs) that relate a soil- or air-concentration to a plant concentration, based on experimental studies that correlate these uptakes with simple chemical properties like vapor pressures, solubilities, and octanol-water partition coefficients. The earliest use of vegetation BCFs (for inorganic contaminants) was for assessing the effects of global radioactive fallout by relating concentrations of radionuclides in plants to concentrations in soil (Ng et al. 1982). Vegetation BCFs have been proposed for organic chemicals for soil and vapor-phase uptake (Briggs et al. 1983; Travis and Arms 1988; Travis and Hattemer-Frey 1988; Bacci et al. 1990, 1992; Sabljic et al. 1990; Trapp et al. 1990; Paterson and Mackay 1991; Schreiber and Schönherr 1992; Hülster and Marschner 1993; McCrady and Maggard 1993; Lorber et al. 1994; McCrady 1994; Paterson et al. 1994; Simonich and Hites 1994a,b; Tolls and McLachlan 1994; and Nakajima et al. 1995).

Surface Waters and Sediments

The behavior of chemicals in surface waters is determined, among other factors, by the rate of physical transport in the water system and chemical reactivity. Physical transport depends to a large extent on the type of water body under consideration (e.g., ocean, sea, estuary, lake, river, or wetland). Schnoor and McAvoy (1981) have summarized important issues related to surface-water transport. At low concentration, contaminants in natural waters exist in dissolved (in the water) and sorbed (to suspended particles) phases. In slow-moving surface-waters, both advection and dispersion are important. In rapidly moving water systems, advection controls mass transport, and dissolved substances move at essentially the same rate as the bulk water. Contaminants that are sorbed to suspended solids (including colloids) can also be entrained in water currents, but they might undergo additional transport processes that alter their effective residence time in surface waters; such processes include agglutination of the suspended particles, sedimentation and deposition of solids, and their scouring and resuspension. Thus, determining the transport of contaminants in surface water requires an understanding of water movement, deposition to the sediment, and resuspension from sediment.

Sediment is the porous layer of solid material and water that forms at the bottom of water bodies primarily as a result of deposition of mineral particles and organic matter. Reuber et al. (1987) note that surface-water sediments have at least two distinct layers. One layer is an active layer characterized by a high degree of chemical and biological activity. The other layer is a deeper, inactive layer in which chemicals are relatively isolated from the water column. Deposition and resuspension of mineral and organic matter to sediments occur continuously in any water body and are an important mechanism for transferring particle-bound contaminants to the sediment layer.

Multimedia Environmental Models

For substances released from waste-incineration facilities, the ambient concentration and deposition fluxes are determined by the partitioning and transport rates of the substances between the different compartments of the environment. Evaluating how chemicals are transported between such compartments requires a model that characterizes multiple environmental media, (i.e., air, soil, vegetation, surface water, sediments, and so forth) in combination. Efforts to assess human exposure to contaminants in multiple media date back to the 1950s when the need to assess human exposure to radioactive fallout and releases led to an assessment framework that included transport both through and among air, soil, surface water, vegetation, and food chains (USNRC 1975, 1977; Hoffman et al. 1979; Moore et al. 1979; Baes et al. 1984a,b; Whicker and Kirchner 1987). Efforts to apply such a framework to nonradioactive organic and inorganic toxic chemicals have been more recent and now are becoming as sophisticated as those extant in the radionuclide field. The first widely used multimedia compartment models for organic chemicals were the "fugacity" models described by Mackay (1991).[1] Fugacity models have been used extensively for modeling the transport and transformation of nonionic organic chemicals in complex environmental systems. Modified fugacity and fugacity-type models have also been used for ionic-organic and inorganic species, including metals. The advantage of the typical multimedia fugacity-type model is the simplicity with which it treats each of the compartments as being well mixed, and allowing for flows and mass transfer between all compartments, and degradation within compartments. Such treatment is clearly an oversimplification but the models, by the judicious selection of compartments to correspond to the penetration depth of the pollutants, can lead to insightful conclusions on the major pathways, reservoirs, and persistence in the environment.

More-recent multimedia models used for assessing releases from incinerators use various approaches. Air dispersion is handled by standard Gaussian plume models, with modification to incorporate wet and dry deposition of materials from the plume. The deposition models are multi-layer transport models, incorporating a well-mixed upper layer in the main plume, an intermediate shear

---

[1] The term "fugacity" is used in thermodynamics to refer to a measure of the tendency of a substance to escape by some chemical process from the phase in which it exists.

layer where the wind-speed increases regularly with height, and boundary-layer near the ground or vegetation surface. Transport of material deposited on the ground is handled largely by compartment models, with pathways of human exposure elaborated to varying degrees, with inter-compartmental transfer rates based on physical modeling, empirical correlations, or fugacity-type approaches. Examples of such models, with descriptions, are given in Lorber et al. (1994); Slob et al. (1993); EPA (1990, 1997b, 1998a).

ASSESSING HUMAN EXPOSURE TO ENVIRONMENTAL CONTAMINANTS

The issue of assessing human exposure to contaminants has been addressed in previous reports of the National Research Council (e.g., NRC 1991b, 1994). Exposure to a substance of concern is defined as contact at a boundary between a human and the environment at a specific concentration for a specific period (NRC 1991b). Human exposure assessment involves measuring or estimating the concentrations of specific substances in each exposure medium, and the time individuals or populations spend in contact with each such medium. Human activity patterns directly affect the magnitude of exposure to substances present in different indoor and outdoor locations. Assessing exposure to contaminants emitted as a result of waste incineration involves characterization of the rates and patterns of incineration emissions, tracking of the emitted material through the environment, and characterizing the amount of human contact with the material. In addition to incineration, other sources (for example, motor vehicles, coal-fired power plants, industrial manufacturing facilities, and some naturally-occurring sources) contribute to the total concentration of contaminants to which humans are exposed. Sexton et al. (1994) and Pirkle et al. (1995) discuss data bases that are available to help establish total exposure concentrations. Incineration facilities add some incremental amount to the total ambient concentrations in the environment for many pollutants, such as nitrogen oxides, sulfur dioxide, particulate matter, volatile organic compounds. For selected pollutants, such as dioxin, incinerators might collectively contribute major fractions of observed ambient concentrations as discussed later in this chapter. A particular incinerator, however, might be the dominant source at a particular location for concentrations of nitrogen oxides, sulfur dioxide, or particulate matter, and may, but not necessarily, be the dominant source for the dioxins.

Exposures to a substance of concern might be dominated by contacts through a single environmental pathway or they might reflect contacts through multiple pathways. Table 4-2 shows some of the pathways of exposure. All possible routes by which contaminants enter the body of an exposed person must be considered—inhalation, ingestion of food or drink, and absorption through skin because such patterns directly affect the magnitude of exposures to substances present in different indoor and outdoor environments.

TABLE 4-2   Examples of Pathways Linking Ambient Airborne Contaminants to Human Exposure

Exposure route	Pathway from ambient air
Inhalation	• Gases and particles in outdoor air • Gases and particles transferred from outdoor air to indoor air
Ingestion	• Fruits, vegetables, and grains contaminated by transfer of atmospheric chemicals to plant tissues • Meat, milk, and eggs contaminated by transfer of contaminants from air to plants that are consumed by animals • Fish contaminated by atmospheric deposition of chemicals directly from air to surface water and by deposition from air to soil with run-off transport to surface water • Mother's milk contaminated by mother's exposure through multiple pathways • Meat, milk, and eggs contaminated through inhalation by animals • Soil contaminated by deposition • Water used for washing or recreation contaminated by deposition
Dermal contact	• Soil contaminated by deposition

Exposure Pathways

Models have been developed for the multimedia transport of pollutants, and the uptake by the food chain, leading to estimates of human daily intake for various scenarios of human activity. The number of processes modeled is large and the uncertainty in the calculated results, particularly for some of the more-complex pathways, is correspondingly large. Such models might yield estimates of total exposure that can be an order of magnitude in error,[2] but such an uncertainty is the norm in risk estimates (and is generally far smaller than the variability in exposures between individuals, and the uncertainty in toxicity values). The models nevertheless are extremely useful in identifying the major pathways of exposure, the major reservoirs for contaminants (e.g., PCBs in sediments), and the approximate residence time in the environment of the contaminants. Inhalation is the most direct path for exposure to pollutants emitted from incinerator stacks and dispersed into the atmosphere. For the pollutants of greatest concern (see Chapter 5), however, the combination of long-range transport, deposition, and uptake of the pollutants by the food chain appears to be the most important mode of exposure.

---

[2] See Chapter 8 for a discussion of important sources of error. The degree of aggregation in the models, uncertainty in the input parameters, and different activity patterns contribute to uncertainty in the results.

Application of various models suggests the major pathways of exposure. The following examples highlight the need to examine indirect pathways of exposure. For cadmium, illustrative of many metals, the major pathway of exposure is usually through the consumption of garden fruits and root vegetables, which absorb cadmium primarily through uptake via the root system. For dioxin (as modeled by the 2,3,7,8-TCDD congener and used to illustrate a highly lipophilic organic compound), the major pathway of human exposure is primarily through meats, especially beef, and dairy products. For mercury, an important exposure pathway is human consumption of fish contaminated by atmospheric deposition of the metal directly from air to surface water and by deposition from air to soil with run-off transport to surface water. Such examinations should include consideration of the pathway by which fish taken for food eventually reaches the table. Subsistence fishers and others who fish for themselves may be at higher risk than the rest of the population that consumes fish, which may have no excess risk whatever.

Uncertainty and sensitivity analyses performed on the calculation of the human exposure to dioxin from municipal-waste combustion indicate that the major uncertainties in the estimated exposure are due to the uncertainties in the deposition rate and ambient air concentrations, and, to a much lesser degree, in the transfer of substances through a terrestrial food chain (Cullen 1995).

ENVIRONMENTAL DYNAMICS OF AND POSSIBLE EXPOSURES TO VARIOUS SUBSTANCES

The following subsections separately discuss the environmental dynamics (transport and fate), and possible human exposures, to particulate matter, cadmium, arsenic, lead, dioxins and furans, carbon monoxide, and hydrogen chloride that are emitted from waste-incineration facilities, as well as other sources. After those subsections, studies are discussed that estimated environmental concentrations of contaminants contributed by waste-incineration facilities. In addition, Chapter 5 provides similar types of estimates for illustrations of health-effects considerations regarding particulate matter, dioxins, lead, and mercury.

Particulate Matter

The transport characteristics of particles depend on their size. Fine and coarse particles in ambient air differ in their chemical composition, solubility, acidity, sources and formation processes, atmospheric lifetime, infiltration indoors, and transport distances (See Table 4-3). Most airborne particles are quite small (less than 0.1 μm in diameter), but most of the particle volume (and mass) is found in particles with diameters greater than 0.1 μm (Whitby 1978). The size distribution of airborne particles is often multimodal. Distributions of particles measured in outdoor air in the United States are almost always bimodal with a

TABLE 4-3 Characteristics of Fine Particles[a] Versus Coarse Particles

Characteristic	Fine Particles	Coarse Particles
Solubility	Significant fraction soluble, hygroscopic, and deliquescent; some portions insoluble.	Largely insoluble and non-hygroscopic.
Major Sources	Combustion and atmospheric transformation of gases. High-temperature processes, smelters, steel mills, incinerators, etc.	Resuspension of soil tracked onto roads and streets. Suspension from disturbed soil, e.g., farming, mining. Resuspension of industrial dusts. Construction, coal and oil combustion, incineration, and ocean spray.
Time Suspended in Air	Days to weeks	Minutes to hours
Travel distance	100s to 1,000s km	1 to 10s km

[a] Fine particles are defined as particles with an aerodynamic diameter less than 2.5 μm.

Source: Adapted from Wilson and Suh 1997.

minimum between 1.0 and 3.0 μm. Fine particles are usually defined as those having an aerodynamic diameter less than 2.5 μm. Fine and coarse particles generally have distinct sources and formation mechanisms, although there may be some overlap. Fine particles are usually formed from gases in two ways: (1) nucleation (i.e., formation of new particles from low vapor-pressure substances present in vapor form, produced either from combustion or from chemical reaction of gases) and (2) condensation of gases onto existing particles. Particles formed from nucleation also coagulate to form relatively larger particles, although particles normally do not grow above 1.0 μm in aerodynamic diameter by these processes. Particles formed as a result of chemical reaction of gases are termed secondary particles because the direct emission from a source is a gas (e.g., SO_2 or NO) that is subsequently converted to a low vapor-pressure substance (e.g., sulfuric acid, nitric acid) that subsequently nucleates or condenses. Examples include sulfates, some low-volatile organics, and ammonium salts. Such transformations can take place locally, during prolonged stagnations of ambient air, or during transport over long distances, and are affected by moisture, sunlight, temperature, and the presence or absence of fogs and clouds. In general, particles formed from these types of secondary processes will be more uniform in space and time than those that result from primary emissions. Particles directly emitted by sources, referred to as primary particles, are also found in the fine-particle fractions (the most common being particles less than 1.0 μm in aerodynamic diameter from combustion sources).

In contrast to fine particles, most of the coarse-particle fraction of ambient aerosol originated as particles emitted directly to the atmosphere, and some combustion-generated particles, such as fly ash and soot, might also be found in the coarse fraction.

Every particle in the atmosphere tends to settle to the ground through the effects of gravity, but the tendency to settle is opposed or abetted by other effects including electrostatic and aerodynamic forces. The net effect is that particles deposit to the ground at velocities that depend primarily on their particle diameter and density. For coarse particles, controlled primarily by gravity, the deposition velocity is proportional to the square of the particle diameter. For very fine particles, deposition is controlled more by electrostatic and other effects than by gravity, so that they deposit more rapidly than would be expected from gravity and their size alone. The result is that fine particles with aerodynamic diameters between 0.1 and 1.0 μm have the minimum deposition velocity of particles. Such fine particles will remain suspended for much longer times (on the order of days to weeks for fine particles as opposed to minutes to hours for coarse particles) and will travel much farther (i.e., hundreds to thousands of kilometers) than the coarse-particle fraction particles (i.e., kilometers to tens of kilometers) (Watson et al. 1995).

Fine particles originating outdoors infiltrate into homes and buildings to a greater degree than do coarse particles (Lioy et al. 1990). Indoor particulate matter (PM) levels are especially important because most people spend the majority of their time indoors, and thus a large amount of their exposure to PM may occur while inside. About 50-90% of the indoor fine particles are of outdoor origin (Clayton et al. 1993; Thomas et al. 1993). Spengler et al. (1981) found that for the Harvard Six City study, long-term mean infiltration of outdoor-origin $PM_{2.5}$ (particulate matter of aerodynamic diameter 2.5 μm or less) was 70% for homes without air conditioning and 30% in homes with air conditioning. Koutrakis et al. (1992) using New York State data of homes without smoking or fireplaces found that 60% of the $PM_{2.5}$ mass was from outdoor sources. Thus, ambient particles penetrate indoors and are available to be breathed into the lungs.

Because they can be transported long distances, penetrate indoors readily, reach deep into the lung, and are the particles most enriched in toxic compounds, it is the fine particulate matter which is of the greatest human-health concern when considering particulate matter or its precursors emitted as a result of waste incineration. The materials which are preferentially concentrated in the fine-particle fraction include volatile metals, such as cadmium and lead, and many low-volatility organic chemicals that adsorb to particle surfaces.

EPA (1998b,c) reports measured concentrations of ambient $PM_{2.5}$ in the United States that range from 13.5 to 37 μg/m^3 in urban areas and 3.1 to 21.6 μg/m^3 in nonurban areas. Also, EPA (1998b,c) estimates that the mean PM_{10} concentration in ambient air in 1997 was 24 $\mu g/m^3$; the 10th percentile value was about 16 $\mu g/m^3$ and the 90th percentile was 32 $\mu g/m^3$.

Cadmium

Cadmium is released into the environment by human activities such as mining and smelting operations, fuel combustion, waste disposal and application of phosphate fertilizer or sewage sludges (Elinder 1985). Cadmium can be present in waste input to an incinerator in the form of the metal (e.g., as cadmium plating), salts, and alloys (e.g., some solders and batteries). It forms a number of salts, including cadmium chloride ($CdCl_2$), cadmium sulfate ($CdSO_4$), and cadmium sulfide (CdS). Cadmium and its salts can be vaporized during waste incineration and emitted to the air as chlorides, oxides, or in elemental form. Such vapors rapidly condense onto particles, either those emitted simultaneously or ambient particles. Because the condensation is to particle surfaces, the fine-particle fractions (with higher surface area per unit mass of particle) become relatively enriched in cadmium.

The sources, sinks, and distribution of cadmium in many ecosystems have yet to be fully evaluated, and cadmium transfer rates between the different compartments of the environment are only poorly known. Aspects of the global cycle of cadmium have been summarized by Nriagu (1980). The major natural sources of cadmium to the active parts of the environment are from mobilization of cadmium from the large reservoir that exists in the lithosphere. The major sink for cadmium that enters the active compartments is burial in freshwater or ocean sediments.

Behavior in the Environment, Pathways, and Exposure

In aqueous systems, water hardness and pH, determine the speciation of cadmium. In fresh water at typical environmental pH values of 6 to 8, Cd^{+2} is the predominant species (Bodek et al. 1988). In the presence of sulfide ions and under reducing conditions, cadmium sulfide is formed over a wide pH range. The resulting precipitation of cadmium sulfide can serve to control the effective solubility of cadmium in natural waters.

In aqueous environments, cadmium will partition between the aqueous and solid phases (e.g., between water and soil particles in soil). This partitioning is described by a distribution, or sorption, coefficient, K_d (in units of L/kg), that is the concentration ratio, at equilibrium, of a chemical species attached to solids or particles (mol/kg) to the chemical concentration in a solution, mol/L, with which the particles have contact. Several mechanisms define this partition relationship—including cation exchange, adsorption, speciation, co-precipitation, and organic complexation. Bodek et al. (1988) have reviewed and compared a number of sorption models for cadmium in soil-water and sediment-water systems. They report that, in soils, estimated K_d values range from 1 to 9,000, with a typical value (at low water concentrations) on the order of 1,000; and that, in sediments, estimated K_d values range from 1 to 160,000, with a typical value (at low water concentrations) on the order of 6,000.

Plants are contaminated with cadmium via two routes—one is uptake of cadmium in soil through the roots and the other is deposition of cadmium from air onto leaf surfaces with translocation to other plant parts. Cadmium residues in plants are typically less than 1 mg/kg (IARC 1993).

The plant-soil partition coefficient, K_{ps}, expresses the ratio of contaminant concentration in plant parts in mg/kg (plant fresh mass) to concentration in wet root-zone soil, in units of mg/kg. Root uptake of cadmium as Cd^{+2} in plants is passive and occurs though uptake by roots of cadmium dissolved in water; cadmium is highly mobile in plants and readily translocated to other plant parts (Bodek et al. 1988). Plant-soil partition coefficients have been reported in the range 0.015 to 2.1 mg/kg with a likely value on the order of 0.1 mg/kg (Bowen 1979; Friberg et al. 1979; Nriagu 1980; Baes et al. 1984a).

According to Bodek et al. (1988), airborne deposition is believed to contribute to concentrations of cadmium found in plant leaves. At low concentrations, the ratio of plant-leaf concentration to air concentration, when air and plant environments are in contact, can be estimated based on the balance of gains from wet and dry deposition versus losses by wash-off and plant decay.

Atmospheric emissions of cadmium from human sources are estimated to exceed those from natural sources by about an order of magnitude (IARC 1993). ATSDR (1997a) summarizes references indicating that the mean levels of cadmium in ambient air range from less than 0.001 $\mu g/m^3$ in remote areas to 0.003 to 0.04 $\mu g/m^3$ in the United States. Cadmium metal and cadmium salts exist in ambient air primarily in fine suspended particulate matter. When inhaled, some fraction of this particulate matter is deposited in the lung airways and the rest is exhaled. In urban areas, an individual who breaths 20 m^3 of air will inhale about 0.2 $\mu g/day$ cadmium.

Cadmium enters drinking water directly from pollution sources, deposition from air to surface water, soil runoff to surface water, or leaching from rocks and soils into ground water. The concentration of cadmium dissolved in the open ocean is less than 0.005 $\mu g/L$ (Nriagu 1980; IARC 1993). The concentration of cadmium in drinking water is generally reported to be less than 1 $\mu g/L$ but it may increase up to 10 $\mu g/L$ as a result of industrial discharge and leaching from metal and plastic pipes (Friberg et al. 1974; ATSDR 1997a). An individual who consumes 2 L of water daily with a cadmium concentration of 1 $\mu g/L$ will have an intake of 2 $\mu g/d$.

For aquatic organisms, the bioconcentration factor (BCF) provides a measure of chemical partitioning between tissue and water and has units of mol/kg (fish) per mol/L (water). Bodek et al. (1988) report both ocean- and freshwater-fish bioconcentration factors in the range 200-50,000 L/kg, with 2,000 L/kg being a typical value in this range of reported values.

Humans can be exposed to soil contaminants through soil ingestion and through dermal uptake following soil contact with skin. For metal contaminants such as cadmium, the amount of intake via these pathways is typically less significant than the amount resulting from inhalation, water intake, and food-consumption pathways (McKone and Daniels 1991). Levels of cadmium in soil vary widely. In nonpolluted areas, concentrations in top soil are about 0.25 mg/kg (ppm) (EPA 1985); whereas in polluted areas, levels of up to 800 mg/kg have been measured (IARC 1993).

Indoor dust may be contaminated by deposition of particles from the air (originating from an emission source, or from suspension of contaminated soil), or by tracking of contaminated soil from outside. Friberg et al. (1974) report that the concentration of cadmium in the dust within houses was related to cadmium concentrations on air particles more than to soil concentrations.

Food is the main source of cadmium for non-occupationally exposed individuals. The gastrointestinal uptake of cadmium from food is generally less efficient than from water or by the lungs, because cadmium binds to food constituents (IARC 1993). The average daily intake of cadmium through food varies among individuals and by geographical area. An assessment using a Total Diet Study estimates the daily dietary intake of cadmium to be almost 15 μg/day (Gunderson 1995). Chaney et al. (1999) report that when zinc is present with cadmium at a ratio that is typical of geological materials (i.e., 100:1); zinc inhibits plant uptake, transfer to edible tissues, and absorption of cadmium in the intestine. However, when cadmium is present without zinc, food-chain mobility is much greater.

Arsenic

During waste incineration, arsenic (As) can be mobilized and emitted to the air as various inorganic compounds or in elemental form. Arsenic has valance states of -3, 0, $+3$, or $+5$, and is generally found in waters as H_3AsO_4, $H_2AsO_4^{-1}$, and $HAsO_4^{-2}$, as well as $H_2AsO_3^{-1}$, and $H_2AsO_4^{-1}$. The principal arsenic-bearing minerals include arsenopyrite (FeAsS), niccolite (NiAsS), cobaltite (CoAsS), tennantite ($Cu_{12}As_4S_{13}$), enargite (Cu_3AsS_4), and native arsenic.

By 1990, 70% of U.S. consumption of arsenic became attributable to the wood preservative industry and 20% to agricultural uses (ATSDR 1998b). Arsenic is also used in glass, nonferrous alloys, and electronics. Arsenic is released into the environment by human activities including arsenical pesticide and preservative use, metal smelting, waste incineration, and coal combustion.

Behavior in the Environment, Pathways, and Exposure

The metal arsenic is insoluble in water (Weast et al. 1986). Trivalent arsenic compounds are quite soluble at ambient temperatures. Pentavalent arsenic

(AS_2O_5) is more soluble than trivalent. In aqueous systems, arsenic forms anions in solution, and thus it does not form complexes with simple anions such as Cl^- (Bodek et al. 1988). Anionic arsenic complexes behave like ligands in water. In aerobic aqueous systems, the arsenic acids $H_2AsO_4^{-1}$ and $HAsO_4^{-2}$ are the dominant species in the pH range 2-11. Below pH 2, the arsenious acid, $H_2AsO_3^{-1}$ can be the dominant species in reducing conditions and above pH 12, the arsenious acid, $HAsO_3^{-2}$ appears.

Bodek et al. (1988) have reviewed and compared a number of sorption measurements and sorption models for arsenic in soil-water and sediment-water systems. They report that, in soils and sediments estimated K_d values range from 15-5,500 mol/kg.

As a likely result of uptake in water, arsenic is absorbed by roots and readily translocated to other plant parts (Bodek et al. 1988). Plant-soil partition coefficients have been reported in the range 0.01-0.04 mg/kg on a dry-mass basis (Baes et al. 1984a), which corresponds to the approximate range of 0.002 to 0.008 on a fresh-mass basis.

According to Bodek et al. (1988), airborne deposition is believed to contribute to some of the concentrations of arsenic found in plant leaves. At low concentrations, the ratio of plant leaf concentration to air concentration when air and plant environments are in contact, can be estimated based on the balance of gains from wet and dry deposition versus losses by wash-off and plant decay. Bodek et al. (1988) report both ocean and freshwater fish bioconcentration factors in the range 10-500 L/kg.

Arsenic metal and arsenic compounds (with the exception of arsine gas) have low volatility and exist in air primarily incorporated as fine suspended particulate matter. When inhaled, some fraction of this particulate matter is deposited in the airways of the lung and the rest is exhaled. Mean concentrations of arsenic in ambient air are estimated to be usually in the range of less than 0.001-0.003 $\mu g/m^3$ in remote areas and in the range of 0.02-0.03 $\mu g/m^3$ in urban areas (ATSDR 1998b). Air concentrations of arsenic near nonferrous metal smelters were reported to reach 2.5 $\mu g/m^3$ (Schroeder et al. 1987). Arsenic exposures can occur through contact with water, food, soil, and house dust in ways similar to those discussed for cadmium.

ATSDR (1998b) indicates that food is typically the greatest source of arsenic exposure for the general population. NRC (1999b) reports that one of the most comprehensive studies of arsenic in food was published in 1993 (Dabeka et al. 1993). The food groups containing the highest mean arsenic compounds were fish (1.66 $\mu g/g$), meat and poultry (0.024 $\mu g/g$), bakery goods and cereals (0.024 $\mu g/g$), and fats and oils (0.019 $\mu g/g$). The average daily dietary ingestion of total arsenic by Canadians was estimated to be 38.1 μg (48.5 μg for adults). Adams et al. (1994) estimates that food contributes 93% of total intake of arsenic in the United States, and seafood contributes 90% of that 93%.

Mercury

Mercury is a naturally occurring metal that combines with other elements, such as chlorine, sulfur, or oxygen, to form inorganic compounds. It occurs commonly in nature as the sulfide, cinnabar. Also, mercury combines with carbon to form organic compounds. Mercury is found in trace amounts in fossil fuels. Mercury is released into the environment by a number of human activities, which include incineration, fossil-fuel combustion (especially coal burning), pulp and paper manufacture, paint manufacturing and applications, preparation of amalgams for dental work, laboratory usage, battery applications, disposal of fluorescent lighting, cement manufacturing, fungicides, and medical application. In addition, soil degassing during natural fires, volcanic activity, and biogenic sources are important contributors to global emissions to the atmosphere (Nriagu 1990).

EPA (1997c) reported that U.S. annual anthropogenic emissions of mercury in 1994-1995 was 158 tons. The emissions of mercury from resource recovery facilities were found to decline with time over the period covered, 1985-1992. This is a reflection of the decline in mercury consumption in the United States, from 2,241 tons in 1980 to 793 tons in 1990, to 463 tons in 1995, with a proportionally larger decrease in consumer products which might end up in municipal solid-waste (mercury consumption for battery production in the United States decreased from 1,058 tons in 1980 to 117 tons in 1990). EPA (1997b) estimated that waste incineration contributed about 33% of the national mercury emissions in 1994-1995.

Behavior in the Environment, Pathways, and Exposure

Mercury (Hg) emissions from incinerators into the air are mostly in the form of either elemental mercury vapor or mercuric chloride ($HgCl_2$). Mercury transport may be long range, with the distance of transport dependent on the rate of conversion of elemental mercury to the soluble mercuric ion. Mercury is washed from the atmosphere by rain (wet deposition), but also deposits (dry deposition) in both vapor and particulate form. Models of the fluxes of mercury have been developed for different watersheds (Porcella 1990; Fitzgerald and Clarkson 1991; Lindquist 1991; Hudson et al. 1994). Some of the mercury deposited in lakes or soils is converted to methylmercury by organisms in soils and water bodies. Methylmercury is of major importance because of its high lipid solubility. The bioaccumulation of mercury in fish is such that the concentration of mercury in fish is much higher than in the ambient water. A significant fraction of mercury input to water bodies might be taken up by the fish. For example, Porcella (1990) estimated that of the 1.5 g/year depositing into a seepage lake in Wisconsin, 0.2 g/year were taken up by the fish. Hall et al. (1997) reported that food was the predominant source of mercury uptake in fish. A major path of human exposure to mercury is from eating fish from contaminated water bodies. The ability to model the pathways of mercury from the source of emissions to human uptake is constrained at present by the difficulty of calculating the interconversion of the elemental mercury and mercuric ion in the atmosphere (which determines the atmospheric lifetime of the mercury), and of determining the rate of methylation of mercury (which determines its uptake by the biota).

Because of the long range transport of mercury, regional average concentrations are uniform within a factor of two to three. The values reported for Wisconsin by Fitzgerald et al. (1991) are representative of continental values. These are a gas-phase concentration of 1.57 ng/m^3 and a particulate concentration of 0.022 ng/m^3. Mean values for the air over a forested watershed in Tennessee were 5.5 ng/m^3 (Schroeder and Fanaki 1988), with the corresponding particle bound concentrations of 0.03 ng/m^3. The particle-bound values can be seen to be less than 1% of the vapor values. However, particulate-mercury concentrations are greater in precipitation than in ambient air (ATSDR 1999).

Values are reported for concentration in rain of 10.3 ng/L, and a wet deposition rate of 6.8 $\mu g/m^2$-year by Fitzgerald et al. (1991) for Wisconsin. Values reported by Glass et al. (1991) of 18 ng/L for rain and 15 $\mu g/m^2$-year for Minnesota are within a factor of a little more than two of those reported in Wisconsin, and are supportive of the view that the volatile elements are uniformly distributed over wide areas.

Concentrations in freshwater fish are typically in the range of 0.1-1 $\mu g/g$ fish. For example, trout from Lake Ontario had average values that declined from 0.24 $\mu g/g$ in 1977 to 0.12 $\mu g/g$ in 1988 (Borgmann and Whittle 1991). Fish from the Savannah River had concentrations of 0.10 to 0.72 $\mu g/g$ (Winger et al. 1990). Although the concentration in fish and the water do not always correlate well because of interfering factors (such as age of the fish, pH, and the different bioavailability of various forms of mercury) the concentration in fish is of the order of a million times that in the water. The biological concentration factor of methylmercury in fish in a freshwater lake was 3 million L/kg (Porcella 1994).

Potential sources of general population exposure to mercury include inhalation, ingestion of drinking water and foodstuffs, and exposure through dental and medical treatments. Food, particularly fish consumption, is the major environmental path of exposure for mercury. Studies of the dietary intake conducted by the Food and Drug Administration show an average daily intake for adults of 0.03 $\mu g/kg$ of daily weight remarkably independent of age and sex, or 2.1 $\mu g/day$ for a 70-kg adult (Cramer 1994). Using a terrestrial food chain model, Travis and Blaylock (1992) estimated an average daily intake of 6.3 $\mu g/day$ for adults with over 50% coming from fish intake. Other studies of dietary intake of mercury are presented in ATSDR (1999).

Lead

Waste products that contain lead include storage batteries, ammunition waste, solder, pipes, and other metal products; consumer electronic products; solid waste and tailings from lead mining; items covered with lead-based paint; and solid wastes generated by mineral ore processing, iron and steel production, and copper and zinc smelting. The general population can be exposed to lead in ambient air, foods, drinking water, soil, and dust (ATSDR 1997b). Table 4-4 presents estimates of environmental lead concentrations in remote, rural, and urban areas.

Metals processing is the major source of lead emissions to the atmosphere. The arithmetic mean concentration of lead in ambient air in 1997 is estimated by EPA to have been 0.04 $\mu g/m^3$. EPA's estimate of the 95th percentile concentration is 0.12 $\mu g/m^3$ and of the 5th percentile concentration is 0.01 $\mu g/m^3$ (EPA 1998b,c). By comparison, the National Ambient Air Quality Standard (NAAQS) for this pollutant is 1.5 $\mu g/m^3$ as an annual average. The highest ambient air concentrations of lead are found in the vicinity of ferrous and nonferrous smelters, battery manufacturers and other stationary sources of lead emissions (EPA 1998b,c). EPA estimates that less than 1% of the public water systems in the United Sates have water entering the distribution system with lead levels above 5 $\mu g/L$. Those systems are estimated to serve less than 3% of the population that receives drinking water from public systems (EPA 1991a). EPA also estimates that lead levels between 10 and 30 $\mu g/L$ can be found in drinking water as a result of plumbing corrosion and subsequent leaching of lead (EPA 1989).

Atmospheric deposition is an important source of lead found in soils. The strong absorption of lead to organic matter in soil tends to limit the bioavailability of lead and thus it tends not to bioaccumulate in aquatic and terrestrial food chains. Lead can be added to food crops through uptake from soil, direct deposition onto crop surfaces from the atmosphere, during transport to market, food processing, and kitchen preparation.

ATSDR (1997b) reports that data from Phase 2 of the National Health and Nutrition Examination Surveys (NHANES) III (conducted during October 1991 to September 1994) indicate that the overall geometric-mean blood-lead level of the population aged 1 year or younger was 0.11 $\mu mol/L$ (2.3 $\mu g/dL$). Among those aged 1-5 years, approximately 4.4% had blood-lead levels of 10 $\mu g/dL$, representing an estimated 930,000 children with levels high enough to be of concern.

Dioxins and Furans

Dioxins and furans refers collectively to polychlorinated dibenzodioxins (PCDDs) and polychlorinated dibenzofurans (PCDFs). Those chemical compounds are generally classified as halogenated aromatic hydrocarbons (HAHs).

TABLE 4-4 Environmental Lead Concentrations in Remote and Rural Areas and Urban Areas[a]

	Remote and Rural Lead Concentration, $\mu g/g^b$	Reference	Urban Lead Concentration, $\mu g/g^b$	References
Air	0.05	Lindberg and Harriss 1981	0.3	Facchetti and Geiss 1982; Galloway et al. 1982
Fresh water	1.7×10^{-5}	Elias et al. 1982	0.005-0.030	EPA 1986a, Vol. II
Soil	10-30	EPA 1986a, Vol. II	150-300	EPA 1986a, Vol. II
Plants	0.18^c	Elias et al. 1982	950^d	Graham and Kalman 1974
Herbivores (bone)	2.0^d	Elias et al. 1982	38^d	Chmiel and Harrison 1981
Omnivores (bone)	1.3^d	Elias et al. 1982	67^d	Chmiel and Harrison 1981
Carnivores (bone)	1.4^d	Elias et al. 1982	193^d	Chmiel and Harrison 1981

[a] Values can be highly variable, depending on organism and habitat location
[b] Except $\mu g/m^3$ in air
[c] Fresh weight
[d] Dry weight

Source: NRC 1993.

Chlorinated and brominated dibenzodioxins and dibenzofurans are tricyclic aromatic compounds with similar chemical and physical properties. There are 75 congeners of chlorinated dibenzo-*p*-dioxins. 2,3,7,8-Tetrachlorodibenzo-*p*-dioxin (TCDD) is the most widely studied of these compounds. TCDD and chemically similar compounds are collectively called dioxins; TCDD serves as the reference compound for this class of compounds (EPA 1994b,d), but it represents a small portion of incineration emissions of PCDDs and PCDFs.

Sources

Historical records of the concentrations in sediment cores show that in the Great Lakes area the levels of dioxins started to rise greatly in the mid 1930s (Czuczwa and Hites 1986), a time corresponding to the growth in the production of chlorinated organic chemicals. The levels in the sediments began to decline in the 1970s, as particulate emission controls began to be imposed. Although such observations are consistent with combustion sources being a major source (particularly incinerators without heat recovery which were phased out in the 1970s), unidentified sources still might be dominant, because attempts at mass balances suggest that the observed deposition rates are greater than can be accounted for by known sources (Brzuzy and Hites 1996).

In addition, there are substantial differences between the homologue distribution of dioxins found in the environment and those emitted by incinerators—differences that cannot be explained by models of the environmental fate of dioxins from combustion sources. Other sources which may be important include the burning of wood treated with pentachlorophenols, secondary copper smelting, fireplaces, and motor vehicles.

Behavior in the Environment, Pathways, and Exposures

The 2,3,7,8 chlorinated dioxin and furan congeners appear to be resistant to natural degradation, bioaccumulate in many organisms and, possibly, biomagnify to the highest levels of the food chain (ATSDR 1998a). The dominant transformation processes affecting their fate are surface photolysis and gas-phase diffusion or volatilization with subsequent photolysis (Yanders et al. 1989). Models (ATSDR 1998a) of the behavior of TCDD show that it is transported primarily through the air and distributed regionally, and that it accumulates primarily in soils. The decomposition rates, both photochemical and bacterial, decrease with extent of chlorination so that the higher chlorinated PCDDs persist longer.

ATSDR (1998a) reports that most of the measurements of CDDs in air tend to be very close to current detection limits. CDDs are found at the greatest concentrations in urban air with octachlorinated dioxin (OCDD) being the most prevalent congener (up to 0.100 ppq), heptachlorinated dioxins (HpCDDs) being the next most common congener, and 2,3,7,8-TCDD being the least common congener (0.014 ppq). Concentrations of all CDDs are highest in the air near industrial areas. Rural areas usually have very low or undetectable levels of all CDDs. In urban and suburban areas, concentrations of CDDs may be greater during colder months of the year when furnaces and wood stoves are used for home heating. Furst et al. (1990), Gilman et al. (1991), Theelen (1991), and Schaum et al. (1994) estimate that about 90% of human exposure to dioxin occurs via contaminated food, including human milk. Hattemer-Frey and Travis (1989) estimated that meat and dairy products accounted for 98% of the total intake of TCDD. The primary mechanism by which dioxin-like compounds enter the terrestrial food chain is suspected to be via atmospheric deposition. Support for this hypothesis includes studies that have measured dioxin compounds even in the most remote areas throughout the world, where atmospheric transport and deposition is the only plausible mechanism (ATSDR 1998a). Deposition can occur directly onto plant surfaces or onto soil. Soil deposits can enter the food chain via direct ingestion (especially applicable to children) or soil ingestion by food animals (cows, for example). Dioxin compounds become available to plants by volatilization and vapor absorption, root uptake or particle resuspension, and adherence to plant surfaces.

The multi-compartment fugacity model of Mackay, combined with the terrestrial food chain model, has been used to calculate the exposure to dioxins in Southern Ontario, Canada (Paterson et al. 1990). The tetra chlorinated, hexachlorinated, and octachlorinated dioxins were treated separately because of their different physico-chemical properties. Air, soil, water, and sediment were considered in the analysis. One major conclusion drawn from the calculations is that the soil compartment is the principal reservoir for the dioxins. From the concentrations and fluxes in the different compartments, the human exposure models are developed using the correlation of Travis and Hattemer-Frey (1988) on the assimilation of the dioxins in dairy products, beef, and vegetation. The results for the tetrachlorinated dioxins are shown in Figure 4-2. For the system modeled, fish consumption is shown to be the major pathway of exposure. The results obtained from the model are compared with experimental observations in Table 4-5, which show that the calculated concentrations in the different compartments are generally within one to two orders of magnitude of the measured values, with the calculated values being low. The authors indicated that these discrepancies could be a result of either the measurements being reported for more polluted areas, and therefore being high, or due to not accounting for all dioxin sources in the model input parameters. Despite their uncertainty, the models provide valuable insight as to which pathways to exposure are important and in which compartments in the environment dioxins accumulate.

The correlation between total exposure and local emissions is expected to be low because the pollutants are dispersed over wide areas, and the foods that are implicated as the major path of exposure will contain a fairly high portion of

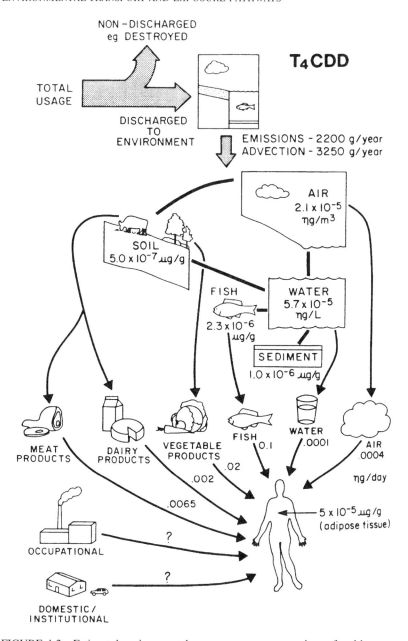

FIGURE 4-2   Estimated environmental compartment concentrations of and human exposures to T$_4$CDD by various pathways in Southern Ontario. Emission and advection rates
are provided at the top of the figure. Transfer rates are indicated along the arrows. Source:
Paterson et al. 1990. Reprinted with permission from *Emissions from Combustion Processes: Origin, Measurement, Control;* copyright 1990, CRC Press, Boca Raton, Fla.

TABLE 4-5   Comparison of Predicted and Measured PCDDs for Southern Ontario

Medium	T_4CDD^g	H_6CDD^g	O_8CDD^g
Air (ng/m^3)			
Estimated	2×10^{-5}	2.2×10^{-5}	3.8×10^{-5}
Reported[a]	1×10^{-5}	6×10^{-4}	5×10^{-4}
Water (ng/l)			
Estimated	5.7×10^{-6}	2.7×10^{-4}	3.8×10^{-4}
Soil (mg/kg)			
Estimated	4.8×10^{-8}	1.6×10^{-7}	7.9×10^{-7}
Reported[a,b]	$(<1-9) \times 10^{-3}$	1.7×10^{-4}	3.5×10^{-3}
Sediment (mg/kg)			
Estimated	1×10^{-6}	8.7×10^{-7}	1.4×10^{-5}
Reported[d]	2.6×10^{-5}	1×10^{-5}	5.6×10^{-4}
Vegetation (mg/kg)			
Estimated	4.8×10^{-8}	1.6×10^{-7}	7.9×10^{-7}
Reported[b,e]	4×10^{-7}	$<1 \times 10^{-8}$	$<1 \times 10^{-6}$
Reported[f]	$<2 \times 10^{-8}$	$<3 \times 10^{-7}$	1×10^{-6}
Meat (mg/kg)			
Estimated	2.2×10^{-8}	7.4×10^{-8}	3.5×10^{-7}
Reported[b,e]	$<1 \times 10^{-6}$	$<1 \times 10^{-6}$	1×10^{-6}
Reported[f]	$<2 \times 10^{-6}$	$<1 \times 10^{-6}$	$(3-24) \times 10^{-6}$
Dairy (mg/kg)			
Estimated	7.7×10^{-9}	2.6×10^{-8}	1.3×10^{-7}
Reported[e]	2.5×10^{-6}	$<1 \times 10^{-6}$	3.6×10^{-5}
Reported[f]	$<1 \times 10^{-7}$	$<1 \times 10^{-7}$	1×10^{-6}
Fish (mg/kg)			
Estimated	2.3×10^{-6}	1.3×10^{-5}	1.8×10^{-5}
Reported[b]	$<1 \times 10^{-6}$		
Human Adipose Tissue (mg/kg)			
Estimated	4.8×10^{-5}	5.3×10^{-5}	3.9×10^{-5}
Reported[c]	5×10^{-6}	7.2×10^{-5}	5.6×10^{-4}

[a] Cantox Inc "Health Hazard Evaluation of Seicfic PCDDS, PCDFS, and PAH in Emissions from the Proposed Petrous/S.C. Resource Recovery Incineration and From Ambient Background Sources", Petrous/S.C. Operations Ltd., Oakville, Ontario (February, 1988)

Travis, C. C., and H. A. Hattemer-Frey, "Human Exposure to 2,3,7,8-TCDD", *Chemosphere*, 16:2331-2342 (1987)

[c] Stanley, J. S., Boggess, K. E., Onstot, J., Sack, T. M., Remmers, J. C., Breen, J., Kutz, F. W., Carra, J., Robinson, P., and Mack, G. A., "PCDDs and PCDFs in Human Adipose Tissue from the EPA FY82 NHATS Repository", *Chemosphere*, 15:1605-1612 (1986)

[d] Astle, J. W., Gobas, F. A. P. C., Shiu, W. J., and Mackay, D., "Lake Sedimentation in Historic Records of Atmospheric Contaminants by Organic Chemicals", pp. 57-77 in *Sources and Fates of Aquatic Pollutants*, R. A. Hites and S. J. Eisenreich, Eds., American Chemical Society, Washington, D. C ., 1987.

[e] Davies, K. "Concentration and Dietary Intake of Selected Organochlorines, including PCBs, PCDDs, PCDFs in Fresh Food Composites Grown in Ontario, Canada", *Chemosphere*, 17:263-276 (1989)

[f] Ontario Ministry of Agriculture and Food, Ontario Ministry of the Environment, Toxics in Food Steering Committee. Polychlorinated Dibenoz-p-dioxins and Polychlorinated Dibenzofurans and other Organochlorine Contaminants in Food, Toronto. (1988).

[g] "T_4CDD" refers to tetra-chlorinated dioxin, "H_6CDD" refers to hexa-chlorinated dioxin, and "O_8CDD" refers to octa-chlorinated dioxin.

Source: Paterson et al. 1990. Reprinted with permission from *Emissions from Combustion Processes: Origin, Measurement, Control;* copyright 1990, CRC Press, Boca Raton, Fla.

products imported from other regions. Hattemer-Frey and Travis (1989) concluded that emissions from municipal solid-waste incinerators do not substantially increase human exposure to CDDs and CDFs above normal background levels.

Carbon Monoxide

Outdoor sources contributing most carbon monoxide (CO) emissions are motor vehicle exhaust, industrial processes, nontransportation fuel combustion, and natural sources such as wildfires. Wood stoves, cooking, tobacco smoking, and space heating are major sources of indoor CO emissions (EPA 1998b,c).

In the atmosphere, CO is transported, by diffusion and eddy currents to the troposphere and stratosphere, where it is oxidized to carbon dioxide by hydroxyl radicals. In soil, microorganisms oxidize it to carbon dioxide. The oceans are reservoirs for CO.

Background levels of CO are quite low, under 1 ppm. Although annual average CO levels are usually below 9 ppm (the 8-hour National Ambient Air Quality Standard), there is great variability in urbanized areas, especially through the day, as a result of traffic patterns. It is unlikely that waste combustion is a major source of CO exposure to the general public.

Hydrogen Chloride

Hydrogen chloride exists as an aqueous acid aerosol in the environment. Anthropogenic and natural chlorine compounds have the same dispersion characteristics, and transport and mixing are similar. Dissociation of hydrogen chloride to form free chlorine atoms can occur by photochemical reaction or by reaction of hydrogen chloride with OH. These free chlorine atoms react by well-established pathways to destroy ozone in the stratosphere.

Gaseous chlorine compounds are removed by rainfall and adsorption onto particles. Chlorine is probably removed from the atmosphere by interaction with particles or water cloud processes. Background levels of particle phase chlorine are primarily due to ocean spray, while volcanic gases account for most gaseous hydrogen chloride. Ambient levels of particle phase chlorine compounds are about 3 $\mu g/m^3$ in U.S. cities, while the total chlorine are about 10-100 $\mu g/m^3$.

ENVIRONMENTAL POLLUTION CONCENTRATIONS ASSOCIATED WITH WASTE INCINERATION

At large distances, the only emissions from incineration facilities which need to be considered are those which contribute significantly to the total regional emissions. At short distances, a site-specific evaluation of the contribution of waste incineration to the local concentration of individual pollutants is needed.

TABLE 4-6   Qualitative Contribution to Air Emissions from All Sources, Dispersion Scale, and Routes of Exposure of Waste Incineration Pollutants

Pollutant	Contribution to U.S. Emissions by Incinerators	Geographic Scale of Dispersion	Major Pathways of Exposure
CO	Minor[a]	Local	Inhalation
HCl	Minor	Local	Inhalation
Dioxins	Major	Regional	Food (especially meat, dairy products, and fish)
PAH	Minor	Regional	Food
PCBs	Minor	Regional	Food
Mercury	Major	Regional	Food (especially fish, meat, and dairy products)
Cadmium	Intermediate	Local, Regional	Food
Lead	Intermediate	Local	Variable, mainly food, such as cereals, vegetables
Particles, fine	Minor	Regional	Inhalation
Particles, coarse	Minor	Local	Inhalation

[a] Major refers to >20%, Minor refers to < 1%.

Source: Adapted from Koshland 1997.

The fraction of any pollutant transported over long distances depends on its environmental residence times in various environmental media. For organic compounds, the controlling factor is often an atmospheric residence time governed by a reaction rate with OH. Values of these for many of the air toxics in combustor effluents are provided by Koshland (1997). For inorganic elements which occur in the condensed phase or as soluble vapors, the atmospheric residence time is determined by the deposition velocity and rainfall events. Table 4-6 summarizes the scale of dispersion and the exposure routes for the pollutants for which the possible health impacts of waste incineration are presented in Chapter 5.

Estimates of Ambient Air Concentrations of Various Pollutants from Waste Incineration

Estimates of air pollutant concentrations associated with hazardous-waste incinerators and cement kilns that burn hazardous waste are provided here to give some idea of the contribution of an incineration facility to air quality. The estimates are not intended to represent the full range of air concentrations that might be associated with incineration facilities currently operating. These estimates of atmospheric concentrations were developed from the databases used by

EPA to support the Maximum Achievable Control Technology (MACT) standards as proposed in April, 1996 for hazardous-waste incinerators and cement kilns using hazardous waste as fuel. An approximate estimate of the impact of the proposed MACT on ambient air quality is also presented for illustrative purposes only. The emission rates were all summarized and analyzed for the shape of their statistical distributions. In all cases, there was a reasonable fit to a lognormal model. These distributions were multiplied together along with the probability distribution for unit air concentrations from Cullen (1995) using a Monte Carlo routine. The atmospheric concentrations were calculated by using 5,000 trials for each chemical. Results (rounded to two significant figures) are summarized in Tables 4-7 through 4-10.

The primary uncertainty associated with this analysis is in Cullen's unit air concentration distribution. This was apparently developed for a municipal solid-waste incinerator in Bridgeport, CT, for a radial area located between 0.4 km and 30 km from the stack—a distance that included the point of maximum ambient concentration. Thus, these results will reflect ambient air concentrations expected from the incinerators and cement kilns considered in this exercise only to the extent that the emission characteristics and the local meteorology of the Bridgeport, CT, incinerator are similar to others. It must be emphasized that a person could be exposed to higher or lower concentrations depending on where he or she was located with respect to any particular combustor, and depending on characteristics of that particular combustor.

In these calculations, the particulate-matter concentrations reflect total rather than respirable particles. It is not possible to generalize, with any certainty, regarding respirable particulate-matter concentration, because there is a lack of data concerning particle-size distributions in emissions. EPA (1992a) states that approximately 62% of the mass of the particles escaping a fabric filter for a municipal solid-waste incinerator are less than or equal to 12 μm in diameter and approximately 54% of the particles are less than 2.3 μm in diameter. These factors could be applied to the concentrations given for total particulate to yield an estimate of respirable particulate.

This analysis should be viewed as illustrative rather than comprehensive. One primary source of uncertainty is the assumption that one single probability distribution of dispersion coefficients applies to all incinerators in the database. The committee was unable to be more comprehensive because of the lack of any database that linked emissions estimates with dispersion estimates for individual facilities.

Post-MACT data were taken from information presented at the U.S. EPA Boiler and Industrial Furnace and Hazardous Waste Incinerator Technical Meeting, Kansas City, KS., March 7, 1996. The database presented at this meeting included mass flow (in pounds per year) in addition to gas flow in dry standard cubic feet per minute (dscfm) for 20 commercial hazardous-waste incineration units and 30 cement kilns using hazardous waste as fuel. It should be kept in

TABLE 4-7 Estimated Ambient Air Concentrations of Various Pollutants Associated with Hazardous-Waste Incinerator Emissions Prior to Compliance with MACT (All concentrations are annual averages in units of micrograms per cubic meter.)

Pollutant	Number of Samples	Mean	Standard Deviation	Median	Range
Mercury	718	5.5×10^{-4}	7.2×10^{-3}	7.2×10^{-6}	1.9×10^{-10} to 0.4
Lead	789	2.2×10^{-2}	0.2	2×10^{-4}	2×10^{-8} to 7.2
PCDDS/F TEQ[a]	182	4.1×10^{-9}	9.4×10^{-9}	5.6×10^{-10}	7.8×10^{-13} to 8.8×10^{-8}
Particulate Matter	633	8.9×10^{-2}	0.31	1.5×10^{-2}	8.6×10^{-6} to 7.7

[a] PCDD/F is the toxic equivalent of the sum of the polycholorinated dibenzodioxins and polychlorinated dibenzofurans.

TABLE 4-8   Estimated Ambient Air Concentrations of Various Pollutants that may be Associated with Hazardous-Waste Incinerator Emissions After Compliance with MACT (All concentrations are annual averages in units of micrograms per cubic meter.)[a]

Pollutant	Mean	Standard Deviation	Median	Range
Mercury	8.7×10^{-5}	4.6×10^{-5}	7.9×10^{-5}	9.3×10^{-6} to 4.6×10^{-4}
PCDDS/F TEQ[b]	5.8×10^{-10}	3.0×10^{-10}	5.2×10^{-10}	5.3×10^{-11} to 2.5×10^{-9}
Particulate matter[b]	0.20	0.10	0.18	2.2×10^{-2} to 0.96

[a] Based on the MACT emission limits that were proposed in April 1996.

[b] PCDD/F is the toxic equivalent of the sum of the polycholorinated dibenzodioxins and polychlorinated dibenzofurans.

mind that an individual facility might have more than one unit. Data are presented in Tables 4-7 through 4-10 for mercury, PCDDS/F TEQs, and particulate matter. Lead is not presented in Tables 4-8 and 4-10 because the proposed MACT standards did not include an explicit standard for lead.

Due to the uncertainties associated with the air modeling as mentioned above, the results shown should not be used in an absolute sense; however, they might be useful in a relative sense to detect trends. In this calculation, the air concentrations attributable to cement kilns are somewhat higher than those for hazardous-waste incinerators, even after imposition of MACT. This is likely due to a combination of factors. The main factor appears to be that the MACT standard is on a mass per volume basis (e.g., 0.2 µg PCDDS/F TEQ/dscm) and the cement kilns have a much greater air flow rate and, consequently, volume than hazardous-waste incinerators. (The average air flow rate for cement kilns is about 3,000 dscm/min. The average air flow rate for hazardous-waste incinerators is about 900 dscm/min.) A higher flow rate indicates that the typical mass emission of pollutants from cement kilns would be higher than hazardous-waste incinerators when a common concentration-based emission limit is used. The average difference in estimated ambient concentrations may be an artifact of using a single dispersion distribution for all facilities, however, the emission characteristics of cement kilns may on average be sufficiently different from hazardous-waste incinerators to cancel the effect. The committee cautions that comparisons between facilities based on these calculations is not a valid exercise.

TABLE 4-9 Estimated Ambient Air Concentrations of Various Pollutants Associated with Emissions from Cement Kilns that Burn Hazardous Waste Prior to Compliance with MACT (All concentrations are annual averages in units of micrograms per cubic meter.)

Pollutant	Number of Samples	Mean	Standard Deviation	Median	Range
Mercury	790	3.2×10^{-3}	2.1×10^{-2}	1.2×10^{-4}	5.6×10^{-9} to 5.2×10^{-1}
Lead	957	1.6×10^{-1}	5.0×10^{-1}	1×10^{-2}	7.7×10^{-7} to 6.9
PCDDS/F TEQ[a]	135	7.6×10^{-8}	5.2×10^{-7}	4×10^{-9}	1.5×10^{-11} to 1.6×10^{-5}
Particulate Matter	217	0.73	1.6	0.28	4.5×10^{-3} to 30

[a] PCDD/F is the toxic equivalent of the sum of the polychlorinated dibenzodioxins and polychlorinated dibenzofurans.

TABLE 4-10   Estimated Ambient Air Concentrations of Various Pollutants That Might Be Associated with Emissions from Cement Kilns That Burn Hazardous Waste after Compliance with MACT (All concentrations are annual averages in units of micrograms per cubic meter.)[a]

Pollutant	Mean	Standard Deviation	Median	Range
Mercury	2.5×10^{-4}	1.9×10^{-4}	2.0×10^{-4}	1.3×10^{-5} to 2.2×10^{-3}
PCDDS/F TEQ[b]	1.6×10^{-9}	1.3×10^{-9}	1.6×10^{-9}	9.2×10^{-11} to 1.2×10^{-8}
Particulate matter	0.6	0.45	0.47	3.2×10^{-2} to 4.8

[a] Based on the MACT emission limits that were proposed in April 1996.

[b] PCDD/F is the toxic equivalent of the sum of the polycholorinated dibenzodioxins and polychlorinated dibenzofurans.

Results from Environmental Monitoring Studies Around Incineration Facilities

Mathematical models and calculations have utility as tools for prediction and correlation of measurements in the environmental sciences. But uncertainties often remain after modeling because of the complexity of the environmental problem being modeled and the necessity to make assumptions throughout the modeling process.

Monitoring is an alternative to modeling environmental pollutants in the vicinity of incinerators. Also it provides a means of assessing a model's reliability. Monitoring can be used to assess local concentrations directly and, thus, avoid some uncertainty. Although environmental monitoring studies have been conducted around waste incinerators, some of the toxicants released by incinerators persist on a regional scale, rather than only on a local scale. Modeling and monitoring complement each other. Monitoring is useful for calibrating and validating models. Models are useful for interpolating and extrapolating monitoring data over space and time.

Ambient air is the most-common environmental medium that is monitored. For example, EPA (1991b) carried out a detailed study of ambient air quality in the vicinity of a municipal solid-waste (MSW) combustor in Rutland, VT. This facility burned 240 tons of waste a day; an electrostatic precipitator and wet scrubber were used to control particulate emissions and acid-gas emissions, respectively. In the investigation, air-dispersion modeling was conducted to determine locations for ambient monitoring and environmental sampling analyses for metals, polychlorinated dibenzo-*p*-dioxins (PCDDs) and polychlorinated diben-zofurans (PCDFs), and total particulate matter. EPA (1991b) found no correla-tion between the amount of waste burned and ambient-air particle concentra-tions. It also found that the proportions of the different compounds in ambient air did not resemble those in the stack gas. It was concluded that the incinerator was not the primary source of PCDDs and PCDFs in the ambient air surrounding the facility.

Other studies have reported similar findings. Hunt et al. (1991) performed ambient-air monitoring for particulate- and vapor-phase PCDDs and PCDFs in the vicinity of the Bridgeport, CT, waste-to-energy facility. Measurements were taken before and after the plant became operational. The results showed little difference in ambient concentrations of chlorinated dibenzodioxins and furans between the preoperational phase (0.097 pg/m^3) and the postoperational phase (0.088 pg/m^3). Stubbs (1993) examined trace metals and air-quality measures in the vicinity of the greater Vancouver, British Columbia, municipal incinerator and concluded that startup and operation of the plant had no measurable effect on air quality. A detailed study of ambient air in the vicinity of a greater Detroit plant that burned refuse-derived fuel was undertaken. The study evaluated many potential chemicals of concern (PCDDs and PCDFs, respirable particles, metals, polycyclic aromatic hydrocarbons [PAHs], polychlorinated biphenyls [PCBs], chlorobenzenes, chlorophenols, and inorganic acids) at four monitoring sites over a period of 2.5 years. One of the monitors was installed at the expected point of maximal effect, as predicted by air-dispersion modeling. The results were ana-lyzed with two-sample tests on means, multiple regressions, and principal-com-ponents analysis. All statistical procedures showed that there was no observable effect of the facility on the measured concentrations of any of the chemicals studied.

Single environmental media other than air have also been evaluated. Eitzer (1995) analyzed bovine-milk samples for chlorinated dibenzodioxins and furans from farms near a municipal solid-waste resource-recovery incinerator in Con-necticut, and found no statistically significant differences between preoperation-al and postoperational concentrations. The facility was designed to incinerate up to 620 tons per day of municipal solid waste and was equipped with a spray dryer and fabric filter for emission controls. The preoperational phase consisted of 17 samples from 5 farms. The postoperational phase included 12 samples from the same farms. Student's T-tests showed no statistically significant differ-ences at a 95% level of confidence between preoperational and postoperational results for any individual congener (the mean and standard deviation presented in Table 4-11 show that estimates have high uncertainty). Similar results were obtained for furans.

Ramos et al. (1997) analyzed bovine milk samples from 12 dairy farms in Spain and 23 samples of pasteurized bovine milk for PCDDs and PCDFs. They found that the levels of dioxins in the milk samples from farms located in rural

TABLE 4-11   Dioxins and Furans in Connecticut Cow's Milk

Congener	Preoperational (fg/g) Mean ± SD (n=17)	Postoperational (fg/g) Mean ± SD (n=12)
2,3,7,8-TCDD	17 ± 24	15 ± 27
1,2,3,7,8-PeCDD	6.4 ± 9.0	6.8 ± 12
1,2,3,4,7,8-HxCDD	31 ± 43	26 ± 35
1,2,3,6,7,8-HxCDD	32 ± 26	23 ± 14
1,2,3,7,8,9-HxCDD	15 ± 20	9.3 ± 15
1,2,3,4,6,7,8-HpCDD	94 ± 120	120 ± 140
OCDD	770 ± 1,500	1,700 ± 3,400

Source: Adapted from Eitzer 1995.

areas without specific dioxin sources (background levels) were slightly lower than those found in milk from the vicinity of potential dioxin emission sources (a waste incinerator, and chemical and metallurgical facilities) and similar to milk near to a paper production facility. In contrast to the conclusions of studies presented above, the authors concluded that the waste incinerator seems to be the emission source with the highest influence on the bovine milk gathered in its vicinity. The average dioxin concentrations found in pasteurized commercial milk were lower than those found in raw milk and were comparable to those found in retail milk from other countries.

McLaughlin et al. (1989) measured dioxins and furans in the soil near a municipal solid-waste incinerator in Hamilton, Ontario. This program was initiated because airborne emissions exceeded the Provincial guidelines for these chemicals. Fourteen soil samples including three control sites and the calculated point of maximum effect were analyzed. The incinerator had been operating for 10 years when the samples were obtained. All samples contained some dioxin and furan congeners. OCDD was the congener most-frequently detected. The range of OCDD was from less than 1.3-3,500 parts-per-trillion (ppt) for the study area and 810-3,200 ppt for the controls. There was no concentration gradient or deposition pattern that was consistent with the direction of prevailing winds or the location of the maximum ground-level concentration. On the basis of these data, the authors concluded that stack emissions from the incinerator have not accumulated in surface soil in the vicinity of the plant.

Schuhmacher et al. (1998) determined concentrations of PCCDs and PCDFs in 24 soil samples collected near a municipal solid-waste incinerator (Tarragona, Catalonia, Spain). Principal Component Analysis and hierarchical cluster analysis were used to compare these soil samples with a set of 10 additional samples collected outside the influence of the plant. The authors concluded that no remarkable PCDD or PCDF contamination was found, and soils in the vicinity of the incinerator provide patterns of PCDDS/Fs quite similar to those obtained in soils collected far from the influence of that facility.

Deml et al. (1996) measured dioxins and furans in blood and human milk of persons living in the vicinity of a municipal solid-waste incinerator in Germany. The facility had been in operation for 13 years and combusted 350,000 tons of waste per year. Blood samples were obtained for 43 persons who had been living in the study area for at least 10 years and 3 persons who had lived there for 8 years. The dioxin and furan concentrations in blood for the study group ranged from 3-19 picograms/gram (TEQ), compared to a control-group concentration of 10-48 pg/g (TEQ). Similar results were found for dioxins and furans in mother's milk. These authors concluded that living in the vicinity of this facility does not result in a higher body burden for dioxins and furans. Based on previous discussions in this report about the emissions of individual incinerators, such a finding is not surprising.

Kurttio et al. (1998) studied concentrations of mercury in hair of people in the proximity of a hazardous-waste-treatment plant that contains an incinerator. A baseline survey of the surrounding population and environment was conducted prior to the plant's operation in 1984; ten years later, investigators studied the same subjects. In 1984 and 1994, the median hair mercury concentrations were 0.5 mg/kg and 0.8 mg/kg, respectively. The researchers concluded that mercury exposure increased as distance from the facility decreased; however, the increase in exposure was minimal and, on the basis of current knowledge, did not pose a health risk.

Bache et al. (1991) analyzed metals and PCBs in vegetation around an incinerator operating without air-pollution controls. The results showed that, of six metals and PCBs considered, only lead was statistically significantly higher than background. The mean of 9 upwind samples for lead was 2.1 mg/kg of vegetation with a standard deviation of 1.2 mg/kg. The downwind sample values depended on distance from the stack. The closest sample was 30 mg/kg. Samples declined to a value that was within the 95% confidence interval of the background data at a distance of 650 meters from the stack. However, Carpi et al. (1994) found increased concentrations of mercury (206 parts-per-billion (ppb), compared with a control value of 126 ppb) in sphagnum moss within 1.6 km of a municipal solid-waste incinerator in New Jersey.

Collett et al. (1998) analyzed levels of cadmium and lead in air and surface-soil samples collected in an area around the Baldovie municipal-waste incinerator in Scotland. They reported that the spatial distribution of lead levels in soils showed a marked variation downwind from the Baldovie incinerator in comparison with the background level for the area. However, the lead levels remained well within the typical range of lead in rural, unpolluted, British soils. The authors compared the observed levels of lead in local soils with the predicted downwind long-term ground-level lead distribution in air and found that atmospheric emissions of lead originating from the Baldovie incinerator directly determine concentrations of lead in soils within a radius of 5 km of the incinerator. However, in the case of cadmium, the authors found neither a marked nor extensive contamination of the sampled area around the incinerator; the levels were within the typical range of cadmium levels in rural, unpolluted, British soils.

There are two dominant mechanisms through which plants can accumulate metals from the atmosphere. Materials in the vapor phase may be directly taken up into plants. Both vapor and particles may be washed out of the atmosphere by precipitation (wet deposition), and dry deposited directly on leaves. The increased concentration of mercury in the Carpi et al. (1994) study and the increased lead in the Bache et al. (1991) study were probably a consequence of wet deposition rather than vapor uptake. The significance of this conclusion is that wet deposition typically occurs in the immediate vicinity of a source, whereas vapor-phase uptake can occur on a regional basis. Yasuhara et al. (1987) showed that incinerators were not important contributors of dioxins and furans to local soil or sediment samples.

Thus, single-medium studies indicate that important dioxin and furan concentrations could not be detected in bovine milk, soil, or vegetation, but increases in lead could be found in soil and vegetation and increases in mercury could be found in moss and human hair samples collected near incinerators.

In addition to the single-medium studies, there have been several multimedia studies around incinerators. Laidlaw Environmental Services, Ltd. (LESL) has operated a hazardous-waste treatment, storage, and disposal facility in Sarnia, Ontario, for over 25 years. A component of this facility is a liquid-injection hazardous-waste incinerator that treats 120×10^6 L of waste per year. Emissions are controlled with a secondary combustion chamber, spray dryer, and fabric filter. Both LESL (Ecologistics 1993a,b) and the Ontario Ministry of Environment and Energy have conducted multimedia monitoring in soil and vegetation around the facility, including locations determined by air modeling to have the greatest potential concentration. Most organic chemicals were not detected in either study area or control locations. For example, in 1992, OCDD, typically the most-common congener, was not detected in soils at detection limits ranging from 0.0052-0.050 nanograms/gram (ng/g). The results obtained in those studies were evaluated by comparison of samples obtained from the study area to those obtained from a control area, and by comparison of concentrations in soil to those considered to be typical by the MOEE (1989). A comparison of the metals of the greatest potential toxicological significance to the "upper limits of normal" developed by MOEE is given in Table 4-12. On the basis of these studies, it was concluded that the facility was not a major source of metals or PCDDs and PCDFs in the environment.

Stubbs and Knizek (1993) analyzed vegetation in the vicinity of the greater Vancouver refuse incinerator and concluded that trends in soil and vegetation trace elements in the study area in the study period indicated little or no change due to the startup and operation of the facility. The results are given in Table 4-13. They also concluded that the facility had no measurable effect on trace-element or PAH concentrations in soil, vegetation, or vegetative growth in the

TABLE 4-12 Metals in Soil Near a Hazardous-Waste Incinerator Compared with Background

Element	Upper Limit of Background (mg/kg)	Number of Samples	Range of Concentrations (mg/kg)
Arsenic	10	21	3.6-6.2
Cadmium	3	21	0.3-2.3
Chromium	50	21	15-39
Lead	150	21	16-28
Mercury	0.15	5	0.03-0.10

Sources: Adapted from MOEE 1989; Ecologistics 1994.

vicinity of the facility. Fruin et al. (1994) presented the results of a multimedia monitoring study (ambient air, soils, and sediment) in the vicinity around a hazardous-waste incinerator operated by 3M (Minnesota Mining and Manufacturing) in Minnesota. The incinerator is a rotary kiln with a secondary combustion chamber, heat-recovery equipment, and five pollution-control devices. The study focused on particulate matter in ambient air and metals in soils. The study found that the incinerator contributed less than 1% of the total suspended particulates and the respirable particulates (PM_{10}) to the total concentration in the ambient environment. A total of 180 soil samples in the potential impact zone of the incinerator were also analyzed for 21 metals. The results for the metals of greatest potential toxicological significance in nearby agricultural land, along with ranges for background presented by the authors, are included in Table 4-14. On the basis of the particulate and metals data, the authors concluded that there were negligible contributions of those combustion products from the incinerator to local concentrations.

In one of the largest studies of its kind, the Texas Natural Resource Conservation Commission (TNRCC 1995) evaluated environmental media (air, soil,

TABLE 4-13 Metals in Soil Near the Greater Vancouver Incinerator

Element	Preoperational (ppm)	Postoperational (ppm)
Arsenic	0.11	0.09-0.12
Cadmium	1.48	0.67-1.17
Nickel	2.47	1.14-1.57
Lead	1.79	1.74-3.13
Selenium	0.08	0.05
Fluorine	3.58	2.52-4.67
Sulfur	0.19	0.16-0.21

Source: Adapted from Stubbs and Knizek 1993.

TABLE 4-14   Metals in Soil Near the 3M Incinerator

Element	Background Range (ppm)	Facility Range (ppm)
Arsenic	(not given)	< 10-13
Cadmium	Trace-1	< 0.07-1.0
Chromium	1-1,500	5.2-12
Lead	< 5-700	8.1-29
Mercury	10-3,400	7.0-18

Source: Adapted from Fruin et al. 1994.

and vegetation) in a community that contains two cement kilns that use hazardous waste as fuel as well as a cement kiln and secondary steel mill that use conventional fuel. Over 940 ambient-air samples were analyzed for suspended particles, 188 for respirable metals, and 135 for volatile organic compounds, and 175 soil samples were analyzed for various potential pollutants. Other analyses were also conducted of acidic gases and chemicals on vegetation. As an example of their results, the dioxin and furan concentrations in soils potentially affected by the facilities ranged from 0.3 to 17.9 ppt (TEQ) with a median of 1.4 ppt (TEQ). This may be compared to the range for unaffected background of 0.8-3.2 ppt (TEQ) with a median of 1.8 ppt (TEQ).

Lorber et al. (1998) examined PCDD/PCDFs in ash from, and soil and air around the Columbus, Ohio, municipal waste-to-energy facility. This facility was estimated to emit as much as 1,000 g of TEQ per year for its approximate 11 years of operation (compared with a 1994 estimate of total emissions of 9,300 g of TEQ per year from all sources in the United States). The effect of the incinerator was detected in air and soil samples. Two air samples known to be downwind of the operating facility had TEQ concentrations approximately 5-fold higher than background. Approximately 2% of the total emitted dioxins were estimated to be present in the soil within about 3 km of the incinerator. However, the authors concluded that despite the magnitude of the emissions, soil and air concentrations in the urban area of Columbus did not exceed urban air and soil concentrations of dioxins found around the world.

Limitations of the studies cited above include reflection of a nonrandom set of facilities; inconsistency of methods, and problems with sampling and analytical techniques, detection limits, number and location of samples, duration of studies, contaminant contribution from other emission sources, and quality assurance and quality control. Therefore, no definitive conclusions can be drawn about waste incineration in general. However, taken as a whole, the weight of the evidence contained in those studies suggests the following. First, in principle, measurement of contributions of substances within various environmental media is a feasible method for assessing environmental emissions from incinerators. Second, the results are consistent with the hypothesis that emissions of dioxins and furans are more important on a regional than a local scale, whereas the emissions of some metals, such as mercury, are important close to the stack and on a regional scale. Third, it appears that incinerators with similar waste streams, operating conditions, and emission controls to those studied are unlikely to now be a major contributor to local ambient concentrations of the chemicals of concern noted in this report.

CONCLUSIONS

- Although releases to the environment from incineration facilities occur mainly by air emissions, multiple potential pathways in the environment to humans exist. Variations in the physical properties of substances of concern and the extent to which they persist in the environment result in differences in the types of pathways. Results of environmental monitoring studies around incineration facilities indicate that the specific facilities studied are unlikely to be major contributors to local ambient concentrations of the substances of concern noted in this report. However, methodological limitations of the studies do not allow for general conclusions to be made about waste incineration's contributions to environmental concentrations of those contaminants. They also do not allow for characterization of total human exposure.

- The air concentration of gases (including carbon monoxide and acid gases) and fine particles can be estimated by dispersion models to data on stack emissions, stack height, and local meteorological conditions. For the case of heavy metals carried by the particles, the deposition fluxes of the particles and the transfer from the soils into the food chain through vegetable produce and animal fodder needs to be taken into account. Models for the dispersion and the uptake into the food chain are available for use in risk assessments.

- Mercury and dioxins and furans are classes of pollutants where historically incinerators are estimated to have contributed a significant portion of the total national emissions. These classes of pollutants are characterized by their long-range air transport, persistence, and relative uniformity of deposition fluxes on regional bases. Whereas one incinerator might only contribute a small fraction of the total environmental concentration of these compounds, the sum of the emissions of all the incineration facilities in a region could be considerable. Because a number of older incinerators had been closed down and replaced by modern low-emitting units, it is now uncertain how much incineration contributes to the environmental concentrations of these compounds.

- Computational models for the environmental fate and transport of mercury and the dioxins and furans provide useful information for assessing the major exposure pathways for humans, but are unable to provide overall estimates of environmental contributions from an individual facility to better than within a factor of 10. The models suggest that fish consumption is potentially the major pathway of human exposure to mercury and that meats, dairy products, and fish are potentially the major pathways of exposure to dioxins and furans.

- Because the food chain is potentially the primary path for exposure to dioxins and furans and toxic metals such as mercury, the correlation between total exposure and local emissions is expected to be low. Because of the persistence of these pollutants and their long range transport, not all relevant sources contributing to exposure pathways are local. In addition, the foods that are implicated as the major pathway of exposure could include a fairly high portion of products imported from other regions.

- Detailed information is required on the distribution of contaminants in the environment once they are released from waste-incineration facilities. Better assessments are needed of transport, accumulation, and physical and chemical transformations of contaminants through all potentially important exposure pathways, including air, food, soil, and water.

- Exposures to dioxins and furans, mercury, and other heavy metals are best assessed by monitoring food consumption. Drinking water may be an important pathway in some circumstances. Models can be used to establish the chemicals' pathways to humans, regional distribution, and persistence in the environment.

RECOMMENDATIONS

- Environmental assessment and management strategies for emissions from individual incineration facilities should include an appropriate regional-scale framework for assessing the collective dispersion, persistence, and potential long-term impacts of incinerator emissions on human health.

- Better material balance information—including measurements of source emissions to air and deposition rates to soil, water, and vegetation—are needed to determine the contribution of waste-incineration facilities to environmental concentrations of persistent chemicals. The variation of these emissions over time needs to be taken into account for the short-term to determine if any important emission increases occur at an incineration facility, and for the long-term to measure changes due to replacement of less-efficient incinerators with modern, lower-emitting units.

- In characterizing potential health effects of waste-incineration emissions, all environmental media should be assessed. Also, all possible exposure routes by which contaminants enter the body of a person should be considered, including inhalation, ingestion of food or drink, and absorption through skin. Such an approach is consistent with EPA's guidelines for health risk assessments.

5

Understanding Health Effects
of Incineration

To understand the possible health effects attributable to waste-incineration emissions, information is needed on contributions made by incineration to human exposures to potentially harmful pollutants and the responses that might result from such exposures. As discussed in this chapter, various tools have been used in attempts to evaluate effects of incineration. Of these tools, all of which contribute to our understanding, risk assessment methods have provided the most-detailed information for regulatory decisionmakers. Although past regulatory risk assessments have suggested that the risks posed by emissions from a well-run incinerator to the local community are generally very small, the same may not be true for some older or poorly run facilities. Some of the available assessments, however, may now be considered inadequate for a complete characterization of risk, for example, due to their failure to account for changes in emissions during process upsets, or because of gaps in and limitations of the data or techniques of risk assessment available at the time. There are limitations in the data and techniques of risk assessment, for example, in considering the effect of potential synergisms between chemicals within the complex mixtures to which humans are exposed, or the possible effects of small increments of exposure on unusually susceptible people. In addition, there are important questions not typically addressed by the usual risk assessment for single facilities such as the collective effect of pollutants emitted from multiple units; regional-scale effects of persistent pollutants; and the effects on workers in the facilities themselves.

This chapter examines the tools used to evaluate the potential for health effects from incineration facilities, and discusses some of the results obtained with those tools. The two primary tools are environmental epidemiology and risk assessment, both of which have been the subject of National Research Council reports (e.g., NRC 1991a, 1994, respectively). In addition, environmental monitoring studies provide immediately useful estimates of ambient concentrations, while biomarker studies hold some promise for future application. The first section of the chapter discusses these tools, and their strengths and limitations relative to one another.

There have been few epidemiologic studies in populations characterized as exposed to contaminants emitted by incineration facilities. Thus, there is a lack of evidence of any obvious health effects related specifically to incinerator exposure. That is, there have been few anecdotal reports that indicated any particular concern for incinerators (as opposed to air pollution in general, for example) or that generated testable hypotheses. Moreover, as discussed later in this chapter, it would be difficult to establish causality given the small populations available for study, the possible influence of factors such as variations in the susceptibility of individuals and emissions from other pollution sources, and the fact that effects might occur only infrequently or take many years to appear. The second section of the chapter summarizes what data are available, and discusses what conclusions can be drawn from those data.

The main information on potential health effects that might arise in populations potentially exposed to substances emitted by incineration facilities comes from risk assessments of individual chemicals emitted by incinerators, combined with monitoring of emissions from incinerators. Such assessments typically indicate that, of the many agents present in incinerator emissions and known to be toxic at high exposures, only a few are likely to contribute the majority of any health risks and such health risks are typically estimated to be very small. This chapter examines the toxic effects of such agents. It also illustrates ways to compare the expected ranges of environmental concentrations attributable to incineration with concentrations known to be toxic, and in the context of total exposures.

The toxic agents were selected for discussion on the basis of the current state of knowledge of the nature of emissions from incinerators and the results of various risk assessments. They are particulate matter (PM), carbon monoxide (CO), acidic gases (i.e., NO_x, SO_2, HCl) and acidic particles, certain metals (cadmium, lead, mercury, chromium, arsenic, and beryllium), dioxins and furans, polychlorinated biphenyls (PCBs), and polyaromatic hydrocarbons (PAHs). The emissions of most of those substances were considered in Chapters 3 and 4.

Particulate matter, CO, lead, and acidic gases and acidic particles have been under regulatory scrutiny for the longest period. Typically, there are well-defined statutory limits on their emission rates or allowable ambient concentrations or increments in ambient concentrations under federal or state statutes. In many risk assessments, such materials have been evaluated solely by comparisons with such statutorily defined limits, limits that have been designed to reduce certain risks from these pollutants below acceptable values. Although there are occupational-exposure limits for most of the other metals and organic compounds listed above, there are no well-defined ambient or emission standards under federal or some state regulations; however, in risk assessments, those materials are typically found to contribute to the majority of the estimated risk, either in contribution to lifetime cancer risks or in contribution to potential noncancer effects. Historically, risk assessments have identified the dioxins and furans as the principal contributors to estimated risks posed by most incinerators with arsenic often next. However, estimates of relative contributions of pollutants to total risk depend on incinerator emission characteristics, populations potentially exposed, potential routes of exposure, and, to some extent, the amount of information that has been collected.

In addition, this chapter discusses "at-risk" populations (populations that might be at increased risk due, at least in part, to pollutants emitted from incinerators). The chapter ends with the main conclusions on understanding health effects of waste incineration reached by the committee and presentation of research needs.

TOOLS FOR EVALUATING HEALTH EFFECTS

Whenever searching for small or subtle health effects of exposures to environmental contaminants, it is best to use a variety of approaches and to critically compare their results. The primary tools that have been used include epidemiologic studies and risk assessments. These are separately discussed in detail below, although it should be realized that there can be a good deal of overlap between the approaches. Environmental monitoring, biomarkers of exposure or effect, and life-cycle assessment are other commonly used tools that produce data which often confirm, support, or enhance the findings obtained during the conduct of epidemiologic or risk-assessment investigations. Exposure assessment plays an important role in may of those approaches.

Such approaches are used to evaluate multiple environmental media (air, surface water, soil, groundwater, sediments, and any other media that might be distinguished), multiple exposure pathways, many scenarios for exposure, multiple routes (inhalation, ingestion, and dermal), multiple chemicals, multiple population groups, and many health end points.

However, the approaches currently used to assess the effects of waste incineration are typically site-specific and facility-specific and so fail to address two important questions regarding a facility or site:

- To what extent does an incineration facility alter the environmental concentrations of substances of concern or alter the existing magnitudes of human exposure to those substances?
- What are the overall local and regional contributions of waste incineration to human exposures?

Epidemiologic Studies

Epidemiologic studies are conducted to test hypotheses about the occurrence (usually prevalence or incidence) of a health outcome, to measure the strengths or sizes of relationships between such outcomes and quantifiable factors (e.g., the magnitude of exposures) or qualifiable factors (e.g., exposure status), or to generate testable hypotheses about such relationships. The methodology, strengths, and weaknesses of environmental epidemiologic studies have been discussed in previous NRC reports (NRC 1991c, 1997). As discussed there, the principal strengths of epidemiologic studies are:

- The people studied include those likely to have been exposed to the material of interest. For incinerator emissions, there is no extrapolation necessary from single chemicals to the complex mixtures to which humans are actually exposed.
- Humans themselves are studied in actual exposure conditions—there is no extrapolation from different animal species or different conditions.
- Individual and group variability in both exposure and sensitivity are necessarily taken into account.

The principal challenges to be addressed by epidemiologic studies in establishing causality include:

- Identifying suitably exposed populations of sufficient size.
- Identifying effect modifiers and/or potentially confounding factors.
- Identifying biases (including reporting biases) in data collection (e.g., Neutra et al. (1991) present an interesting case study of this problem).
- Measuring exposures.
- Measuring effects that are small, might occur only infrequently, or take many years to appear.

Risk Assessments

Risk assessment is the use of procedures to estimate the probability that harm will arise from some action such as the operation of a facility. The procedures used to perform risk assessments vary widely, from a snap judgment to the use of complex analytic models. However, risk assessments of incineration or incineration facilities have become more structured and formalized, following the four-step paradigm described in previous NRC reports (NRC 1983, 1994).

In the case of a particular incinerator, the first step, hazard identification, might begin with enumeration of the chemicals present in emissions and suspected of posing health hazards (and this alone might be an expensive proposition in unusual specific cases). The emissions have to be quantified, the potential health effects identified, and the conditions under which a chemical might cause those effects defined. The attempt to obtain emission-rate estimates might take the form of direct measurements, which are limited by the sensitivity of the measuring methods, the variability over time of emission rates, the cost of such measurements, and the inaccuracies affecting all such field work. Alternatively, similar measurements from other, comparable facilities might be used as bases to estimate emissions. The result is generally a list of chemicals with their expected average emission rates and sometimes a measure of the variability of the emission rates with time—for example, how short-term emission rates might differ from the long-term average. In many cases, there may be a list of the emission rates that are identified as maximums by the owner or operator of the facility.

After developing a list of chemicals identified as potentially of concern, a dose-response assessment is used to evaluate quantitatively the relation between exposures and toxic responses. Ideally, this assessment would consider all the particular conditions of exposure, including the complete mix of other potential contaminants from incineration, and exposures to the same and different chemicals from other sources. In practice, dose-response assessments are limited, by the regulatory milieu of most risk assessments, to the use of cancer potency-slope estimates or unit risks[1] (for the evaluation of cancer risks) and reference, doses[2] (for the evaluation of noncancer risks) published in the Integrated Risk Information System (IRIS)[3] or other regulatory documents by the Environmental Protection Agency (EPA) or the Agency for Toxic Substances and Disease Registry (ATSDR).

Most of the effort of individual risk assessments has gone into the evaluation of exposure, which is the third step in the risk-assessment paradigm. As discussed in Chapter 4, exposure assessment involves an estimation or measure-

---

[1] Cancer potency-slope estimates or unit risks. The human cancer potency-slope is the incremental increase in lifetime cancer risk per incremental unit of lifetime average dose (generally by ingestion, occasionally by other routes of exposure). The estimates of cancer potency-slope is obtained by assuming that the dose-response curve may be linear at low doses, and extrapolating to low dose from higher experimental doses. In many cases, there is an additional extrapolation from laboratory animals to humans. The unit risk is the incremental increase in lifetime cancer risk per incremental unit of air concentration of an airborne carcinogen. It is estimated using methods similar to those used for cancer potency-slope, but with slightly different assumptions adopted for inter-species extrapolation.

[2] The reference dose is a long-term average dose rate that is expected to result in no noncancer health effects in humans. It is obtained from experimental results in humans or animals by a relatively well-defined procedure that incorporates safety factors to account for all the defined extrapolations performed.

[3] IRIS. EPA's (1992b) Integrated Risk Information System (IRIS) is a database of human health effects that might result from exposure to various substances found in the environment. IRIS is accessible via the Internet at http://www.epa.gov//iris.

ment of the concentration of specific substances in each environmental medium, and the time individuals or populations spend in contact with the substances. The network of exposure pathways becomes more and more complex as more-remote regions are incorporated. Food contaminated near an incineration facility might be consumed by people close to the facility or far away from it. Thus, local deposition on food might result in some exposure of populations at great distances, due to transport of food to markets. However, distant populations are likely to be more exposed through long-range transport of pollutants and low-level, widespread deposition on food crops at locations remote from a source incineration facility. To be most useful, exposure assessments need careful definition of the scenarios to which the assessments apply. Within such scenarios, the distribution of individuals or populations exposed need to be accounted for, and other variabilities and uncertainties incorporated (EPA 1992c). In order to dovetail with the dose-response assessments, care must be taken in the exposure assessment so that doses can be evaluated in the correct way. Potential doses can be expressed as the average rates at which material crosses the epithelial layer of an exposed individual (by inhalation or ingestion) or enters the outer layer of skin (e.g., through dermal contact) per unit of body weight per day (EPA 1992d; DTSC 1992a,b). However, such measures do not necessarily correspond to the does-response measures (e.g., carcinogenic potency-slope, unit risk, and reference doses), which typically relate response to exposures rather than doses. In the absence of such exact correspondence, exposure-dose relationships may become crucial.

The final step of the risk-assessment paradigm, risk characterization, involves integrating the results of exposure assessment, dose-response assessment, and hazard assessment in such a way as to "develop a qualitative or quantitative estimate of the likelihood that any of the hazards associated with the agent of concern will be realized in exposed people" (NRC 1994). Risk-assessment results are generally expressed as lifetime cancer risks (calculated by taking the sum—over the pollutants of interest—of the products of lifetime average exposure to each pollutant and its potency slope) or as summary hazard indices (the sum over various chemicals of the ratio of estimated dose of each chemical to its reference dose). In the case of lead, projected blood-lead concentrations are used. A complete risk characterization should also contain a full discussion of the uncertainties associated with the estimates of risk.

Risk assessment of waste incineration facilities can involve the following aspects:

- Measurement or estimation of emission rates from specific facilities.
- Modeling designed for tracking the flow of substances of concern through the environment.
- A large body of information on toxicity of many emitted substances, in particular of dose-response information.

- Characterization of the expected effect of new incinerators, or of what might happen in the future with any incinerator.

Such risk assessments are congruent with most regulatory schemes—the principal inputs to risk assessments are also characteristics of incinerators that are usually regulated, for example, emission rates.

The lack of complete data leads to uncertainties involved and the problem of communicating such uncertainties. Those uncertainties arise from the following:

- The lack of complete emission data, especially for nonstandard operating conditions.
- The problem of dose-response assessment at low doses, and in particular of low-dose, cross-species, inter-route, and temporal dose-pattern extrapolation.
- The lack of toxicity data on most products of incomplete combustion.
- The lack of physical and chemical information on relevant characteristics of substances of concern.
- The use of unverified models of transport of substances in the environment, due to incomplete knowledge as to how such transport occurs.
- The variability of all aspects of the assessment, due to variations in physical conditions (e.g., topography, temperatures, rainfall, soil types, and meteorological conditions), characteristics of people (e.g., eating habits, residence times, age, and susceptibility), and so on, leading to wide ranges of exposures and risks for different people.
- The possibility of errors and omissions in the assessment (e.g., omission of an important pathway of exposure).

Because of the variability and uncertainty, most risk assessments have not been designed to quantify actual health risks; rather they have been designed solely for regulatory purposes to yield upper-bound estimates of health risks that may be compared to regulatory criteria.

Other Tools

Environmental monitoring and biological markers of exposure or effect are two tools often used in conjunction with epidemiologic or risk assessment investigations. These tools aid in identifying or confirming pollutants that may give rise to adverse health effects. Life-cycle assessment (LCA) has been used to evaluate the resource consumption and environmental burdens associated with a product, process, package, or activity throughout its lifetime over large geographic regions. LCA can be used in conjunction with risk assessments to assess effects over a broad scale—from the time of introduction of a chemical into the environment to its destruction.

Environmental Monitoring Studies

In principle, it is desirable to measure concentrations of certain pollutants directly from the incinerator in the surrounding environment. Such monitoring is most commonly of the ambient air, but soil, water, sediments, vegetation, and foods have at times been monitored for some of the emitted pollutants.

Environmental monitoring is principally useful because it directly measures the concentrations of certain materials from a particular incinerator, in some cases in the media of immediate interest (e.g., dioxins in vegetation and cows' milk). No health effects are measured. For use in evaluating health effects, however, environmental monitoring suffers from several disadvantages, because:

- There is usually a problem in distinguishing the contribution of the incinerator to environmental concentrations.
- Monitoring measurements are limited both in space and in time while concentrations are often highly variable in both time and space.

For these reasons, environmental monitoring is usually most useful in confirming, calibrating, or disproving the modeling efforts used in risk-assessment methodology.

Biologic Markers (Biomarkers) of Exposure or Effect

There is now considerable interest in the use of biologic markers of exposures or effects in epidemiologic studies of the health risks posed by some occupational and environmental exposures (NRC 1989a,b, 1992a,b, 1995). Some of these studies are relevant to likely exposures to substances emitted from incinerators—for example, measurements of specific congeners of PCDDs and PCDFs in blood and adipose tissues of exposed workers (Schecter et al. 1994), analyses of chlorophenol and pyrene metabolites in blood and urine of incinerator workers (Angerer et al. 1992), analysis of selected DNA adducts in blood samples of incinerator workers and measurement of various indexes of metal exposure in workers (Malkin et al. 1992).

Such studies are likely to be generally useful for evaluating exposures to specific materials that might be present in incinerator emissions or evaluating the presence of effects that might be associated with incinerator emissions. However, no biomarker of exposure or effect associated uniquely with incinerator emissions has been identified, nor is any such biomarker likely to be identified, inasmuch as incineration emissions as a class do not (so far as is now known) have components that are peculiar to them nor that cause unique effects.

Thus, although the use of biomarkers might add substantially to the accuracy of measurement of exposures and effects in epidemiology, it is not likely to reduce substantially other major sources of uncertainty that are entailed in the application of epidemiology to incinerator emissions.

RESULTS OF EPIDEMIOLOGIC STUDIES OF INCINERATOR-EXPOSED POPULATIONS

This section discusses the findings from epidemiologic studies of incinerator-exposed populations, including the few studies of human populations in the vicinity of incinerators and the more-detailed health studies of workers in these facilities. In general, information is rather sparse on the relationship between human exposure to pollutants released to the environment by incinerators and the occurrence of health effects.

Studies of Local Populations

In one of the earliest epidemiologic studies of populations in the vicinity of waste incinerators, Zmirou et al. (1984) obtained data on the use of medications for respiratory illnesses over a 2-year period among residents of a French village at distances of 0.2, 1, and 2 km from a refuse incinerator. Medication use was determined by examining prescription forms filed by the residents after each purchase. The purchase of respiratory medications (bronchodilators, expectorants, antitussants, and so on) decreased as the distance of the residences from the incinerator increased, and the relationship was statistically significant. However, the prevalence of other possible confounding risk factors for respiratory illness, such as socioeconomic and geographical situation, were not accounted for in this study, and no causal associations can be inferred.

After reports of illness and neurologic symptoms in workers employed at the Caldwell Systems, Inc. hazardous-waste incinerator in western North Carolina and health complaints of nearby residents, the Agency for Toxic Substances and Disease Registry (ATSDR) performed a cross-sectional study in the surrounding community for the prevalence of self-reported respiratory, musculoskeletal, neurologic, irritative, and other symptoms (ATSDR 1993a). A higher prevalence of self-reported respiratory symptoms, but not of respiratory or other diseases, was found in the target population than in a nearby comparison population. Prevalence data were adjusted for age, sex, and cigarette smoking. Members of the population close to the incinerator were almost nine times more likely to report recurrent wheezing or cough, and they were almost twice as likely as those living further from the site to report respiratory symptoms (after adjustment for smoking, asthma, and environmental concern). Other symptoms—including chest pain, poor coordination, dizziness, and irritative symptoms—were also statistically significantly greater in the population close to the incinerator. However, the investigators noted that neither the prevalence of physician-diagnosed diseases (as reported by subjects) nor hospital admissions for these diseases differed between the target and comparison populations, and they pointed out that the retrospective nature of the study (the incinerator operated from 1977 to 1988, and the cross-sectional study was conducted in 1991) limited interpretation of the findings. One of the major concerns was recall bias associated, in part, with the greater than 2-year gap between the shutdown of the incinerator and the conduct of the symptom survey. Another factor was the large amount of adverse publicity that the incinerator received before shutdown. Although the investigators attempted to control for recall bias by stratifying their results according to the respondents' expression of environmental concern, they concluded that they were only partially successful, inasmuch as the higher rate of self-reported symptoms from the population close to the incinerator was not associated with any difference in physician-diagnosed disease rates or in hospital-admission rates between the two communities. The investigators also acknowledged that they had no direct measures of community exposure to incinerator-emitted pollutants, which had ceased more than two years before the study, and thus could not estimate differences in exposures among individuals within the population close to the incinerator. Thus, this study is of limited utility in evaluating the effect of incinerator exposures, but emphasizes the necessity of controlling for various types of bias.

Wang et al. (1992) tested the lung function of 86 primary-school children living in Taiwan near a wire-reclamation incinerator and compared the results with those in 92 schoolchildren in a school in a "nonpolluted city." All children had been inhabitants of their districts since birth and had similar socioeconomic backgrounds. Air pollution in the incinerator district was considerably greater than that in the comparison city. SO_2 concentrations were 18.1 and 2.1 parts per billion (ppb), respectively, and NO_2 concentrations were 12.6 and 2.1 ppb. Questionnaire responses yielded no differences in the prevalence of respiratory symptoms among children in the two areas. However, the prevalence of children with abnormal forced expiratory volume in 1 second (FEV_1) was statistically significantly greater in the incinerator community (17.5% vs. 3.2% with abnormal test results). Two groups of children with no reported respiratory symptoms were tested later for bronchial hyperactivity—26 children in the target population and 26 children in the comparison population. A positive methacholine-challenge test was found in 9 of the former and only 1 in the latter group. The authors concluded that "the high level of air-pollution" in the population close to the incinerator was associated with a detrimental effect on lung function in primary-school children; however, they did not obtain data that would allow them to ascribe the measured air pollution to emissions from the incinerator, nor did they characterize other sources of air pollution in the target population. Thus, this study appears to demonstrate that higher concentrations of air pollutants alter pulmonary function in children, but does not directly allow any inference about the contribution of incinerators as opposed to other pollutant sources to either environmental concentrations or health effects in particular.

Gray et al. (1994) studied the prevalence of asthma in children living in two regions of Sydney, Australia, where incinerators burned sewage sludge and in one comparison community within the same metropolitan area. They measured respiratory illness in the previous year by questionnaire, airway hyperactivity by histamine-inhalation tests, and atopy by skin tests in 713 children 8-12 years old in the two regions and in 626 children of the same age in a comparison community without an incinerator. All children attending public and parochial schools within a 5-km radius of each of the study communities were selected for the study. Measurements of SO_x, NO_x, H_2S, O_3, and particulate matter during the study period showed no differences among the three regions. The prevalence of current asthma, atopy, symptom frequency, or asthma of any category of severity was not statistically different between incinerator and comparison regions. Results of tests of baseline lung function and of airway hyperactivity also did not differ among the three groups of children. The authors pointed out that their study was not designed to measure short-term acute effects of pollutant exposures. They also noted that the prevalence of asthma symptoms and atopy in this population of Sydney children, including those from the incinerator and comparison communities, was comparable with that in four other populations of children studied in Australia, and they concluded that emissions from high-temperature sewage-sludge incinerators appeared to have no adverse effect on the prevalence or severity of childhood asthma.

Shy et al. (1995) reported on the first year of a 3-year study of three incinerator communities and three comparison communities in southwestern North Carolina. The study was designed primarily to assess the acute respiratory effects of living in the neighborhood of an incinerator. Of the incinerators, one was a biomedical-waste incinerator, one a municipal-waste incinerator, and the third an industrial furnace fueled by liquid waste. Comparison neighborhoods were pair-matched to the incinerator communities on density and quality of housing and were upwind of and at least 3 km from the incinerators. In each neighborhood, 400-500 households were surveyed by telephone for sociodemographic characteristics, including prevalence of such respiratory risk factors as smokers in the home, and the prevalence of acute and chronic respiratory symptoms. No differences in respiratory-symptom prevalence were found between the subjects living near to either biomedical-waste incinerator or municipal-waste incinerator and their comparison communities. Several chronic respiratory symptoms were reported to have a higher prevalence in the liquid-waste combustor community than in its comparison group, but this difference did not persist when the symptom prevalence in the liquid-waste combustor community was compared with the pooled prevalence of symptoms in the three comparison communities.

Concentrations of particulate matter, including PM_{10} and $PM_{2.5}$, and of acidic gases, including SO_2 and HCl, were monitored in each of the study areas and did not differ measurably between target and comparison communities, either on a daily-average or monthly-average basis. Results of baseline lung-function tests also did not differ statistically significantly between target and comparison communities. Subjects with a history of recent wheeze or other asthma-like symptoms and nonsmoking subjects with no history of respiratory symptoms were recruited from each study community to record twice-daily peak expiratory-flow rates, acute respiratory symptoms, and (among asthmatics) use of asthma medications for 35 consecutive days during each year of study. None of the paired communities showed a difference in peak expiratory flow rates, adjusted for age, sex and height, or in the incidence of acute respiratory symptoms over the 35-day recording period during the first year of study.

A chemical mass-balance analysis of particle sources during the period of the study estimated that a maximum of 3% of the particle mass in ambient air could be attributed to emissions from the biomedical-waste incinerator on days when the prevailing wind was blowing directly from the incinerator toward the air-monitoring station less than 1 km away. On days when the prevailing wind was in other directions, the contribution of the incinerator to the particle mass measured at the monitoring station was less than 1%. Shy et al. (1995) concluded that data from the first year of study were compatible with the null hypothesis of no difference in acute or chronic respiratory symptoms or lung function between paired target and comparison communities and that particle and acid-gas emissions from the three incinerators contributed trivial quantities to the ambient-air concentrations in the adjacent neighborhoods.

Thus, the few community-based epidemiologic studies reported to date have yielded no evidence that acute or chronic respiratory symptoms are associated with incinerator emissions. However, that conclusion is based on only two community studies, that of Gray et al. (1994) in Sydney, Australia, and that of Shy et al. (1995) in North Carolina. In both measures of air quality, specifically of particles and gases, showed no difference between the incinerator and comparison communities. The lack of difference in concentrations of commonly measured air pollutants found in these studies does not rule out the possibility of differences in concentrations of unmeasured pollutants of concern (such as PCDDs and PCDFs) that may be present in incinerator emissions as well as in background pollution. Thus, such measurements do not directly show that there can be no excess of respiratory effects due to incinerators. However, the absence of differences in the prevalence of asthma among exposed children in the Sydney study and the absence of differences in the incidence of acute respiratory symptoms or in lung function in the North Carolina study are at least suggestive that unmeasured pollutants from well controlled incinerators are not causing overt short-term effects on the respiratory system.

An excess of lung-function abnormalities was found in the schoolchildren study of Wang et al. (1992) in Taiwan, in which the target population had considerably higher measured concentrations of ambient SO_2 and NO_2. This supports the conclusion that if incinerator emissions result in violation of air-quality standards, the adverse health effects attributable to the excesses can be expected.

After reports of a cluster of cases of cancer of the larynx near an incinerator of waste solvents and oils in Lancashire, UK, Elliott et al. (1992) analyzed the incidence of cancers of the larynx and lung in areas adjacent to all 10 licensed incinerators of waste solvents and oils in Great Britain that began operation before 1979. Exposures and cancer risks were assessed at the aggregate, or "ecological" level. No data were obtained that would allow linking of individual exposure to cancer risk. Postal-coded cancer-registration data were available for 1974-1984 in England and Wales and for 1975-1987 in Scotland. Standardized observed-to-expected incidence ratios were calculated for each postal-code area stratified by distance from the incinerator, within 3 km and 3-10 km away. Expected values were based on national rates and were stratified by region and a measure of socioeconomic status. None of the observed-to-expected incidence ratios within 3 km or 3-10 km away differed statistically significantly from unity for the two cancers. When data were further evaluated over a range of geographic circles up to 10 km away to test for trend, there was no evidence of higher risk closer to the incinerators. The authors noted that, owing to the restricted number of years available for analysis, their model assumed a lag of only 5-10 year between the beginning of incinerator operation and a potential effect on cancer incidence and that this lag is recognized to be short in light of the epidemiology of most cancers. An additional 10-year follow up of cancer incidence in these populations would be more informative, in that, as the authors note, Fingerhut et al. (1991) observed an excess cancer mortality associated with TCDD workplace exposures only after 20 years of followup. They concluded that the observed cluster of laryngeal cancer at the Lancashire site was unlikely to be attributable to residential proximity to the incinerator.

In a second, more-comprehensive study of cancer incidence in over 14 million people living near 72 municipal solid-waste incinerators in Great Britain for the years 1974-1986, Elliott et al. (1996) studied cancer incidence in relation to residential proximity to the incinerators. All postal-code areas within 7.5 km of one of the municipal incinerators in England, Wales, and Scotland—except those brought into operation after 1975—were divided into eight concentric bands on the basis of distance from the incinerator. The observed cancer incidences in all residents within the 7.5-km study area and in residents within each of the 8 bands were compared with expected numbers of cancers based on national cancer-incidence rates obtained directly from the Small Area Health Statistics Unit database and adjusted for age, sex, region, and a "deprivation score." The deprivation score was an attempt to take into account the prevalence of unemployment, overcrowding, and social class of the head of household; this score was previously found to strongly correlate with cancer rates across Great Britain. Statistically significantly greater numbers of cancers—for all cancers combined and for cancers of the stomach, colon and rectum, liver, and lung—were observed for the entire study area; within the eight geographic bands, the excess of observed over expected numbers increased slightly closer to the incinerators.

However, on further analysis, the authors concluded that those results were likely to be largely explained by residual confounding by the deprivation score. When they compared the ratios of observed-to-expected cancers during the preincinerator period—that is before startup of a site—with postincinerator ratios and assumed a 10-year lag between year of startup and cancer incidence, the authors found that observed-to-expected ratios were somewhat larger during the preincinerator period, particularly for stomach and lung cancers. They also observed that the deprivation score was higher with increasing proximity to incinerators. A review of the histologic coding of liver-cancer cases revealed substantial disagreement between the cancer-registry and death-certificate databases. The authors concluded that the excess cancer cases in areas closest to the incinerators could be accounted for by the higher prevalence of unemployment, overcrowding, and lower social class in these areas, and that these factors were not fully controlled in the analysis but that further investigation, including histologic review of cases, should be done.

In a spatial analysis of risk as a function of distance from various sources of pollution (shipyard, iron foundry, incinerator, and city center) in Trieste, Italy, Biggeri et al. (1996) concluded that air pollution is a moderate risk factor of lung cancer. This is consistent with a study conducted in Rome, Italy (Michelozzi et al. 1998) which reported that mortality from laryngeal cancer declined with distance from the sources of pollution. In contrast, a 10-year follow up study conducted in Finland reported increased mercury exposure as the distance decreased from a hazardous-waste incinerator; however, "the increase in exposure was minimal and, on the basis of current knowledge, did not pose a health risk (Kurttio et al. 1998)."

Studies of Incinerator Workers

Motivated by findings of Pani et al. (1983) that airborne particles collected in the working areas of a municipal refuse incinerator were mutagenic, Scarlett et al. (1990) compared the frequency of urinary mutagens, measured by the Ames assay, in a sample of 104 refuse-incinerator workers in 7 incinerator plants with that in 61 water-treatment plant employees in 11 municipal facilities. When urinary-mutagen frequency was adjusted for age, cigarette-smoking, fried-meat consumption, alcohol use, and use of a wood stove in the home, the frequency of urinary mutagens in incinerator workers was found to be a factor 9.7 times as high as the comparison group of water-treatment plant workers when the assay was performed without microsomal activation and 6.3 times as high with microsomal activation. Mutagens were present in urine of workers at 4 of the 7 incinerators and only 1 of the 11 water-treatment plants.

Two years later, the investigators restudied workers at the same incinerators and water-treatment plants to evaluate the consistency of their earlier results (Ma et al. 1992). Three urine samples, collected at about 1-wk intervals, were obtained from 37 incinerator workers in four facilities and from 35 water-treatment plant workers in eight facilities. When the first urine samples were compared, incinerator workers had positive mutagen assays four times more often than water-treatment workers; the difference was statistically significant. Although the frequency of mutagens was higher among incinerator workers for the second and third urine samples, the differences from frequencies in the water-treatment workers were no longer statistically significant. With microsomal activation, the proportions of incinerator workers who had positive mutagen assays declined in the three urine samples—from 21.6% to 15.2% and then 8.3%. The authors speculated that the trend might be explained in two ways. One is that incinerator workers began to take measures to reduce their exposures. The other is that exposures to mutagenic substances in incinerator plants was highly variable. The authors pointed out that the presence of mutagens in the urine does not establish that mutations are taking place in the cells of these workers, but they did recommend that measures be taken to reduce occupational exposures of incinerator workers to potential mutagens in their work environments.

Angerer et al. (1992) measured concentrations of various organic substances in the blood and urine of 53 workers at a municipal-waste incinerator in Germany and 431 men and women "who belong to different subgroups," also in Germany. No information is provided in the report on the extent of industrial-hygiene controls in the incinerator facility. Statistically significantly higher concentrations of urinary hydroxypyrene, 2,4- and 2-5-dichlorophenol, and 2,4,5-trichlorophenol, and of plasma hexachlorobenzene (HCB) were found among incinerator workers, whereas the controls had higher concentrations of urinary 4-monochlorophenol and tetrachlorophenol. No statistically significant differences between the two groups were found for blood benzene (after stratification on cigarette-smoking), plasma polychlorinated biphenyls, or urinary 2,4,6-trichlorophenol or pentachlorophenol. Urinary hydroxypyrene was measured because it is a metabolite of pyrene and has been shown to be a good indicator of internal dose of PAHs. Plasma PCBs and HCB and urinary chlorophenols were measured because these chemicals, when combusted, are precursors of dioxins and furans, and because they are easier to measure in biological material than the dioxins and furans. The lack of consistent findings between the incinerator and comparison groups for PCBs, HCB, and chlorophenols means this study provides no conclusive evidence on the exposure, absorption, or metabolism of combustion precursors of the PCDDs and PCDFs, and so allows no inference about exposures to PCDDs and PCDFs. However, the higher concentrations of hydroxypyrene might indicate that incinerator workers had higher exposures to PAHs.

Schecter et al. (1994) measured polychlorinated dioxins and dibenzofurans in pooled samples of blood from 85 workers at a relatively old incinerator in New York City and pooled blood from 14 matched controls in the same city. Higher concentrations of several of the dioxin and furan congeners, except TCDD, were found in the blood of incinerator workers. The authors comment that the findings document exposure and bioavailability and suggest a hazard to workers. After the findings were presented, personal protective measures were put into place for the workers at this facility. Because the samples from all workers were pooled, it was not possible to evaluate whether concentrations of congeners were related to the probable extent of occupational exposure, duration of employment, or to potentially confounding exposures; analysis of these variables could have given greater confidence that the findings were attributable to the occupational environment rather than to other sources of the organic pollutants.

In 1992, staff of the National Institute for Occupational Safety and Health (NIOSH) performed environmental sampling to investigate employee exposure to PCDDs, PCDFs, metals, and other substances at three New York City munic-ipal-refuse incinerators (NIOSH 1995). Six area samples from working zones and five bulk fly-ash samples were collected and analyzed for PCDD and PCDF congeners, eight personal-breathing-zone samples and nine area samples were collected for metals during cleaning operations, and 10 samples were collected for respirable dust and silica. Airborne PCDD and PCDF concentrations for four of the six area samples from working zones exceeded the National Research Council guideline of 10 pg/m^3 (one sample by a factor of 80); all four were collected during cleaning operations. The breathing-zone samples approached or exceeded the NIOSH and Occupational Safety and Health Administration criteria for arsenic, cadmium, lead, and nickel. Area samples collected near work locations exceeded relevant evaluation criteria for aluminum, arsenic, cad-mium, cobalt, lead, manganese, and nickel. One of 10 samples exceeded the NIOSH recommended exposure limit for respirable quartz by 50%. The air-borne concentrations of aluminum, arsenic, cadmium, lead, and nickel during some periods of the cleanout of the electrostatic precipitator and of PCDDs and PCDFs during cleaning of the lower chamber were high enough to exceed the protection capabilities of the air-purifying respirators worn by the workers dur-ing these operations. On the basis of this evaluation, NIOSH staff concluded that working in cleanout operations at the incinerators poses a health hazard.

Malkin et al. (1992) analyzed blood samples from 56 high-pressure plant tenders working at three New York City incinerators. The duties of these work-ers—involving precipitator, upper- and lower-chamber, and undercarriage clean-ing—were judged to be those with the highest potential exposure to lead. Blood samples were also obtained from a control group of 25 high-pressure plant ten-ders working at heating plants, where maintenance of boilers was involved. Al-though the average blood-lead concentration (11.0 µg/dL) of the incinerator workers was not high relative to concentrations associated with clinical abnor-malities, they were statistically significantly higher than the average (7.4 µg/dL) in the comparison workers. When the variation in blood lead among incinerator workers was analyzed with multiple-regression modeling (incorporating age and cigarette smoking), workers who did not always wear protective devices or who cleaned the combustion chambers more times in the last year had statistically significantly higher blood lead. None of the known health effects of lead exposure was evaluated in this study. The results suggest that the presence of lead in combustion-chamber fly ash can increase the blood-lead concentrations of incinerator workers.

Only two morbidity or mortality studies of waste-incinerator workers have been reported. Bresnitz et al. (1992) evaluated 86 male workers among 105 active employees at a Philadelphia municipal incinerator. The workers were divided into potential high- and low-exposure groups of 45 and 41, respectively, on the basis of a worksite analysis performed by an independent industrial hygienist. Eight workers had at least one measurement in blood or urine indicating excessive exposure to heavy metals, but these elevations were unrelated to exposure category. Although 34% of the workers had evidence of hypertension, the prevalence of this condition was unrelated to exposure group. None of the biochemical measurements of blood or serum were clinically significant, and, except for hematocrit and serum creatinine, the differences between the two exposure groups were not statistically significant.

Gustavsson (1989) studied the mortality experience of 176 waste-incinerator workers in Sweden. Compared with national and local death rates standardized for age and calendar year, there was an excess of deaths from lung cancer and ischemic heart disease. Analysis of duration of exposure supported the conclusion that the excess of deaths from ischemic heart disease was attributable to occupational factors, whereas lung-cancer deaths were too few to make such an inference.

In summary, workers in the incinerator industry have not been extensively studied for morbidity and mortality risks. A Swedish study found an excess of deaths from lung cancer and ischemic heart disease among a sample of 176 incineration workers. The few available studies reviewed here yield evidence that some workers are exposed to amounts of organic compounds and metals (including dioxins, furans, and lead) that result in increased tissue concentrations. The health consequences of the exposures have not been evaluated through systematic followup of these workers.

A recent report of a retrospective mortality study of a cohort of 532 male subjects employed at two municipal-waste incineration plants in Rome, Italy (Rapiti et al. 1997) revealed an increased risk of gastric cancer. The authors concluded that these findings indicate the need to further investigate the role of cancer as a result of occupational exposure to hazardous waste.

Studies of Animal Populations

Lloyd et al. (1988) studied rates of twin births in cows ("twinning") in an area of central Scotland surrounding two waste incinerators, one a municipal-waste incinerator and the other a chemical incinerator. The study of twin births was prompted by the anecdotal observation of a dramatic increase in twinning among the dairy cattle in the region. The authors noted that some polychlorinated hydrocarbons have estrogenic and fertility-related properties and that endogenous or exogenous estrogens might affect the frequency of twinning. Two postal-code sectors downwind of the incinerators were considered to be areas of primary risk, and this classification was supported by finding comparatively high concentrations of polychlorinated compounds in surface soils in these sectors. Twinning rates in the upwind and more-distant postal-code sectors were 3-13 per 1000 births; the highest rates, 16 and 20 per 1,000, were observed in the two downwind sectors. The incidence of identical twins in cows is rare, but fraternal twins can occur in up to 5% of births, depending on the breed. Delay in mating or artificial insemination can contribute to twinning, as can repeated breeding and artificial insemination. The incidence of twinning is also increased once a cow has given birth to a first set of twins (Hafez 1974). The authors noted that genetic factors in twinning remain to be investigated in this population.

In a second study of the same area, Williams et al. (1992) analyzed the male-to-female ratio in calves at birth by postal-code sector and found an excess of female births downwind of an incinerator. Because of suggestions that pollution from the incinerators might have increased during later years, the data were grouped into two periods, 1975-1979 and 1980-1983. Statistically significantly lower male-to-female ratios were observed in one of the two downwind sectors during both periods, but not in the other downwind sector. By using computer mapping and smoothing techniques to analyze twinning rates in enumeration districts within each postal-code sector, the authors were able to show a persistent excess of female births, compared with other districts, along a northeast-southwest axis from the incinerators, which was consistent with the prevailing wind patterns in the area. Because many factors can alter sex ratios, and these factors were not enumerated in this study, the authors considered it premature to attribute causality to the reported associations.

RESULTS FROM RISK ASSESSMENT STUDIES

There have been hundreds of risk assessments performed on incinerators of various types in many parts of the country. These assessments have taken various forms and followed various protocols. Among the more-detailed have been the assessments for Dickerson County (Brower et al. 1990), and more recently, the Waste Technologies Incinerator (EPA 1997b), but there is no convenient listing or compilation of such assessments or their results. There is no standard way for publishing these risk assessments, and few receive peer review. Although most such assessments are in the public domain, obtaining them is difficult, and there are still many that are likely to have remained private.

Most of these risk assessments are based on methodology that was first introduced in the evaluation of nuclear power plants (NRC 1977). It should be emphasized that these risk assessments were performed to evaluate the risks to the local population; workers's risks were generally not evaluated, nor was the regional impact considered, and not all facilities have been assessed for risk. Experience with them indicates that:

- For modern, well-controlled incinerators, risk estimates for cancer effects even for the most-highly exposed persons (not workers), are generally small to negligible (for example, lifetime cancer risk estimates below 1 in 100,000).
- At least some older, poorly controlled incinerators—had they continued to operate—would likely have resulted in cancer risk (above 1 in 10,000 lifetime risk).
- The principal contributors to risk estimates tend to be dioxins and furans (through food chain routes), arsenic, HCl, mercury, lead, and particles.
- Experience in performing such assessments is extremely important, particularly if new chemicals are inserted into models not designed for them.

Risk assessments have as one of their bases an evaluation of the health effects observed for the materials examined in risk assessment. A fundamental tenet of risk assessment is the ability to perform extrapolations, including extrapolations of dose-response results for health effects observed at different concentrations, in differing exposure circumstances, and even in different species. It is considered, however, that uncertainty is minimized by using the minimum amount of extrapolation possible. The examples in the following section were chosen to illustrate the ranges of data available for the various chemicals.

Observed Health Effects of Materials Present in Incineration Emissions

This section summarizes, for selected pollutants of concern, the adverse health effects that have been documented in humans and animals. These pollutants are known to be produced and released into the environment during the operation of various waste incinerators. The chemicals selected for discussion in this section are particulate matter, CO, acidic gases (NO_x, SO_2, and HCl) and acidic particles, (e.g., as H_2SO_4 or NH_2HSO_4), some metals (cadmium, lead, mercury, chromium, arsenic, and beryllium), and organic compounds—dioxins and furans and some other products of incomplete combustion (PCBs and PAHs). Human health effects have been observed for some of these agents at extremely high concentrations in various exposure circumstances; but such effects have not been observed as a direct result of exposure to emissions from a waste incinerator (as demonstrated in the following sections). PM health effects can apparently occur at concentrations previously considered acceptable. For lead, health effects occur at blood concentrations that are not far above background blood concentrations, but these correspond to ambient air concentrations greater than current standards for lead.

Particulate Matter

Particulate Matter (PM) consists of a mixture of materials. The numbers of particles and their chemical composition can vary within specific particle-size fractions from location to location and over time, depending on the types of source emissions and atmospheric conditions. Concern about airborne particulate matter in recent years has been driven largely by epidemiologic studies that have reported relatively consistent associations between outdoor particulate-matter levels and adverse health effects. However, assessing the specific health risks resulting from exposures to airborne particulate matter, and distinguishing these effects from those produced by gaseous copollutants, involves substantial scientific uncertainty about the influence of copollutants and weather, about whether some particulate-matter fractions (size or chemical) might be more-highly associated with health risks, and about the nature of dose-response relationships between particulate matter and health (NRC 1998, 1999c).

Most available epidemiologic evidence of PM effects have employed direct or indirect metrics of PM mass, irrespective of particle composition or emission source (e.g., see Dockery and Pope 1994).

The most-clearly defined effects associated with exposure to PM have been sudden increases in the number of illnesses and deaths occurring day to day during episodes of high pollution. The most notable of those episodes occurred in the Meuse Valley in 1930, in Donora in 1948, and in London in 1952. During the December 1952 episode, 3,000-4,000 excess deaths were attributable to air pollution, with the greatest increase in death from chronic lung disease and heart disease (United Kingdom Ministry of Health 1954). The death rate increased most dramatically in those older than 45 years and among those with pre-existing respiratory illnesses (such as asthma). Collectively, studies of those and other early episodes left little doubt that airborne PM contributed to the morbidity and mortality associated with very high concentrations of urban aerosol mixtures dominated by combustion products (e.g., from burning coal) or their transformation products (such as aerosols containing sulfuric acid).

The 1982 EPA PM criteria document concluded that the available studies collectively had indicated that mortality was substantially increased when 24-hr airborne-particle concentrations exceeded 1,000 $\mu g/m^3$ (as measured by the black smoke method) in conjunction with SO_2 concentrations over 1,000 $\mu g/m^3$ (the elderly and persons with severe pre-existing cardiovascular or respiratory disease were mainly affected).

The period since the 1982 criteria document (and its 1986 addendum) has seen many reports of time-series analyses of associations between human mortality and acute exposures to PM at or below the pre-1997 U.S. 24-hr standard

(PM$_{10}$ at 150 μg/m^3). As a result, EPA moved to institute a more-stringent U.S. short-term PM mass concentration limit of 65 μg/m^3 for fine particles (PM$_{2.5}$, the mass of particles below 2.5 μm in diameter), and an annual PM$_{2.5}$ limit of 15 μg/m^3. On May 14, 1999, a panel of the U.S. Court of Appeals for the District of Columbia Circuit remanded the new standards for PM$_{2.5}$.

Numerous investigators have reported statistically significant positive associations between relative risk for death and various indexes of PM in many cities in the United States and other countries. The elderly (over 65), particularly those with pre-existing respiratory disease, were found to have higher risks than younger adults (Thurston 1996). Studies suggest that children are also at increased risk from the adverse health effects of air pollution. During the London fog episode, the second highest increase in mortality (after older adults) was in the neonatal age group (relative risk, (RR) = 1.93 for children less than 1 year) (United Kingdom Ministry of Health 1954). More recently, Saldiva et al. (1994) found acute exposure to air pollution in Sao Paulo, Brazil to be significantly associated with respiratory mortality in children less than 5 years of age, although the effect could not be definitively associated with a specific pollutant. Also, Bobak and Leon (1992) and Woodruff et al. (1997) both found long-term averages of air-pollution, including PM, to be associated with increased post neonatal (ages 1 to 12 months) mortality. Thus, air pollution exposure has been associated with increased mortality, with the very young and the elderly being indicated as being especially at risk.

Published summaries of PM reports have converted all results to a PM$_{10}$-equivalence basis and provided quantitative comparisons (Ostro 1993; Dockery and Pope 1994; Thurston 1996). Other summaries have used total suspended particles (TSP) as the reference PM metric (Schwartz 1991, 1994a) and considered many of the same studies included in the PM$_{10}$-equivalence summaries. (Other air pollutants were generally not addressed in deriving the coefficients reported by these summaries.) The results suggest about a 1% change in acute total mortality for a 10-μg/m^3 change in daily PM$_{10}$. Such a change represents a seemingly small increment in risk from exposure to this pollutant, but it must be remembered that peak PM$_{10}$ concentrations are commonly about 100 μg/m^3 above concentrations for an average day, that large populations are affected by this ubiquitous pollutant, and that this reported RR is for total mortality (with even higher RRs being found in studies of more affected specific causes, such as respiratory disease, and for sensitive populations, such as the elderly). Also, the implied increments in lifetime risk from small increments in exposure to particles are very high compared with typical values of regulatory interest. In the reviews cited above, the highest PM$_{10}$-associated relative risks for death were indicated for the elderly and for those with pre-existing respiratory conditions; both constitute populations that appear to be especially sensitive to acute exposures to air pollution.

Aggregate population-based cross-sectional studies using averages across various geopolitical units (cities, metropolitan statistical areas (MSAs), and so on) have examined the relation between mortality and long-term PM exposure. Those community-based studies sought to define the characteristics of a community that are associated with its overall average health status, in this case annual mortality. For example, Ozkaynak and Thurston (1987) analyzed 1980 total mortality in 98 MSAs, using data on PM_{15} and $PM_{2.5}$ from the EPA inhalable-particle monitoring network for 38 of these locations. They concluded that the results suggested an effect of particles on mortality that decreased with increasing particle size.

Prospective cohort studies have considered the effect of PM exposure on the relative survival rates of individuals, as modified by age, sex, race, smoking, and other individual risk factors, finding that PM exposure can lead to substantial shortening of life in the general population. That type of analysis has a substantial advantage over aggregate population-based studies, in that the individual analysis allows stratification according to such important risk factors as smoking. Abbey et al. (1991) described a prospective cohort study of morbidity and mortality in a population of about 6,000 white, non-Hispanic, nonsmoking, long-term California residents who were followed for 6-10 years beginning in 1976. TSP and ozone were the only pollutants considered. In a followup analysis, Abbey et al. (1995) considered exposures to SO_4^{2-}, PM_{10}, and $PM_{2.5}$, as well as visibility (extinction coefficient). In these analyses, no significant associations with nonspecific mortality (i.e., from all natural causes) were reported, and only high concentrations of TSP or PM_{10} were associated with respiratory symptoms of asthma, chronic bronchitis, or emphysema. However, a more recent analysis using an additional 5 years of follow-up on this cohort and improved PM_{10} exposure estimates did predict significant PM-mortality associations among men in this cohort, who reportedly spent significantly more time outdoors than women (Abbey et al. 1999). Dockery et al. (1993) analyzed the mortality experience in 8,111 adults who were first recruited in the middle 1970s in 6 cities in the eastern portion of the United States. The subjects were white and 25-74 years old at enrollment. Dockery et al. (1993) reported that "mortality was more strongly associated with the levels of fine, inhalable, and sulfate particles" than with the other pollutants. Pope et al. (1995) analyzed 7-year survival data (1982-1989) for about 550,000 adult volunteers obtained by the American Cancer Society (ACS). They took great care to control for potential confounding factors on which data were available. For example, several different measures of active smoking were considered, as was the time exposed to passive smoke. The adjusted total-mortality risk ratios for the ACS study, computed for the cities' range of the pollution exposures, were as follows: 1.15 (95% confidence interval, 1.09-1.22) for a 19.9 $\mu g/m^3$ increase in sulfates and 1.17 (95% confidence interval, 1.09-1.26) for a 24.5 $\mu g/m^3$ increase in $PM_{2.5}$. Analysis of life-tables indicate that these effects are associated with more than a 1-year shortening of expected lifespan for the entire population (WHO 1995).

Dockery and Pope (1994) have reviewed the effects of PM_{10} on both respiratory mortality and morbidity. They considered five primary health end points: mortality, hospital use, asthma attacks, respiratory symptoms, and lung function. They concluded that there was a coherence of effects across the end points, with most end points showing a 1-3% change per 10 $\mu g/m^3$. A later analysis by Thurston (1996) indicated that those PM-effect estimates are reduced somewhat if the influences of copollutants are addressed.

Hospitalization data can provide an especially useful measure of the morbidity status of a community during a specified period. Hospitalization data on respiratory-illness diagnosis, or more specifically for chronic obstructive pulmonary disease (COPD) and pneumonia, give a measure of respiratory status. Both COPD and pneumonia hospitalization studies show moderate but statistically significant relative risks, in the range of 1.06-1.25, associated with an increase of 50 $\mu g/m^3$ in PM_{10}. Table 5-1 presents results of several studies of short-term exposure-response relationships of fine-particle sulfates, $PM_{2.5}$, and PM_{10} with different health-effect indicators, as developed by the World Health Organization. The data provide quantitative estimates of the effect of PM (per unit of increment) for each outcome considered.

Acidic Gases and Acidic Aerosols

Nitrogen Oxides

Nitric oxide (NO) is the major nitrogenous pollutant emitted from incineration facilities. Although NO itself is not thought to result in any deleterious health effects at the concentrations surrounding combustion sources, it is readily oxidized in the ambient environment to nitrogen dioxide (NO_2), which is the most biologically significant of the nitrogen oxides. NO_2 exerts its health effects via two primary pathways. One pathway is directly through interactions with the respiratory system when breathed. The other pathway is indirectly through the photochemical formation of atmospheric ozone, a secondary pollutant with much greater respiratory effects than NO_2 itself. Collectively, nitrogen oxides are often assessed as a group known as NO_x.

NO_2 is water-soluble and, when breathed, is efficiently absorbed in the mucous lining of the nasopharyngeal cavity and lung, where it converts to nitrous acid, HNO_2, and nitric acid, HNO_3, which can then react with the pulmonary and extrapulmonary tissues. NO_2 has been shown in occupational settings to be rapidly fatal at extremely high concentrations (i.e., 150,000 ppb and above) because of pulmonary edema, bronchial pneumonia, or bronchiolitis fibrosa obliterans (NRC 1977, Ellenhorn and Barceloux 1988), but these exposures are 10,000 times in excess of ambient concentrations found near sources such as

TABLE 5-1 Results of Several Studies of Short-Term Exposure-Response Relationship of Sulfates, $PM_{2.5}$, and PM_{10} with Different Health-Effect Indicators

Health-Effect Indicator	Estimated Change in Daily Average Concentration Needed for Given Effect, $\mu g/m^3$		
	Sulfates	$PM_{2.5}$	PM_{10}
Daily mortality			
5% change	8[a]	29[a]	50[b]
10% change	16	55	100
20% change	30	110	200
Hospital admissions for respiratory conditions			
5% change	8[c]	10[d]	25[e]
10% change	16	20	50
20% change	32	40	100
Bronchodilator use among asthmatics			
5% change	f	f	7[g]
10% change	f	f	14
20% change	f	f	29
Symptom exacerbations among asthmatics			
5% change	f	f	10[h]
10% change	f	f	20
20% change	f	f	40
Peak expiratory flow (mean population change)			
5% change	f	f	200[i]
10% change	f	f	400
20% change			f

[a] Based on estimates for St. Louis from Dockery et al. 1992.

[b] Based on Pope et al. 1992; Dockery et al. 1992; Schwartz 1993; Kinney et al. 1995; Ito et al. 1995. Relative-risk estimates per 100 $\mu g/m^3$ from these studies were 1.16, 1.16, 1.11, 1.04, and 1.05, respectively; statistically significant estimates only.

[c] Based on Thurston et al. 1994; Burnett et al. 1994. Relative-risk estimates per 10 $\mu g/m^3$ from these studies were 1.11 and 1.03, respectively.

[d] Based on Thurston et al. 1994.

[e] Based on Schwartz 1994b,c,d. Relative-risk estimates per 100 $\mu g/m^3$ from these studies were 1.40, 1.20, and 1.10.

f No data.

[g] Based on Pope et al. 1991; Roemer et al. 1993. Relative-risk estimates per 10 $\mu g/m^3$ from these studies were 1.12 and 1.02, respectively.

[h] Based on Roemer et al. 1993; Pope et al. 1992. Relative-risk estimates per 10 $\mu g/m^3$ from these studies were 1.05 and 1.05, respectively.

[i] Based on Pope et al. 1991, 1992; Roemer et al. 1993. Coefficients from these studies ranged from −0.028 to −0.041 L/min per $\mu g/m^3$; mean baseline peak flow ranged from 260 to 300 L/min.

Source: WHO 1995.

incinerators. Ambient concentrations of NO_2 vary with motor-vehicle traffic density in most U.S. cities, and annual average concentrations range from about 4 to 34 ppb (EPA 1998b,c). Potential acute effects of concentrations above 100 ppb NO_2 can include reduced pulmonary function, inflammation of the lung, and altered host defenses, especially among asthmatics (e.g., Samet and Utell 1990). The concentrations required to produce those effects can be reached indoors when unvented gas stoves or kerosene heaters are present, but are generally above the concentrations that occur in the ambient air (Klaassen et al. 1995). However, studies of healthy subjects exposed to NO_2 from 75 min to 3 hr at up to 4,000 ppb have generally failed to show lung-function alterations (Bascom et al. 1996). Even in susceptible people, such as those with pre-existing respiratory disease, effects at concentrations less than 1,000 ppb are not consistently detected. Concern with respect to present-day ambient concentrations of NO_2 is focused primarily on increases in airway responsiveness of asthmatic people after short-term exposures and increased occurrence of respiratory illness among children associated with long-term exposures to NO_2 (EPA 1993).

Hydrogen Chloride

The irritating properties of hydrogen chloride (HCl) prevent the study of more than transient voluntary exposure at concentrations that are likely to cause serious health effects, so there is a paucity of human data that can be used to evaluate the health effects of exposure to HCl at high concentrations (NRC 1991c). In humans, HCl acts primarily as an irritant of the upper respiratory tract, eyes, and mucous membranes, generally at concentrations over 5 ppm (NRC 1991c). Concentrations of 50-100 ppm are considered barely tolerable (Stokinger 1981). Bleeding of the nose and gums and ulceration of the mucous membranes have been attributed to repeated occupational exposure to HCl mist at high (unspecified) concentrations (Stokinger 1981). Etching and erosion of teeth have been reported in workers exposed to acids in battery, pickling, plating, and galvanizing operations (ten Bruggen Cate 1968); these workers were exposed to various mineral acids, including HCl (0.1 ppm), in combination with other acids, primarily sulfuric acid.

The LC_{50} values for HCl in rats, mice, and guinea pigs are 4,700 ppm, 2,600 ppm, and 2,500 ppm, respectively, for a 5-min exposure (Machle et al. 1942; Darmer et al. 1974). Results of studies in which mice were exposed to HCl vapors or aerosols indicate that vapors and aerosols have comparable toxicity (Darmer et al. 1974). As in humans, HCl was extremely irritating to the eyes, mucous membranes, and skin. In addition, rats and mice had scrotal ulceration and corneal erosion and clouding. Gross examination of animals that died during or shortly after exposure revealed moderate to severe emphysema, atelectasis, and pulmonary edema. No deaths were reported in mice or rats exposed to HCl at 410 and 2,078 ppm, respectively, for 30 min (Darmer et al. 1974).

No pathologic changes were observed in experimental animals exposed to HCl at 33 ppm for 6 hr/day, 5 days/week for 4 weeks. Exposure of rats and mice at 50 ppm for 6 hr/day, 5 days/week for 90 days resulted in statistically significant decreases in body weight, whereas no change was observed in hematologic characteristics, serum chemistry, and urinalysis. Histologic examination revealed dose-related minimal to mild rhinitis at 10, 20, and 50 ppm. Exposure of rats at 10 ppm and higher for 6 hr/day, 5 day/week for life resulted in laryngeal hyperplasia in 22% of the test animals, compared with 2% of control animals, and tracheal hyperplasia in 26% of the test animals, compared with 6% of controls (Sellakumar et al. 1985).

Mortality in the progeny of rats exposed to HCl at 300 ppm on day 9 of pregnancy was $31.9 \pm 9.2\%$, compared with $5.6 \pm 3.7\%$ in controls ($p < 0.01$). The progeny of rats exposed at 300 ppm either for 12 days before pregnancy and of rats exposed on day 9 of pregnancy showed disturbances in kidney function, as measured by diuresis and proteinuria (Pavlova 1976).

Baboons exhibited signs of irritation, such as coughing and frothing at the mouth, during a 5-min exposure to HCl at 810 ppm, but not at 190 ppm (Kaplan 1987). Severe irritation and dyspnea occurred at higher concentrations (16,750 and 17,290 ppm). Dyspnea persisted after exposure, followed by death several weeks later from bacterial infections. Baboons exposed at 500 ppm for 15 min also exhibited signs of irritation (increased respiratory rates) but did not develop hypoxia, did not show changes in respiratory function, and were able to perform escape tasks (Kaplan et al. 1988).

Studies have demonstrated notable differences between primates and rodents in responses to HCl exposure. Exposure of rats and mice to HCl concentrations of 560 ppm for 30 min and less than 50 ppm for 10 min, respectively, produced dose-related decreases in respiratory frequency (Barrow et al. 1979; Hartzell et al. 1985). Baboons exposed to HCl at up to 17,000 ppm for 5 min, however, exhibited increases in respiratory frequency that could be interpreted as a compensatory mechanism in response to hypoxia (Kaplan et al. 1988). Given their greater similarity to humans in the respiratory tract and its function, baboons would probably be more-appropriate animal models than rodents for extrapolation of HCl effects to humans (NRC 1991c).

It has been postulated that a toxic gas or vapor adsorbed on ambient particles of suitable size, perhaps including dust, could be carried to the bronchioles and alveoli, where more-serious damage could occur. Such an effect has been looked at to some extent by the Air Force (Wohlslagel et al. 1976) and found not to be significant in the case of hydrogen fluoride and HCl mixed with alumina particles. However, more recent studies provide evidence that strongly acidic aerosols can constitute a portion of PM that is especially associated with acute respiratory health effects in the general public (Thurston et al. 1992, 1994).

Acidic Aerosols

Most historical and present-day evidence suggests that there can be both acute and chronic effects of the strongly acidic component of PM, i.e., the hydrogen ion (H^+), concentration when it is below pH 4.0 (Koutrakis et al. 1988; Speizer 1999). Increased hospital admissions for respiratory causes were documented during the London fog episode of 1952, and this association has now been observed under present-day conditions. Thurston et al. (1992, 1994) have noted associations between ambient acidic aerosols and summertime respiratory hospital admissions in both New York state and Toronto, Canada, even after controlling for potentially confounding temperature effects. In the 1994 report, statistically significant independent H^+ effects remained even after the other major copollutant, in the regression model, ozone was considered. H^+ effects were estimated to be largest during acid-aerosol episodes ($H^+ \geq 10$ µg/m^3 as sulfuric acid or H^+ at ≈ 200 nmol/m^3), which occur roughly 2 or 3 times per year in eastern North America. The studies provide evidence that present-day strongly acidic aerosols might represent a portion of PM that is contributing to the significant acute respiratory health effects noted for PM in the general public.

Results of recent symptom studies of healthy children indicate the potential for acute acidic PM effects in this population. Although the "6-city Study" of parent diaries of children's respiratory and other illness did not demonstrate H^+ associations with lower respiratory symptoms except at H^+ above 110 nmol/m^3 (Schwartz et al. 1994), upper respiratory symptoms in two of the cities were found to be most-strongly associated with high concentrations of H_2SO_4 (Schwartz et al. 1991). Two recent summer-camp (and schoolchildren) studies of lung function have indicated a statistically significant association between acute exposures to acidic PM and decreases in the lung function of children, independent of those associated with O_3 (Neas et al. 1995; Studnicka et al. 1995).

Reported associations between chronic H^+ exposures and children's respiratory health and lung function are generally consistent with adverse effects as a result of chronic H^+ exposure. Preliminary bronchitis prevalence rates reported in the "6-city Study" locales were found to be more-closely associated with average H^+ concentrations than with PM in general (Speizer 1989). Follow-up studies of those cities (and a seventh) that controlled for maternal smoking, education, and race suggested associations between summertime average H^+ and chronic bronchitic and related symptoms (Damokosh et al. 1993). Bronchitic symptoms were observed 2.4 times more frequently (95% confidence interval, 1.9-3.2) at the highest acid concentration (H^+ at 58 nmol/m^3) than the lowest concentration (16 nmol/m^3). Furthermore, in a followup study of children in 24 United States and Canadian communities (Dockery et al. 1996) in which the analysis was adjusted for the effects of sex, age, parental asthma, parental education, and parental allergies, bronchitic symptoms were confirmed to be statistically significantly associated with strongly acidic PM (relative odds, 1.7; 95% confidence interval, 1.1-2.4). It was also found in the "24-city Study" that mean forced vital capacity (FVC) and forced expiratory volume in one second (FEV_1) were lower in communities that had high concentrations of strongly acidic PM (Raizenne et al. 1996). Thus, chronic exposures to highly acidic PM have been associated with adverse effects on measures of respiratory health in children.

Asthmatic subjects appear to be more sensitive than healthy subjects to the effects of acidic aerosols on lung function, but reported effective concentrations differ widely among studies (EPA 1986b). Adolescent asthmatics might be more sensitive than adult asthmatics and might experience small decrements in lung function in response to H_2SO_4 at concentrations only slightly above peak ambient concentrations (for example, less than 100 $\mu g/m^3$ H_2SO_4, or 2,000 nmol/m^3) (Koenig et al. 1983, 1989). Even in studies reporting an overall absence of statistically significant effects on lung function, individual asthmatic subjects appear to demonstrate clinically important effects (Avol et al. 1990). Two studies from different laboratories have suggested that responsiveness to acidic aerosols correlates with the degree of baseline airway hyperresponsiveness (Utell et al. 1983; Hanley et al. 1992).

Studies have also examined the effects of exposure to both H_2SO_4 and ozone on lung function in healthy and asthmatic subjects (Frampton et al. 1995). Two recent studies found evidence that H_2SO_4 at 100 $\mu g/m^3$ potentiates the ozone response, in contrast with previous studies. Animal studies support the hypothesis of a synergism between acidic aerosols and ozone (e.g., Last et al. 1986). Overall, acidic aerosols appear to be a contributing factor in the toxicity of PM at present-day ambient levels, either alone or in conjunction with ozone exposure. Thus, to the extent that incineration emissions increase the acidity (i.e., lowers the pH) of ambient PM, they may be expected to also increase the toxicity of those ambient aerosols.

Carbon Monoxide

Carbon monoxide (CO) is a colorless, odorless, poisonous gas formed during combustion processes as a result of carbon not being completely oxidized to carbon dioxide (CO_2).

CO binds strongly to hemoglobin, with an affinity over 200 times that of oxygen. The binding of CO with hemoglobin is not readily reversible, so it reduces the oxygen-carrying capacity of the blood significantly. CO concentrations above 25 ppm might lead to carboxyhemoglobin (COHb) concentrations of 5%, which has been associated with cardiovascular and respiratory disease and can interfere with pregnancy. Major damage to brain and lung occurs at 50% COHb, and death at 70%.

The body's natural production of CO results in a normal background COHb saturation concentration of 0.4-0.7%. In the nonsmoking population, COHb concentrations of 0.5-1.5% are typical; in those who smoke a pack of cigarettes per day, 5-6% is typical. COHb in newborns of smoking mothers is 1.1-4.3%. A blood COHb concentration of about 5% would be expected after an exposure to CO at 35 ppm for 6-8 hr (Ellenhorn and Barceloux 1988).

COHb of 2-4% has been associated with a decrease in time to myocardial ischemia and angina (Allred et al. 1989), and 2.9% has led to significant reduction in exercise tolerance and onset of angina (Kleinman et al. 1989). Furthermore, tunnel officers who were exposed to CO and who had COHb over 5% had an increased risk of dying from arteriosclerotic heart disease (Stern et al. 1988). Recently, Morris et al. (1995) reported that an increase of 10 ppm in CO in ambient air pollution was associated with a 10-37% increase in the rate of hospital admissions for congestive heart failure among those over 65.

Fetal hemoglobin has a greater affinity for CO than does adult hemoglobin; fetal COHb concentrations are typically 10-15% higher than maternal concentrations. Maternal exposure to CO at 30 ppm will lead to 5% COHb in the mother and 6% COHb in the fetus. Both the mother and the fetus are also more susceptible during pregnancy. CO has been shown to interfere with pregnancy in rats; although control rats had 100% successful pregnancy, the success rate for those exposed to CO at 30 ppm was only 69% (COHb was 4.8%), and for those exposed at 90 ppm, only 38% (Garvey and Longo 1978).

Fetuses, newborns, and pregnant women are especially susceptible to CO. Other high-risk groups include those with pre-existing heart disease and those over 65 years old (Morris et al. 1995). Hemoglobin reaches equilibrium with CO much more rapidly in people with anemia than in normal subjects; thus, a 4-hr exposure to CO at 20 ppm led to a COHb concentration of 4-5% in anemic subjects, but only 2.5% in normal subjects. Overall, CO from incinerators is not considered to be an important health factor (see discussion of "Implications to Human Health").

Metals

Metals associated with incinerator emissions include cadmium, lead, mercury, chromium, arsenic, and beryllium. Results of human and animal studies that examined the health effects of these metals are discussed below. It should be noted that for many of the health effects of concern, exposures are uncertain or unknown and are related not to incinerators but rather to occupational studies or case reports of accidental spills or releases.

Cadmium

The various inorganic forms of cadmium investigated to date have shown similar toxic effects (ATSDR 1997a). All soluble cadmium compounds are cumulative toxicants. Inhalation studies of cadmium-containing aerosols have shown that particle size is a major determinant of toxicity, whereas the chemical form of cadmium is relatively unimportant (Hirano et al. 1989a,b; Rusch et al. 1986). Similarly, oral-exposure studies of inorganic cadmium compounds have shown that absorption of the divalent ion (Cd^{2+}) results from ingestion of all soluble salts and that uptake rates of free cadmium ions and those complexed with proteins are similar.

Except at very high exposures, absorbed cadmium is bound almost totally to the protein metallothionein. The cadmium-metallothionein complex is readily filtered by the glomerulus and is largely reabsorbed in the proximal tubules of the kidney (Foulkes 1978).

The toxic effects of cadmium in humans and animals are similar. The major toxic effects are acute and chronic inflammation of the respiratory tract, renal tubular effects, and lung cancer.

In general, respiratory effects occur after cadmium exposures that are usually seen only in occupational settings, and environmental exposures to cadmium are unlikely to result in acute or chronic respiratory disease. Whereas animal studies have shown that inhaled cadmium can cause lung cancer in rats (Takenaka et al. 1983; Oldiges et al. 1989), human data are less convincing. Thun et al. (1985) reported an exposure-response relationship between cumulative cadmium exposure and lung cancer. On the basis of those findings, EPA has classified cadmium as a group B1 (Probable) human carcinogen by inhalation; a unit risk[4] of 1.8×10^{-3} per $\mu g/m^3$ was calculated (EPA 1992b).

Human and animal data on the neurotoxicity of cadmium are sparse, but there is evidence that neurobehavioral changes appear in adults and children after exposures smaller than those causing renal effects (Marlowe et al. 1985; Struempler et al. 1985; Hart et al. 1989). Animal studies have found behavioral and structural nervous system changes after relatively small oral or parenteral cadmium exposures.

Other toxic effects include cardiovascular effects, hematologic changes, and gastrointestinal changes. These occur after very high exposures after which respiratory and renal changes are also prominent.

ATSDR has estimated inhalation cadmium exposures that pose minimal risk to humans (minimal-risk levels, MRLs) (1997a). An MRL is defined as an estimate of the greatest daily human exposure to a substance that is likely to be without an appreciable risk of noncancer adverse effects over a specified duration of exposure. On the basis of a no-observed-adverse-effects level (NOAEL) of 0.7 $\mu g/m^3$ in a study of workers that reported a prevalence of proteinuria of 9% after a 30-year exposure at 23 $\mu g/m^3$ (Jarup et al. 1988), ATSDR (1997a)

---

[4] In its Integrated Risk Information System (IRIS), EPA defines inhalation "unit risk" as the upper- bound excess lifetime cancer risk estimated to result from continuous exposure to an agent at a concentration of 1 $\mu g/m^3$ in air.

estimated an inhalation MRL at 0.2 $\mu g/m^3$ and an oral MRL at 0.7 $\mu g/kg$ per day. The average daily dietary intake of cadmium by adult Americans is about 0.4 $\mu g/$ kg per day (Gartrell et al. 1986). The average levels of cadmium in smokers approaches the MRL (Nordberg et al. 1985), and therefore smokers are at risk of renal disease from any additional cadmium exposures, including incinerators.

The health effects of cadmium compounds in humans are summarized in Table 5-2.

Lead

Lead has been studied more thoroughly than any of the other pollutants of concern in connection with waste incineration. The public-health importance of lead is due both to its ubiquity in the environment and to the fact that it can affect virtually every organ system in humans and animals. Some effects of lead occur at intakes producing blood concentrations that are low compared with blood concentrations that were considered normal within the past 4 decades (see Table 5-3), and for some no threshold has been demonstrated. In addition, well-defined susceptible subpopulations exist, including fetuses, pre-school-age children, the elderly, smokers, alcoholics, those with nutritional disorders, those with neural or renal dysfunction, and those with genetic diseases that affect heme synthesis (ATSDR 1997b). Direct toxicity to peripheral nerves used to be common among poorly protected lead-exposed workers.

The toxicity of lead and its various inorganic and organic compounds after inhalation, ingestion, or dermal absorption depends on the total body burden and the distribution among various target organs (ATSDR 1997b). The two principal routes of exposure are ingestion and inhalation. About 50-90% of inhaled lead is absorbed by the body, whereas less than half of ingested lead is retained. Children absorb more lead through the gastrointestinal tract than do adults, about 30% compared with less than 10%, although dietary factors are important. Vitamin C, vitamin D, and calcium deficiencies might double or even triple the fraction of ingested lead that is absorbed.

Lead is stored in various body tissues, including blood, kidney, brain, and bone (ATSDR 1997b). Lead in blood has a half-life of about 35 days, in soft tissue about 40 days; and in bone about 20 years. The commonly measured blood-lead concentration is a complex function of prior exposures, showing rapid response to short-term fluctuations in lead intake; while bone lead concentration is more a measure of long-term exposure.

Blood lead has been the most commonly used biomarker of risk (ATSDR 1997b). Many studies have relied on blood-lead measurements as surrogates for biologically relevant lead exposures, doses, or body burdens. Such measurements may, however, be unreliable as indicators of long-term exposures during periods when exposure is changing. Some of those shortcomings can be overcome by measuring bone lead with x-ray fluorescence, which has been used in

TABLE 5-2 Cadmium Compounds: Health Effects in Humans

Type of Toxicity	Route of Exposure	Type of Effects[a]	Reference
Cardiovascular	Inhalation and ingestion	Conflicting findings in relation to cadmium exposures and cardiovascular toxicity; cardiovascular abnormalities reported at very high exposures	Shigematsu 1984; Kazantzis et al. 1988
Cancer	Inhalation	Unit risk of 1.8×10^{-3} per $\mu g/m^3$ based on 1 study	Thun et al. 1985
Gastrointestinal	Ingestion	Ingestion of large amounts causes acute gastric irritation and inflammation; acute doses > 0.07 mg/kg cause nausea, vomiting, and abdominal pain; death reported at 5 g	Nordberg et al. 1973
Hematologic	Inhalation	Anemia reported at very high exposures	Friberg 1950; Kagamimori et al. 1986
Neurologic	Inhalation	Neurobehavioral changes reported at exposures less than those causing renal effects	Marlowe et al. 1985; Struempler et al. 1985; Hart et al. 1989
Renal	Inhalation	Renal tubular damage at renal cortex concentrations of 50-200 $\mu g/g$ wet weight; proteinuria reported at 0.023 mg/m^3	Roels et al. 1983; Jarup et al. 1988; Bernard and Lauwerys 1989; Buchet et al. 1990
Respiratory	Inhalation	Acute toxicity: lung edema, tracheobronchitis, pneumonitis, and fever reported at concentrations > 0.5 mg/m^3; death from acute respiratory failure at higher concentrations Chronic toxicity: impaired lung function and emphysema	Friberg 1950; Townshend 1982; Grose et al. 1987; Greenspan et al. 1988

[a] For many of the health effects of concern, exposures are uncertain or unknown and are related to occupational studies or case reports of accidental spills or releases.

TABLE 5-3 Lead Compounds: Health Effects in Humans

Type of Toxicity	Route of Exposure	Type of Effects[a] (μg/dL blood-lead concentrations)	Reference
Cardiovascular	Inhalation	Cardiac lesions and ECG changes at high concentrations; increased blood pressure reported at 7 μg/dL	Pirkle et al. 1985; Schwartz 1988; Coate and Fowles 1989; ATSDR 1997b
Cancer	Inhalation	Inconclusive in humans	EPA 1988a; ATSDR 1997b
Developmental	Inhalation	No major congenital anomaly; conflicting data on reduced birthweight and gestational age at 15 μg/dL	Bellinger et al. 1984; Needleman et al. 1984; McMichael et al. 1986; Bornschein et al. 1989; Greene and Ernhart 1991; ATSDR 1997b
Gastrointestinal	Inhalation	Colic, abdominal pain, constipation, cramps, nausea, vomiting, anorexia, and weight loss reported at acute high concentrations	Matte et al. 1989; ATSDR 1997b
Hematologic	Inhalation	Inhibition of ALA-D activity reported at 12 μg/dL; increased erythrocyte protoporphyrins reported at 15 μg/dL; reduced hemoglobin synthesis reported at 40 μg/dL for children; frank anemia reported at 80 μg/dL	Pueschel et al. 1972; Chisolm et al. 1985; Roels and Lauwerys 1987; ATSDR 1997b
Hepatic	Inhalation	Abnormal liver-function tests and inhibition of cytochrome P-450 reported at high concentrations	Saenger et al. 1984; ATSDR 1997b

Neurologic	Inhalation	Neurologic and neurobehavioral abnormalities reported at 40-60 µg/dL; decreased nerve-conduction velocities reported at 30-50 µg/dL; encephalopathy reported at 100 µg/dL (adults), 80 µg/dL (children); IQ deficit reported at 6-70 µg/dL; increase in hearing thresholds reported at 4-56 µg/dL; convulsions and death resulting from severe nervous system depression at above approximately 80 µg/dL.	Kehoe 1961; Rummo 1974; Baker et al. 1979, 1983; Hänninen et al. 1979; Rummo et al. 1979; Seppäläinen et al. 1983; Robinson et al. 1985; Lansdown et al. 1986; Fulton et al. 1987; Schroeder and Hawk 1987; Schwartz and Otto 1987; Harvey et al. 1988; Cooney et al. 1989a; Needleman and Bellinger 1989; Pocock et al. 1989; NRC 1993; ATSDR 1997b
Renal	Inhalation	Acute nephropathy—structural and functional changes in proximal convoluted tubules, Fanconi's syndrome (aminoaciduria, phosphaturia, glucosuria, increased sodium excretion, and decreased uric acid excretion)—reported at acute high concentrations Chronic nephropathy (interstitial nephritis and tubular and glomerular disease) reported at 40 µg/dL (adults) and 80 µg/dL (children)	Pueschel et al. 1972; Weden et al. 1979; Buchet et al. 1980; Pollock and Ibels 1986; Maranelli and Apostoli 1987; Ong et al. 1987; Verschoor et al. 1987; Huang et al. 1988; ATSDR 1997b
Reproductive	Inhalation	Increased incidence of miscarriages and stillbirths reported at concentrations of 10 µg/dL or more; adverse effects on testes reported at 50 µg/dL	Assennato et al. 1987; Baghurst et al. 1987; Rodamilans et al. 1988; Hu 1991; ATSDR 1997b

[a] For many of the health effects of concern, exposures are uncertain or unknown and are related to occupational studies or case reports of accidental spills or releases.

epidemiologic studies and found to be useful, in conjunction with blood lead, for assessing body lead burden.

Neurotoxicity is a major health concern with respect to lead exposure. Available data suggest that children are more sensitive than adults and respond to lead at lower doses (Rom 1992). Severe lead encephalopathy, occasionally fatal, occurs in adults with blood lead above 100 μg/dL (Kehoe 1961) and in children with blood lead as low as 80 μg/dL (NRC 1993). Adults have less-severe but overt neurologic and neurobehavioral effects at blood lead as low as about 40-60 μg/dL (Baker et al. 1979, 1983; Hänninen et al. 1979), and decreased nerve-conduction velocities have been reported at blood lead of 30 μg/dL (Seppäläinen et al. 1983).

Studies of neurodevelopmental effects in children have produced less-conclusive findings with respect to identifying a threshold at which effects appear. On the one hand, statistically significant relationships have been found between intelligence quotient (IQ) and blood lead in children whose individual blood-lead ranged from 6 to 46 μg/dL (Schroeder and Hawk 1987) and in groups of children whose average exposures ranged from 5.6 to 22.1 μg/dL (Fulton et al. 1987). Increasing blood lead was associated with decreasing IQ in each of those studies, and no threshold for this effect was observed. Further data on children's IQ at higher lead concentrations suggest a deficit of about 5 points in average IQ in groups of children with mean blood lead of 50-70 μg/dL compared with a control group with mean blood lead of 21 μg/dL (Rummo 1974; Rummo et al. 1979), and a deficit of about 4 IQ points in groups of children with estimated blood lead of 30-50 μg/dL compared with a control group with mean blood lead less than 15 μg/dL (Needleman 1979). Other investigations, however, have failed to find an association between low blood lead and neurobehavioral effects or IQ deficits (Lansdown et al. 1986; Harvey et al. 1988; Cooney et al. 1989a,b; Pocock et al. 1989).

Overall, the data suggest that lead causes neurobehavioral disturbances in children at concentrations below 50 μg/dL, and possibly below 20 μg/dL. No threshold for such effects can yet be demonstrated.

Other well-documented effects of lead exposure at blood-lead concentrations above 40 μg/dL in humans are renal impairment, hematologic effects, cardiovascular effects (including high blood pressure), gastrointestinal and liver abnormalities, and reproductive and developmental effects (ATSDR 1997b). With respect to the latter, no human evidence suggests that low prenatal exposure to lead is associated with any major structural congenital anomaly (McMichael et al. 1986). Studies of prenatal exposure at low concentrations, however, have produced conflicting data with respect to low birth weight and gestational age (Bellinger et al. 1984; Needleman et al. 1984; Bornschein et al. 1989; Greene and Ernhart 1991). Some evidence suggests that lead reduces gestational age, even when maternal blood lead is below 15 μg/dL. Similarly, miscarriages and stillbirths have been reported in exposed women whose blood lead was 10 μg/dL or higher (Baghurst et al. 1987; Hu 1991) and adverse effects on the testes of offspring of women whose blood lead was 40-50 µg/dL have been reported (Assennato et al. 1987; Rodamilans et al. 1988).

The evidence for lead as a human carcinogen is inconclusive, but lead exposures have caused renal tumors consistently in experimental animals under suitable experimental conditions (ATSDR 1997b).

Although lead toxicity has been known since antiquity, there remains considerable debate about safe exposures and about the body burden of lead below which no adverse effects might be anticipated. Several toxic effects appear to have different thresholds of exposure, and some have no clearly defined safe exposure (that is, no identifiable threshold exposure). In that regard, EPA (1986b) and ATSDR (1997b) have expressed concern about the emerging evidence of a constellation of effects that occur at low blood-lead concentrations (10-15 µg/dL, or even lower), including subtle neurologic and neurobehavioral changes, growth and blood-pressure effects, inhibition of aminolevulinic acid dehydrase and pyrimidine-5'-nucleotidase activity, reduction in serum 1,25-dihydroxyvitamin D, and increase in erythrocyte protoporphyrins. The health effects of lead toxicity in humans are summarized in Table 5-3.

Mercury

Adverse human health effects of exposure to mercury are dependent on the particular chemical species of mercury, the magnitude and route of exposure, and the degree to which the mercury is metabolized.

Mercury exists in inorganic and organic forms. Inorganic mercury occurs in 3 valence states: metallic (elemental) mercury (Hg), mercurous salts (Hg^+), and mercuric salts (Hg^{2+}). The most commonly encountered organic forms are alkyl mercury species (notably methylmercury and ethylmercury) that result largely from microbial metabolism of inorganic mercury in the environment.

Metallic mercury is poorly absorbed from the gastrointestinal tract, but inhaled metallic mercury vapor is well absorbed from the lungs (ATSDR 1994). Metallic mercury can oxidize to the mercuric state. Soluble mercuric compounds (Hg^{2+}) are well absorbed from the intestine and are the most commonly encountered inorganic salts of mercury. Mercurous salts (Hg^+) are absorbed from the intestine, but are unstable in the presence of sulfhydryl groups and convert to either metallic mercury or the mercuric state. Therefore, mercurous compounds can share the toxic characteristics of both metallic and mercuric mercury. Organomercury compounds are absorbed well from the intestine and are less readily oxidized to the mercuric state than are metallic or mercurous compounds.

In general, dermal absorption is the least-likely route of uptake of mercury, although it appears that dermal absorption can be substantial under some circumstances (ATSDR 1994). Metallic mercury is absorbed through the skin, but at much lower rates than by inhalation. Inorganic salts might be absorbed to a greater degree, but quantitative data are lacking.

In humans, metallic mercury and organomercury compounds cross the blood-brain barrier and the placenta, and the major health effects of concern for these compounds are nervous system impairment and fetal toxicity. Inorganic salts of mercury do not cross the placenta or blood-brain barrier readily, so they are typically less toxic to the fetus and produce fewer central nervous system effects. The kidney appears to be the most-sensitive organ after ingestion of inorganic salts. The renal tract is the principal route of excretion of all forms and species of mercury. All mercury compounds have some degree of renal toxicity.

Other organs affected by mercury at higher exposures include the respiratory, cardiovascular, hematologic, gastrointestinal, and reproductive systems. Such toxic effects at high exposures might reflect the high affinity of mercuric mercury for sulfhydryl groups. Results of animal studies support a concern for the neurologic, renal, developmental and reproductive, and respiratory effects of mercury exposure in humans.

Most data concerning human health effects are related to occupational exposures, accidental spills and releases, or the major environmental contamination and consumption of fish contaminated with methylmercury in Minamata, Japan. Most reports of mercury exposures that caused serious human health effects predate 1970 and occurred in workplaces and other settings where exposures were generally high; the health effects observed under such conditions might not be directly pertinent to chronic, low-level exposures generated by waste incineration.

Several recent issues have rekindled interest in this metal. Mercury in dental amalgams has raised concerns about the slow absorption of mercury from dental fillings. Mercury arising from environmental pollution, notably from industrial pollution, waste incineration, other combustors, and natural sources has caused problems with surface water contamination in lakes and streams, and has raised concerns about human-health effects of eating fish with high tissue mercury concentrations (Amdur et al. 1991). The latter subject is particularly relevant to the present discussion. The health effects of mercury compounds in humans are summarized in Table 5-4.

Chromium

Chromium is most commonly encountered in 4 valence states: 0 (metal), II (chromous), III (chromic), and VI. Cr(VI) in the environment is almost always related to human activity (ATSDR 1993b).

Cr(III) is an essential nutrient, forming an organic complex that facilitates the interaction of insulin with cell-membrane receptors. The recommended dietary intake of Cr(III) is 50-200 µg/day (ATSDR 1993b; NRC 1989b).

In general, Cr(VI) compounds are more toxic than Cr(III) compounds, and are better absorbed after inhalation, ingestion, and dermal contact (ATSDR

TABLE 5-4 Mercury Compounds: Health Effects in Humans

Type of Toxicity	Form of Mercury	Route of Exposure	Type of Effects[a]	Reference
Cardiovascular	Metallic	Inhalation	Increased blood pressure and heart rate at acute high concentrations (>0.27 mg/m^3); hypertension, death from ischemic heart disease, and cerebrovascular disorders reported at chronic exposure to high concentrations	Vroom and Greer 1972; Piikivi 1989; Aronow et al. 1990; Barregård et al. 1990; Bluhm et al. 1992; Soni et al. 1992; Taueg et al. 1992; ATSDR 1994
	Mercurous and mercuric salts	Inhalation	Increased blood pressure and acrodynia reported in children	Warkany and Hubbard 1953; ATSDR 1994
Cytogenic	Metallic	Inhalation	No chromosomal damage from occupational exposures (clastogenic for mammalian germ cells)	ATSDR 1994
Death	Metallic and organic	Inhalation	Death reported at exposures of 1.0 mg/m^3	Hill 1943; ATSDR 1994
	Inorganic and organic	Ingestion	Death reported at exposures of 10-60 mg/kg	Troen et al. 1951; Gleason et al. 1957; Bakir et al. 1973; Tsubaki and Takahashi 1986; ATSDR 1994
	Inorganic	Dermal	Death reported at exposures of 93 mg/kg	ATSDR 1994
Developmental	Organic	Inhalation and ingestion	Developmental toxicity at high concentrations	Engleson and Hermer 1952; Amin-Zaki et al. 1976; Marsh et al. 1980, 1981, 1987; ATSDR 1994;
Gastrointestinal	All forms	Inhalation and Ingestion	Acute gastrointestinal toxicity (nausea, diarrhea, abdominal pains, inflammation of oral mucosa, inflammation, and acute ulceration) reported at 20-30 mg/kg; similar effects seen after inhalation	Campbell 1948; Brown 1954; Tennant et al. 1961; Sexton et al. 1978; Fagala and Wigg 1992; ATSDR 1994

(continued)

TABLE 5-4 (Continued)

Type of Toxicity	Form of Mercury	Route of Exposure	Type of Effects[a]	Reference
Hematologic	Metallic	Inhalation	Metal fume fever, syndrome like leukocytosis, decreased red cell counts, and anemia reported at high concentrations	Campbell 1948; Fagala and Wigg 1992; ATSDR 1994;
Neurologic	Metallic and organic	Ingestion, inhalation, and dermal	Nervous system impairment and behavioral abnormalities reported to occur at 0.014-0.076 mg/m^3 (inhalation) and at 0.74 mg/kg (oral)	Al-Saleem 1976; Cinca et al. 1980; Bluhm et al. 1992; Adams et al. 1983; ATSDR 1994; Marsh et al. 1995
Renal	Metallic, inorganic salts, and organic	All routes	Acute renal toxicity (oliguria, hematuria, urinary casts, impaired ability to concentrate urine, and proteinuria), chronic nephrotic syndrome reported at 15-30 mg/kg	Friberg et al. 1953; Afonso and de Alvarez 1960; Murphy et al. 1979; Cinca et al. 1980; Samuels et al. 1982; Jaffe et al. 1983; Piikivi and Ruokonen 1989; Kang-Yum and Oransky 1992; Soni et al. 1992; ATSDR 1994
Reproductive	Metallic and ingestion	Inhalation	Increased rate of spontaneous abortions and complications with labor reported at high concentrations	Mishonova et al. 1980; Cordier et al. 1991; ATSDR 1994
Respiratory	Metallic	Inhalation	Lung edema and fibrosis reported at very high concentrations; airway and parenchyma inflammation, chest pain, cough, breathlessness, and reduction in vital capacity reported at >44 mg/m^3	Tennant et al. 1961; Milne et al. 1970; Bluhm et al. 1992; Soni et al. 1992; Taueg et al. 1992; ATSDR 1994

[a] For many of the health effects of concern, exposures are uncertain or unknown and are related to occupational studies or case reports of accidental spills or releases.

1993b). However, after ingestion, Cr(VI) is reduced to Cr(III) in the stomach, so that the ingestion route is of lesser importance.

Occupational exposures of humans to chromium—mainly Cr(VI)—compounds have caused ulceration and perforation of the nasal septum; respiratory tract irritation; sensitization of the respiratory tract, skin, and mucous membranes; and increased risk of lung cancer. Renal damage, gastrointestinal changes, and hematologic effects have also been described. Skin problems caused by direct contact with chromium compounds include ulceration and allergic sensitization. Among chromate workers, those problems were severe in the past when skin contact was high (Lucas and Kramkowski 1975).

The health effects of chromium compounds in humans are summarized in Table 5-5.

Arsenic

Arsenic is a powerful human toxicant. Exposures to inorganic arsenic compounds—chiefly oxides and oxyacids (arsenates and arsenites)—are the most-common sources of exposure, although organic arsenicals (mainly methyl or phenyl arsenates) have been used widely in agriculture. Organic arsenicals are considered less toxic than the inorganic forms (ATSDR 1998b).

Inhalation exposures of humans to inorganic arsenic compounds have led to acute and chronic respiratory irritation, and to lung cancer (EPA 1988b; ATSDR 1998b). A wide variety of adverse health effects, including skin and internal cancers and cardiovascular and neurological effects, have been attributed to chronic arsenic exposure, primarily from drinking water (NRC 1999b). Direct skin contact has led to local irritant effects and hyperkeratoses.

Little information is available on the effects of organic arsenicals on humans. Results of animal studies suggest that organic compounds can have effects similar to those of inorganic forms (ATSDR 1998b). However, no studies have demonstrated that organic arsenic is carcinogenic in humans (ATSDR 1998b). Reduction in the carcinogenicity of arsenic, particularly at low exposure concentrations, has been linked to its methylation in vivo (Marcus and Rispin 1988). Because most forms of organic arsenic are already methylated, there is good reason to expect organic arsenic would be far less carcinogenic in humans than inorganic forms. ATSDR (1998b) has reviewed animal studies of organic arsenic and concluded that it may have weak carcinogenic potential.

In general, the toxicity of arsenic in all its forms is less in experimental animals than in humans, so animal data on arsenic are considered less reliable predictors of human effects than are animal data for many other substances (ATSDR 1998b). The major health effects of arsenic compounds in humans are summarized in Table 5-6.

TABLE 5-5 Chromium Compounds: Health Effects in Humans

Type of Toxicity	Form of Chromium	Route of Exposure	Type of Effects[a]	Reference
Cancer	Primarily Cr(VI)	Inhalation	Increased risk of lung cancer	Mancuso and Hueper 1951; Bidstrup and Case 1956; Hayes et al. 1979; Hill and Ferguson 1979; Braver et al. 1985; Davies et al. 1991; Pastides et al. 1991; ATSDR 1993b
Dermal	Chromium compounds	Dermal	Ulceration, allergic sensitization reported at high concentrations	Lucas and Kramkowski 1975; ATSDR 1993b
Gastrointestinal	Chromium salts	Inhalation	Stomach cramps, gastric and duodenal ulcers, and gastritis reported at high concentrations	Mancuso 1951; Sassi 1956; ATSDR 1993b
Hematologic	Cr(III) and Cr(VI)	Inhalation	Leukocytosis or leukopenia, decreased hemoglobin, and prolonged bleeding times reported at high exposures	Mancuso 1951; ATSDR 1993b
Renal	Cr(VI)	Inhalation	Proteinuria and renal tubular necrosis reported at high exposures	Franchini and Mutti 1988; ATSDR 1993b
Respiratory	Primarily Cr(VI)	Inhalation	Ulceration and perforation of the nasal septum, respiratory tract irritation, and decreased lung function reported at 0.002 mg/m^3; sensitization of respiratory tract and mucous membranes	Mancuso 1951; Taylor 1966; Bovet et al. 1977; Keskinen et al. 1980; Lindberg and Hedenstierna 1983; Novey et al. 1983; ATSDR 1993b

[a] For many of the health effects of concern, exposures are uncertain or unknown and are related to occupational studies or case reports of accidental spills or releases.

TABLE 5-6 Arsenic Compounds: Health Effects in Humans

Type of Toxicity	Route of Exposure	Type of Effects[a]	Reference
Cancer	Inhalation/inorganic compounds	Lung cancer	Axelson et al. 1978; Enterline and Marsh 1982; Lee-Feldstein 1986; Sobel et al. 1988; ATSDR 1998b; NRC 1999b
	Ingestion/inorganic compounds	Skin cancer; evidence of increased risk of liver, bladder, and kidney cancers	Zaldivar 1974; Chen et al. 1985, 1986; Wu et al. 1989; ATSDR 1998b; NRC 1999b
Cardiovascular	Ingestion/inorganic compounds	Cardiac arrhythmia and ventricular tachycardia or fibrillation reported at high doses; vascular lesions leading to blackfoot disease reported at chronic low exposures	ATSDR 1998b; NRC 1999b
	Ingestion/inorganic compounds	Raynaud's disease	ATSDR 1998b; NRC 1999b
Dermal	Ingestion/inorganic compounds	Various skin disorders	ATSDR 1998b; NRC 1999b
	Dermal	Local irritation, hyperkeratoses	ATSDR 1998b; NRC 1999b
Gastrointestinal	Ingestion/inhalation	Nausea, vomiting, diarrhea, and abdominal pain reported at acute high doses or repeated lower doses	ATSDR 1998b; NRC 1999b

(continued)

TABLE 5-6   (Continued)

Type of Toxicity	Route of Exposure	Type of Effects[a]	Reference
Hematologic	Ingestion/inorganic compounds	Anemia reported ≥ 0.17 mg/kg per day	Kyle and Pease 1965; Harrington et al. 1978; Franzblau and Lilis 1989; ATSDR 1998b; NRC 1999b
Neurologic	Ingestion/inorganic compounds	CNS abnormalities associated with acute and chronic exposures; peripheral neuropathies associated with acute and chronic low exposures (0.02-0.5 mg/kg per day)	Franzblau and Lilis 1989; Goebel et al. 1990; ATSDR 1998b; NRC 1999b
Respiratory	Inhalation/inorganic compounds, most commonly arsenic trioxide	Acute (> 0.1-1 mg/m^3) and chronic respiratory irritation and inflammation in airways and lung parenchyma	Ide and Bullough 1988; Morton and Caron 1989; ATSDR 1998b; NRC 1999b

[a] For many of the health effects of concern, exposures are uncertain or unknown and are related to occupational studies or case reports of accidental spills or releases.

Beryllium

Appreciable human exposures to metallic beryllium or its salts occur almost exclusively in workplace settings. The burning of coal and fuel oil contributes a small inhaled burden to the general public, particularly in urban areas where the median air concentration is about 0.2 ng/m^3 (ATSDR 1993c).

Inhalation is the principal route of exposure (ATSDR 1993c). Granulomatous lung disease is the most common health effect in humans, although beryllium disease is a multisystem disorder (ATSDR 1993c). Dermal contact can cause sensitization and systemic illness, but beryllium compounds are absorbed poorly through the skin (ATSDR 1993c). Absorption of beryllium from the gastrointestinal tract is also poor, and this route of exposure has rarely caused appreciable toxicity.

Epidemiologic data have suggested an increased risk of lung cancer associated with occupational exposures to beryllium. Results of a recent study that accounted for smoking habits and used an appropriate unexposed comparison group showed an increased risk of lung cancer among exposed people (Steenland and Ward 1991). Animal studies in rats and monkeys have also shown that beryllium can cause lung tumors (ATSDR 1993c).

The current workplace exposure limit for beryllium of 0.002 mg/m^3 was established in 1950 to prevent nonmalignant beryllium disease and has been successful in reducing the rate of chronic lung disease (ATSDR 1993c). That limit might not be sufficient to protect against lung cancer. In this regard, EPA (1992b) has estimated the upper bounds for inhalation unit risk of 2.4×10^{-3} m^3/μg, and for ingestion a potency of 4.3 kg-d/mg.

The health effects of beryllium toxicity in humans are summarized in Table 5-7.

Organic Compounds

Dioxins and Furans

Dioxins and furans have been the subject of much controversy and study (e.g., NRC 1994). Their toxic effects are summarized below.

Acute Toxicity

Case studies of acute reactions caused by exposure to TCDD have been documented. Workers exposed to TCDD in a plant explosion were examined; they had a number of acute symptoms "characterized by skin, eye and respiratory tract irritation, headache, dizziness, and nausea" (Suskind and Hertzberg 1984). The acute symptoms subsided within a week but were followed by "acneiform eruption, severe muscle pain affecting the extremities, thorax and shoulders, fatigue, nervousness and irritability, dyspnea, complaint of decreased libido and intolerance to cold" (Suskind and Hertzberg 1984).

TABLE 5-7 Beryllium Compounds: Health Effects in Humans

Type of Toxicity	Form of Beryllium	Route of Exposure	Type of Effects[a]	Reference
Cancer	Beryllium or its salts	Inhalation	Lung cancer	Steenland and Ward 1991; ATSDR 1993c
Dermal/ocular	Beryllium salts	Dermal, ocular	Ocular granulomas and sensitization resulting in papulovesicular dermatitis and ulceration reported at high concentrations	ATSDR 1993c
Respiratory	Beryllium or its salts	Inhalation	Acute tracheobronchoalveolitis reported at high concentrations; chronic beryllium disease— multisystem granulomatous disease—principally involving lungs reported at lower concentrations	Van Ordstrand et al. 1945; Hardy and Tabershaw 1946; Kriebel et al. 1988a,b; Rossman et al. 1988; ATSDR 1993c

[a] For many of the health effects of concern, exposures are uncertain or unknown and are related to occupational studies or case reports of accidental spills or releases.

Dioxin is acutely toxic to experimental animals at sufficiently high doses. The lethal dose of TCDD varies extensively among species and with sex, age, and route of administration (NRC 1994). A symptom known as severe wasting syndrome has been reported in several laboratory animals. Weight loss typically manifests itself within a few days after exposure and is associated with a loss of adipose and muscle tissue (Max and Silbergeld 1987). Typically, at the lethal dose, there is a delayed toxicity, and death usually occurs several weeks after exposure (EPA 1994d).

Chronic Toxicity

Results of epidemiologic studies suggest that chloracne (an acne-like eruption due to prolonged contact with certain chlorinated compounds (NRC 1994), increased gamma-glutamyltransferase (GGT) (a hepatic enzyme that is measured in human serum to evaluate liver toxicity), increased diabetes, and altered reproductive hormone concentrations appear to be long-term, noncarcinogenic consequences of exposure to TCDD (EPA 1994d; NRC 1994). Other effects reported include eyelid cysts, hypertrichosis and hyperpigmentation, actinic keratosis (abnormal distribution of the hair), Peyronie's disease (progressive scarring of penile membrane), cirrhosis, liver enlargement, alteration of liver enzyme concentrations, porphyria (alteration of porphyrin metabolism), and renal, neurologic, and pulmonary disorders (EPA 1994d). But results of other studies suggest possible acute effects and few chronic effects other than chloracne.

A chronic-toxicity study performed by Kociba et al. (1978, 1979) on laboratory rats over 2 years showed urinary disorders in females. Alterations of the liver were found in both males and females. Other specific effects of TCDD toxicity in animals include wasting syndrome, hepatotoxicity, enzyme induction (in particular, the induction of cytochrome P-450 1A1, which is responsible for the activation and detoxification of endogenous and exogenous chemicals), endocrine alterations, decreased vitamin A storage, and decreased lipid peroxidation (NRC 1994). The most-consistent syndrome of TCDD toxicity among all animals is wasting syndrome (NRC 1994).

In humans, several studies documenting blood or adipose-tissue measurements, workplace exposure, and the occurrence of chloracne reported increased cancer rates after a relatively long latency in workers exposed to TCDD at relatively high concentrations (Zober et al. 1990; Fingerhut et al. 1991; Manz et al. 1991). Specifically, an excess of respiratory cancer was reported, as was a suggested increased risk of connective, soft tissue, and lung cancers. However, substantial uncertainties with regard to the database of epidemiologic evidence could influence risk estimates (for example, a large variety of tumor types, uncertainties as to exposure, possible confounding with such known human carcinogens as asbestos, and possible confounding with cigarette-smoking).

Several long-term studies have been performed to determine the carcinogenesis of TCDD in experimental animals. Long-term carcinogenicity bioassays of TCDD have been conducted in rats, mice, and hamsters (Van Miller et al. 1977; Kociba et al. 1978; Toth et al. 1979; NTP 1982a,b; Della Porta et al. 1987; Rao et al. 1988). Exposure has been oral, intraperitoneal, dermal, and subcutaneous. Results of the studies have been summarized in NRC 1994, Table 4-2. Increased tumor rates reportedly occurred at several sites in the body in different studies, although the liver was consistently a site of tumor formation in different studies and different species. In studies in which liver cancer occurred, other toxic changes in the liver also occurred. Other organs in which increased cancer rates were observed in animals exposed to TCDD include the thyroid, adrenals, skin, and lungs. Further animal studies indicate that PCDD and PCDF carcinogenesis may proceed through a receptor-mediated mechanism, although details are unclear (Stone 1995), and that PCDDs and PCDFs are tumor promoters in animal liver and skin assays.

Developmental Toxicity

Alterations in development due to dioxin exposure have also been reported in experimental animals (EPA 1994d), including such structural malformations as cleft palate and hydronephrosis in mice, while other species have shown postnatal functional alterations, some irreversible, including effects on the reproductive system and object-learning behavior (EPA 1994d). The resemblance between some effects observed in adult monkeys and neonatal mice exposed to TCDD and those documented in Yusho or Yu-Cheng infants (for example, subcutaneous edema of the face and eyelids, larger and wider fontanel, and abnormal lung sounds) suggests that particular effects reported in these infants were caused by TCDD-like PCB and chlorinated dibenzofuran (CDF) congeners in the rice oil ingested by the mothers (Harada 1976; Urabe et al. 1979; Hsu et al. 1994).

Reproductive Toxicity

Although there have been no studies concerning the reproductive effects of dioxin-related compounds in humans, the potential exists for dioxin and related compounds to cause reproductive toxicity (Kimmel 1988). A variety of animal studies have shown that TCDD and its structurally related compounds affect female reproduction (Kociba et al. 1976; Barisotti et al. 1979; Murray et al. 1979). The foremost effects seem to be decreased fertility, inability to carry to term, and, in rats, decreased litter size. There are also effects on gonads and the estrous cycle. In males, TCDD and related compounds decrease testis and accessory sex organ weights, cause abnormal testicular structure, decrease spermatogenesis, and reduce fertility.

Neurologic Effects

In 1976, an industrial accident at a chemical manufacturing plant near Seveso, Italy, released kilogram amounts of TCDD into the environment. Neurologic effects were reported to have occurred shortly after exposure to TCDD in some workers and residents of contaminated areas (ATSDR 1998a). Symptoms included headache, insomnia, nervousness, irritability, depression, anxiety, loss of libido, and encephalopathy.

Immunotoxicity

Studies in mice, rats, guinea pigs, and monkeys have indicated that TCDD suppresses the function of some components of the immune system in a dose-related manner; that is, as the dose of TCDD increases, suppression of immune function increases. TCDD suppressed the function of cells of the immune system, such as lymphocytes (affecting cell-mediated immune response) and the generation of antibodies by B cells (affecting humoral immune response). Increased susceptibility to infectious disease has been reported after TCDD administration. In addition, TCDD increased the number of tumors that formed when tumor cells were injected into mice.

The effects of TCDD on the immune system appear to vary among species, although most studies used different treatments and are not completely comparable. However, some species seem more sensitive than others to the effects of TCDD on the immune system. It is not known whether humans would be more or less sensitive than laboratory animals.

Other Products of Incomplete Combustion (PICs)

The remainder of this section will focus on two other products of incomplete combustion—PCBs and PAHs.

PCBs

Most of the data on the adverse health effects of PCBs in humans are derived from occupational studies. Dermal and ocular effects of exposure have been relatively well established in these studies (ATSDR 1998c). There are also reports of respiratory, gastrointestinal, hematologic, hepatic, musculoskeletal, developmental, and neurologic effects, but the evidence is not strong enough to establish cause-effect relationships, in part because PCB concentrations were not measured and because other compounds were present in the work environment. Occupational studies have been inconclusive regarding the association of PCB exposure and cancer risk (ATSDR 1998c).

In studies of women assumed to have consumed PCB-contaminated fish, their offspring were found to have neurobehavioral deficits at birth, some of which persisted through the follow-up period of several years from birth. However, the findings are inconclusive because of various limitations of the studies regarding exposure assessment and the comparability of exposed and nonexposed subjects. Lower birthweight and shortened gestational age were reported in infants born to mothers occupationally exposed to PCBs, but these effects did not follow an exposure gradient (ATSDR 1998c). Estimates of PCB body burdens in populations exposed at concentrations commonly found in the United States indicate that neurobehavioral effects can occur after prenatal maternal exposures (ATSDR 1998c). Evaluations of blood samples from women who miscarried or delivered prematurely showed associations between these effects and concentrations of PCBs. Because of confounding factors, including exposure to DDT and other organochlorine pesticides, the adverse developmental effects reported in these studies cannot be attributed specifically to PCB exposure.

Effects of PCBs observed in experimental animals are generally consistent with the human data. Most of the toxicity studies of PCBs have involved oral exposures, and numerous effects have been documented, including hepatic, gastrointestinal, hematologic, dermal, immunologic, neurologic, and developmental and reproductive effects (ATSDR 1998c). Other effects of oral PCB exposure include weight loss, thyroid toxicity, and liver cancer (ATSDR 1998c). Adverse effects on liver and body weight were observed in the only animal-inhalation study of PCBs.

PAHs

PAHs occur ubiquitously in the environment from both anthropic and natural sources. PAHs occur in the atmosphere most commonly as products of incomplete combustion. They are found in the exhausts from fossil fuels; combustion; industrial processes (such as coke production and refinement of crude oil); gasoline and diesel engines, oil-fired heating, and in cigarette smoke. PAHs are present in groundwater, surface water, drinking water, waste water, and sludge. They are found in foods, particularly charbroiled, broiled, or pickled food items, and at low concentrations in refined fats and oils.

Occupational studies of workers who were exposed to mixtures that contain PAHs (for example, from exposure to coke-oven emissions and roofing tars) for long periods show an excess of cancer, particularly of the lung and skin (ATSDR 1995). Several of the PAHs, including benzo[a]pyrene, the most-studied PAH, have caused tumors in laboratory animals by the inhalation, ingestion, or dermal routes (ATSDR 1995). However, many animal studies involving PAHs have been negative with respect to carcinogenicity. Noncancer adverse health effects with PAH exposure have been observed in animals but generally not in humans with the exception of adverse hematological and dermal effects. In various animal studies, most involving oral exposures of test animals, various PAHs increased mortality, primarily because of adverse hematopoietic effects, including aplastic anemia and pancytopenia (ATSDR 1995). Benzo[a]pyrene induced reproductive toxicity in rodents, but the incidence and severity of the effects depended on the strain of animal and the method of administration (ATSDR 1990). Prenatal exposures of rats and mice to benzo[a]pyrene produced a decrease in mean pup weight during postnatal development and caused a high incidence of sterility in the F_1 progeny of mice (ATSDR 1990). PAHs are a broad and complex category of compounds, with many generally co-occurring and as such, are difficult to characterize merely by evaluating individual components. Nevertheless, the occupational health effects of various mixtures of PAHs have been evaluated in some groups of workers, e.g. coke oven workers and roofers, and occupational criteria and standards for protection of workers have been developed.

POPULATIONS AT RISK

In this section, we discuss sensitive populations and worker populations, which may be at especially increased risk because of exposure to incinerator emissions.

Sensitive Populations

Although not a well-defined term, *sensitive subpopulation* refers to some subset of the population that might suffer much more serious adverse health effects as the result of exposure to a toxic agent than the average population. Identifying high-risk persons is a critical part of the definition of sensitive subpopulations. Variation in sensitivity is due to many factors, some more easily recognized than others. For a specific population, these factors may include variations in underlying health conditions, diet, stages of development, and age, as well as genetic differences (e.g., in metabolic rates). Classic examples include the 1952 London smog and the 1948 Donora, PA, episodes, in which the increased mortality associated with pollution most severely affected the very young and the elderly (UK Ministry of Health 1954). Fetuses exposed during organ development can be extremely sensitive to relatively low exposures to chemicals that cause little or no harm in adults; for example, thalidomide interferes with the fetal development of limbs at doses that are harmless in adults. Lead exposure in utero is linked to adverse central nervous system (CNS) development at blood levels lower than required to produce neurological effects in adults (Kimmel and Buelke-Sam 1994). Because the blood-brain barrier is less well developed in infants than in adults, it should be expected that chemicals, in general, can more readily affect the central nervous system of infants (Kimmel and Buelke-Sam 1994). It is hypothesized that DNA-repair mechanisms are not well developed in fetuses or babies—human fetuses have only 20-50% of the DNA repair enzyme activity of adults; if this hypothesis is true, this population may be particularly sensitive to carcinogens (Kimmel and Buelke-Sam 1994). Much greater absorption of lead through the gastrointestinal tract in children (30%) than adults (6%) has been demonstrated (Ross et al. 1992).

The normal decline of many physiologic functions (such as immunologic responses) with aging might make the elderly more susceptible to various pollutants. Increased mortality due to both pulmonary and cardiovascular disease has been documented when pollution (for example, with particles and CO) has been only slightly increased, even when the levels of pollution remained within EPA guidelines (see earlier discussion in this chapter). However, there are substantial uncertainties about whether the correlations between measured pollution indicators (PM, CO, etc.) and mortality reflect cause and effect.

Some behavior patterns of children result in their receiving greater doses of pollutants than adults who experience the same environment. Running and playing outdoors lead to higher breathing rates and hence potentially greater intake of airborne pollutants; this might also affect adults who are working hard or who exercise regularly. Young children engage in a high degree of hand-to-mouth behavior; videotapes have documented about 40 hand-to-mouth actions per hour among young children (Ross et al. 1992). Thus, contaminated dirt and dust might enter children's systems to a greater degree than adults.

Sensitive populations can include those whose health is already compromised. For example, asthmatics respond to SO_2 at lower concentrations than nonasthmatics (see section on Acidic Aerosols and Gases). African-Americans are more likely than whites to have hypertension and kidney disease and therefore could be more susceptible to pollutants, such as lead, which adversely affect the circulatory and renal systems (see section on Lead). Similarly, some people may be much more sensitive to the effects of some chemical exposures because of pre-existing conditions brought on by exposures to other agents (possibly including other chemicals). For example, people who have experienced a hepatitis B virus infection appear to be at greatly increased risk of cancer due to aflatoxin B exposure, compared with those who have not had hepatitis B infection.

Variability in diet can be fairly extreme (e.g., vegan diets, which excludes all animal products, versus average American fare), resulting in substantial differences in the intakes of some pollutants (NRC 1993). Vegans, for example, should have substantially lower exposure to PCDDs and PCDFs, since the majority of the intake of these materials in the average American diet comes from their presence in animal fats. In addition, dietary deficiencies may also play a role in increasing the variability of uptake of certain pollutants. For example, iron deficiency can result in higher uptake of lead in the diet, while calcium deficiency might affect lead excretion (as observed in animal models) (ATSDR 1997b).

It has been observed that at high enough exposures, some chemicals or exposure situations alter the toxic effects of other chemicals or exposure situations. For example, the effects of high exposure to asbestos are compounded by cigarette smoking, so that the relative risk for lung cancer from the combined exposures is substantially higher than the sum of the relative risks due each separately (ATSDR 1993a). Similar interactions (some of them showing protective effects) have been seen in humans for other combinations of high exposures in occupational settings (with smoking generally being one of the exposures), and in medical situations (drug interactions), and at high exposures in animal models. Nevertheless, people are exposed to mixtures of chemicals for which interactions have not been studied (NRC 1988a, 1996).

Xenobiotic materials can be metabolized by various tissues, particularly the liver, and the rate of metabolism can be altered by exposure to various exogenous chemicals or drugs such as alcohol and tobacco (Petruzzelli et al. 1988). The metabolites produced may be more or less toxic than the parent chemical, and may be more or less easy for the body to further metabolize or excrete, so that the effective toxicity of any material may depend on such factors as the rate of metabolism and excretion. These factors can be highly variable within the human population. The genetic variability of AAH mediated metabolic rates in the liver, for example, is in a range of a factor of several thousand, possibly leading to variability in sensitivity to toxic effects from PAH exposure. Such variability is one possible explanation for the 7.3-fold odds-ratio for a particular genetic difference in PAH metabolism observed for squamous-cell lung cancer cases among Japanese light smokers (Nakachi et al. 1991). Similarly, another genetic factor, extensive-hydroxylator phenotype, has been associated with an increased risk of lung, liver, and bladder cancer in Americans, and in British workers exposed to asbestos or PAHs at high concentrations (Caporaso et al. 1989).

Overall, laboratory animals of different strains have exhibited a difference of a factor of 40 in tumorigenesis in response to carcinogens. NRC (1994) has estimated that the range of susceptibilities of humans to carcinogens is quite large: 1% of the population might be 100 times more susceptible than the average person, and 1% might be only one-hundredth as susceptible.

Worker Populations

Incinerator operators and maintenance workers, and those involved in the collection, transport, and disposal of fly ash and emission control equipment residues, have the potential to be most exposed to toxic substances associated with incineration.

As is true in many other industries, maintenance and cleaning often present the greatest opportunities for exposure to hazardous materials. The residual wastes after incineration can contain high concentrations of metals and dioxins, and firebrick can add crystalline silica to these hazards (Steenland and Stayner 1997). Air-pollution control equipment collects and concentrates certain toxic chemicals, so workers who maintain and clean these devices may be particularly at risk. Two recent studies of four municipal incinerators have documented very high exposures of workers to hazardous waste during the routine cleaning of the incinerator chambers and the electrostatic precipitators (NIOSH 1995; Richey 1995). According to those studies, the incinerators were periodically shut down (monthly to quarterly) to remove accumulated slag from the walls of the burn chamber and to clean fly ash out of the burn chambers and the electrostatic precipitators. To move the waste material to a point where it can be vacuumed out, the slag and fly ash were swept and shoveled, two operations that generated high airborne concentrations of particles containing heavy metals and dioxins at relatively high concentrations.

Dioxin concentrations were measured by NIOSH (1995) during incinerator cleanout operations, albeit with only single samples in various locations. PCDD and PCDF concentrations (measured as TEQs) ranged from 9 to 800 pg/m^3, uniformly high compared with the NRC (1988b) guideline of 10 pg/m^3 (established after a transformer fire). All samples collected during these maintenance operations indicated that workers were exposed at or above the NRC exposure guideline.

Analysis of bulk samples of fly ash from the first incinerator indicated that the dioxin content increased as one moved from the burn and upper (expansion) chamber (TEQ, 3 parts per trillion, or 3 ppt) to the lower (cooling) chamber (TEQ, 7 ppt) to the electrostatic precipitator (TEQ, 900 ppt). The second incinerator had even more dioxin in the one composite sample from the upper and lower chamber—TEQ, 50 ppt. The NIOSH health hazard evaluation was undertaken after a report of increased PCDD and PCDF concentrations in a pooled blood sample from 56 municipal incinerator workers, indicating that municipal incinerator workers suffer higher exposures on the job than the general population (Schecter et al. 1991).

The NIOSH investigation also indicated that exposures to arsenic, cadmium, lead, and aluminum substantially exceeded occupational exposure limits during the clean-out operations.

Richey (1995) reported very high exposures during cleaning operations at two municipal incinerators over a 2-year period. Personal air samples were collected for 39 workers before a vacuum system was introduced to reduce exposures, and for 22 workers while the vacuum system was in use. During normal maintenance operations, without the vacuum, half the 24 samples from those cleaning the incinerator chamber were above the PEL for lead (50 μg/m^3). In addition, the samples collected from two of seven workers drilling boiler tubes and seven of eight workers cleaning out the electrostatic precipitators were above the PEL for lead. The geometric mean exposures were 36 μg/m^3 and 38 μg/m^3 for the incinerator-chamber cleaning and boiler-tube drilling, respectively (dropping to 5.1 and 3.6 μg/m^3 with the use of vacuum), and cleaning the electrostatic precipitator resulted in a geometric mean exposure of 1,300 μg/m^3, dropping to

320 µg/m³ with vacuum use—still over 6 times the PEL. The same proportion of samples were above the PEL for cadmium. The geometric mean concentrations without and with vacuum use were 1.8 and 0.4 µg/m³ during cleaning of the incinerator chamber, 2.5 and 0.2 µg/m³ during boiler-tube drilling, and 64.1 and 18.9 µg/m³ during cleanout of the electrostatic precipitator. The fact that a separate study of 56 incinerator workers found them to have substantially higher blood lead concentrations than a comparison group of high-pressure plant tenders working at heating plants (Malkin et al. 1992) is consistent with the high lead exposures observed and suggests that incinerator workers in general are at risk of measurably increased lead absorption.

THE COMMITTEE'S CONSENSUS JUDGMENTS ABOUT WASTE INCINERATION AND PUBLIC HEALTH

After considering information on incineration operations and emission characteristics (Chapter 3), environmental behavior of pollutants of concern and contributions of incineration to environmental media (Chapter 4), and health-effects information summarized earlier in this chapter, the committee reached consensus judgments on various degrees of concern about incineration and public health on the basis of what is known, and in view of the lack of important information, as described in this report. The lack of such information has contributed to the substantial concern among many communities about possible adverse health effects resulting from incinerators. It is important to note that uncertainty also exists around current estimates of exposures and health effects with respect to other waste management practices.

In developing its consensus judgments about the various degrees of concern, the committee used an approach similar to a preliminary screening assessment intended to err on the side of caution in the face of substantial uncertainty. But in expressing its degrees of concern, the committee is not attempting to judge whether health effects are occurring. That would take a full-scale evaluation and would require much more information than was available to the committee. After additional information is obtained, it is possible that a degree of concern for a particular pollutant might change.

Table 5-8 shows the committee's qualitative consensus judgments of the relative degrees of concern for potential health consequences generally posed by waste incineration facilities. The following three populations are considered in this table: persons who work at the facilities, persons who live in close proximity to the facilities, and those individuals residing farther away who may be exposed to pollutants from multiple distant incinerator facilities. Each population is expected to experience quite different exposures because of different time-activity patterns, distances from the emission sources, or the chemical-specific nature of the pathways through which they may be exposed.

TABLE 5-8 Degrees of Concern for Potential Health Effects of Waste Incineration as Judged by the Committee[a]

Substance	Before MACT Compliance[b]			After MACT Compliance[b]		
	Potential Effects on Workers at a Facility[c]	Potential Effects from a Single Facility on a Local Population[d]	Potential Effects from Multiple Facilities on a Broader Population[e]	Potential Effects on Workers at a Facility[c]	Potential Effects from a Single Facility on a Local Population[d]	Potential Effects from Multiple Facilities on a Broader Population[e]
Particulate matter	Substantial	Substantial	Minimal	Substantial	Minimal	Minimal
Dioxin[f]	Substantial	Minimal	Substantial	Substantial	Minimal	Substantial
Lead	Substantial	Substantial	Moderate	Substantial	Minimal	Moderate
Mercury	Substantial	Moderate	Moderate	Substantial	Minimal	Moderate
Other metals[g]	Substantial	Moderate	Moderate	Substantial	Minimal	Moderate
Acidic gases[h]	Moderate	Minimal	Negligible	Moderate	Negligible	Negligible
Acidic aerosols[i]	Moderate	Moderate	Moderate	Moderate	Minimal	Minimal

[a] The four degrees of concern (substantial, moderate, minimal, negligible) were chosen based on general estimates of the following aspects: incineration emissions; persistence of emitted substances in the environment; mobility through air, soil, water, and food; potential total exposure through routes of inhalation, ingestion, and dermal absorption; and relative toxicity of the individual substances inferred from studies not involving incineration. The degrees are intended to represent the committee's qualitative assessment and consensus judgment of the possibility of health effects to workers and segments of the general public from incineration emissions. The degrees of concern are not derived from direct evidence of adverse health effects observed from incineration use. Also, a particular degree of concern is not intended to imply the extent to which the committee believes health effects are actually occurring. In addition, the selection of a particular degree resulted only from a preliminary screening effort in the absence of sufficient information to characterize all the important parameters accurately. For example, in this preliminary assessment, the committee did not attempt to assign different degrees of concern according to types of waste incinerated, design and operation of the facility, emission controls, or extent of worker protection. The term "substantial" is used to express the committee's highest degree of concern for a particular pollutant and category might change after more specific information is obtained. The term "substantial" is used to express the committee's highest degree of concern about possible exposures that could lead to health effects among workers, a local population, or a broader population. Lower degrees of concern correspond to less possibility that the specific groups are exposed to concentrations associated with adverse health effects.

b Compliance with MACT (maximum achievable control technology) will result in emissions reductions for many of the pollutants listed. However, not enough information is available to assess the potential impacts on broader populations due to residual emissions of dioxin and metals. Because of such uncertainty and the environmental persistence of dioxins and metals, the committee retained the degree of concern assigned to regional populations prior to MACT for those substances.

c Refers to possible effects due to pollutant exposures to workers at an incineration facility. Facility workers have the greatest potential to be exposed to the waste stream as it enters the incineration facility and is prepared for incineration, as well as the substances resulting from incineration (vapors, particles, gases, and solid noncombustibles such as ash).

d Refers to possible effects due to exposures to pollutants emitted from a local incineration facility. People who live, work, and attend schools near an incineration facility may be exposed to higher concentrations (amounts) of combustion products (vapors, particles, and gases) than those farther away.

e Refers to possible effects due to exposures of individuals to pollutants from multiple distant incineration facilities. Emissions into the environment from a source (incinerator) might be carried long distances from the source or may enter a geochemical-biological cycle near the facility, and so be taken up in the food chain. This distribution is determined in large part by the properties of the substance itself and to a lesser extent on the nature of the source. The contribution of many incinerators to the dioxin content of food is of substantial concern to the committee, and the committee has moderate concern about the contribution of many incinerators to the metal content of food.

f The term "dioxin" includes polychlorinated dibenzodiozins (PCDDs) and polychlorinated dibenzofurans (PCDFs).

g Other metals: cadmium, chromium, arsenic, and beryllium.

h Acidic gases: nitrogen oxides, sulfur dioxide, and hydrogen chloride.

i Acidic aerosols: a gas suspension of solid or liquid particles having strongly acidic components.

Selection of Pollutants

Table 5-8 reflects pollutants emitted by incinerators that currently appear to have the potential to cause the largest health effects due to their toxicity and the potential for exposures to occur. Also, pollutants are included that have the potential to be widely distributed in the environment, as well as those that do not have such a potential. Thus, some pollutants might be important locally and some might be more important when considered on a broader scale. Pollutants were also identified for which typical environmental concentrations are near levels at which health effects are expected. In areas where the ambient concentrations are already close to or above environmental guidelines or standards, even relatively small increments of substances can be important.

Workers at a Facility

The committee considered information presented earlier in this chapter on studies of incineration workers and other types of workers who had been exposed to high concentrations of pollutants listed in the table. Studies at municipal solid-waste incinerators show that workers are at much higher risk for adverse health effects than individual residents in the surrounding area. In the past, incinerator workers have been exposed to high concentrations of dioxins and toxic metals, particularly lead, cadmium, and mercury. Workers may be particularly at risk, not only because of emissions from the facility, but even more so if their work involves maintaining and cleaning the air-pollution control devices without proper safeguards. The electrostatic precipitators and bag houses, where potential emissions are captured, present risks to workers handling the concentrated pollutants.

A Single Facility and a Local Population

As discussed in Chapter 4, results of environmental monitoring studies around individual incineration facilities have indicated that the specific facilities studied were not likely to be major contributors to local ambient concentrations of the substances of concern, although there are exceptions. However, methodological limitations of those studies do not permit general conclusions to be drawn about the overall contributions of waste incineration to environmental concentrations of those contaminants. Particulate matter emitted by incinerators is especially important for local populations living in areas with high ambient concentrations of airborne particles.

Multiple Facilities and a Broader Population

The potential effects of metals and other pollutants that are very persistent in the environment may extend well beyond the area close to the incinerator. Persistent pollutants can be carried long distances from their emission sources, go through various chemical and physical transformations, and pass numerous times through soil, water, or food. Dioxins, furans, and mercury are examples of persistent pollutants for which incinerators have contributed a substantial portion of the total national emissions. Whereas one incinerator might contribute only a small fraction of the total environmental concentrations of these chemicals, the sum of the emissions of all the incineration facilities in a region can be considerable. The primary pathway of exposure to dioxins is consumption of contaminated food, which can expose a very broad population. In such a case, the incremental burden from all incinerators deserves serious consideration beyond a local level.

Before MACT Compliance

The committee is aware that incinerator emissions are expected to decrease as a consequence of improved design and operations, modifications of the waste stream, improved emission control devices, and changing waste management practices. In reviewing incineration practices and emissions data, the committee found that the data typically have been collected from incineration facilities during only a small fraction of the total number of incinerator operating hours. Generally, data are not collected during startup, shutdown, and upset conditions—when the greatest emissions are expected to occur. Furthermore, such data are typically based on a few stack samples for each pollutant. Thus, the adequacy of such emissions data to characterize fully the contribution of incineration to ambient pollutant concentrations for health-effects assessments is uncertain.

After MACT Compliance

Implementation of EPA's regulatory requirements for MACT for incineration facilities is expected substantially to reduce emissions from the highest-emitting facilities. For such facilities, MACT would reduce the degree of concern indicated for potential health effects from exposures within local areas. However, on a broader scale, considering multiple facilities and broader populations, implementation of MACT is unlikely to alter the committee's relative degree of concern for the potential health effects due to pollutants such as dioxin and some metals, and the concerns would remain because these pollutants are persistent, widespread, and potent. Furthermore, there would be no change in the committee's degree of concern for potential worker exposures, because MACT alone would be unlikely to change their exposures.

Various Degrees of Concern

The four degrees of concern (substantial, moderate, minimal, negligible) shown in Table 5-8 are intended to convey the committee's qualitative assessment and consensus judgment of the possibility of health effects to workers and the general public from incineration emissions. A degree was chosen for each specific pollutant and category based on general information on incineration emissions; persistence of the pollutant in the environment; mobility through air, soil, water, and food; potential total exposure through routes of inhalation, ingestion, and dermal absorption; and relative toxicity. The term "substantial" is used to express the committee's highest degree of concern about possible exposures that might lead to health effects among workers, a local population, or a broader population. Lower degrees of concern correspond to less possibility that the specific groups are exposed to concentrations associated with adverse health effects. The following sections provide additional discussion about the levels of concern for specific pollutants.

Particulate Matter

Given the possible health effects of typical environmental concentrations of PM and despite considerable scientific uncertainty, the committee has a substantial degree of concern for potential effects on local populations from exposure to PM contributed by high emitting (principally older) facilities. With modern PM control in a well-run facility, the emissions are so much lower that their contributions to local exposures are very low. Even in the most modern facilities, however, there is continued high concern by the committee for potential health effects from exposure to workers without proper safeguards. The handling of additional emission-control residues by workers might even add to their PM exposures and health risks after MACT implementation. On a broader geographical scale, the collective contribution of incineration facilities is comparatively small, and only minimal concern is associated with incineration on this scale, both before and after MACT compliance.

As seen in Figure 5-1, most U.S. metropolitan areas experience PM air pollution in the range at which adverse effects, including immediately increased mortality, have been associated with PM pollution. Any increases in PM concentrations—and especially in the fine particles emitted by combustion facilities, such as incinerators—can be expected to add to any existing PM health-effects burden. Increases in concentrations will be proportional to the PM emission rate by the facility and can be crudely estimated on the basis of "typical" ambient concentration estimates provided for various incinerator types shown in Tables

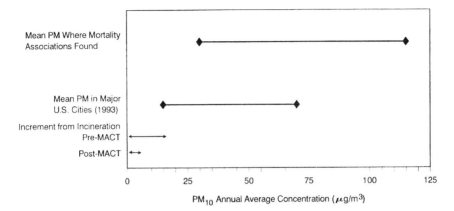

FIGURE 5–1   Comparison of range and mean PM_{10} concentrations in cities in which PM-death associations have been reported, range of mean PM_{10} in U.S. cities in 1993, and range of increment from incineration.

4-8 through 4-11 (see Chapter 4) and the health-effect information presented in Table 5-1. On the basis of these tables, it is seen that the highest PM effect of the uncontrolled incinerators, and especially cement kilns incinerating waste (potentially reaching 30 $\mu g/m^3$ total PM, or about 20 $\mu g/m^3$ PM_{10}), might be projected to produce increases in health effects on the worst days in the highest-effect locations (potentially about a 2% increase in daily mortality and a 4% increase in respiratory hospital admissions on the maximum day in the case of the pre-MACT cement kiln). However, after MACT controls are applied to these plants, such projected air-pollution effects should be reduced by almost a factor of 10. As a result, the local effects of individual post-MACT plants (though still non-zero) would be so small that such projections would represent much less than a 1% increase in risk of acute morbidity or death, even at the most affected receptor on the worst-case day, and it is highly unlikely that such potential effects could be detected by even the most carefully designed epidemiologic study.

Dioxins

The committee has a substantial degree of concern for the potential health effects from exposures of plant workers to highly potent pollutants such as dioxin. There is uncertainty as to whether there is any adequate margin of safety between typical background exposures to dioxins and those with measurable responses that might be related to health. Implementation of MACT controls are unlikely to alter the committee's degree of concern, because MACT is not designed to reduce worker exposures.

On a wider scale, it appears that a portion of dioxins in the environment has been produced by waste incineration and that a portion of the current input into the environment is produced by incineration, but how much is not known. There is substantial evidence that the average concentrations in the biosphere are now decreasing despite past increases in incineration, and it is not clear what effect MACT will have on these average concentrations. The wide dissemination of dioxins throughout the environment including the food supply, results in widespread exposures. Exposure indicators (such as blood and fat concentrations) arising from such exposures are close to the levels that, in some experimental systems, give rise to measurable biologic responses that might be related to adverse health outcomes. Thus, the committee has a substantial degree of concern for the incremental contribution to dioxins emissions from all incinerators on a regional level and beyond. Because the major route of exposure to dioxin is the food chain, the exposure of the local population is not expected to be affected much more by a local incinerator than by one located in another state. The local population shares the widespread increase in dioxin exposure from each incinerator, but experiences minimal additional risk. However, there may be specific individuals who have higher exposures because of their location and activity patterns.

The mechanism of dioxin toxicity is known to be complex. Several acute toxic effects are mediated almost solely (at least in the mouse) by the aryl-hydrocarbon receptor (Fernandez-Salguero et al. 1996), but there are other mechanisms. Studies attempting to elucidate precise mechanisms of action continue, and such studies show detectable effects of dioxin-like materials at concentrations similar to those encountered in the environment although it is unclear to what extent such effects might affect health. Figure 5-2 summarizes some of the dioxin TEQs exposures that are associated with overt toxic effects. Four scales of exposure are shown because no single exposure or dose measure is known to correlate with all toxic effects, and various measures have been used in human and animal studies. The four scales are ambient air concentration, long-term average intake, adipose-tissue concentration, and serum concentration. The scales have been aligned roughly so that the background concentrations—those found in typical U.S. populations—are level (horizontal dotted line), and the range of variation of these typical concentrations is indicated (a question mark indicates little information on the range of variation). On the ambient-air scale are marked the estimated maximal concentrations (worst-case locations) around the worst-case hazardous-waste incinerator and cement kiln, as discussed and depicted in Chapter 4, Tables 4-8 through 4-11.

The average-intake scale indicates average human intakes and the intakes associated with overt toxic effects in animals, and the long-term average intakes found to cause cancer in more than about 10% of laboratory animals.

Adipose-tissue concentrations that correspond in laboratory animals to no overt effects and the tissue concentration roughly corresponding to the concen-

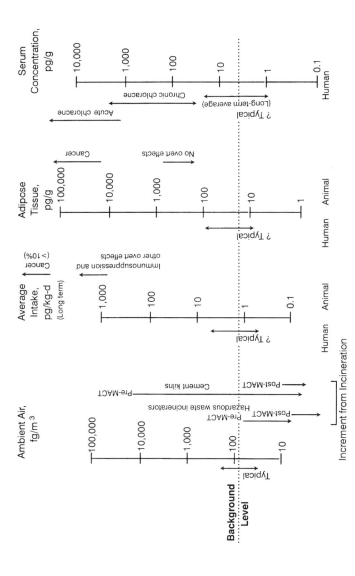

FIGURE 5-2 Dioxin TEQs associated with overt toxic effects and concentrations found in the environment. The typical range of background concentrations is shown by the double ended arrows about the "Background level" starred line, with "?" indicating uncertainty about the range. For the concentration scale on the left, the arrows show the ranges of increments in concentration potentially (as the worst-case location) associated with the labeled incineration sources and conditions. For the three scales on the right, arrows show approximate concentrations associated with the labeled end-points, in animals (to the right of each scale) and humans (to the left of each scale).

trations causing cancer in more than 10% of animals are shown. The ratios between concentrations required to cause cancer in animals and typical background concentrations in humans are different for average intake and for adipose-tissue concentrations, possibly because of differences in the pharmacokinetics of dioxin in animals and humans.

Finally, to indicate the effects of relatively short-term exposure, the serum concentrations in people who have exhibited dioxin-associated chloracne (one effect definitely associated in humans with dioxin exposure) are shown for both very-short-term exposure (e.g., Seveso children) and chronic occupational exposure.

Other Products of Incomplete Combustion

Products of incomplete combustion (PICs) have been defined as organic compounds not originally detected in the waste stream entering the incinerator, but found in incinerator stack-gas emissions (Travis and Cook 1989). PICs can arise as new organic compounds formed during the incineration process itself, might have been present in the original waste stream (but at concentrations below the cut-off level used in analyzing the waste feed), or might have been brought into the incineration system from noncombustion sources (e.g., auxiliary fuel feed, or ambient air introduced into the system). It is hypothesized that most PICs are formed from recombination of molecular fragments outside the combustion zone (Trenholm et al. 1984). Because they are widespread, persistent, and potent, the major PICs of concern are dioxins and furans, which are discussed separately in this section.

Other PICs of potential health concern are PCBs and PAHs. Incinerators are not major emission sources of these on a local or regional scale. Furthermore, in comparison with dioxins and furans, other PICs emitted by incinerators are estimated to have relatively little effect on health, or little is known about their toxicity at the relatively low concentrations emitted.

Lead

Lead at low concentrations can have adverse health effects especially infants and children. Therefore, at the local population level, the committee has substantial concerns regarding contributions to total lead exposure by incinerators operating prior to implementation of MACT controls. Incinerators operated under MACT are expected to emit only a negligible amount of lead locally, so the potential health effects in local populations from lead after the implementation of MACT are seen as minimal. Due to its toxic potential, exposures of incinerator workers to lead is of substantial concern to the committee. Implementation of MACT controls are unlikely to alter the committee's level of concern because MACT is not designed to reduce worker exposures.

Figure 5-3 shows reported effects of lead at various concentrations in the blood. Effects that have been clearly established and are well accepted by the scientific community are indicated by solid lines, effects with less certainty are indicated by dashed lines, and more controversial effects are indicated by dotted lines. For example, frank anemia occurs at blood concentrations of 80 µg/dL or above; reduced hemoglobin synthesis occurs in adults at 50 µg/dL and above, although this effect might occur in children at lower concentrations; loss of hearing acuity occurs above 30 µg/dL, but hearing loss has been measured down to 10 µg/dL; and while the effect of lead on diastolic blood pressure is clear above 50 µg/dL, some studies indicate effects on systolic blood pressure above 30 µg/dL, and effects below 10 µg/dL are seen in some studies. Several effects have no apparent threshold (for example, the effects on children's cognitive function, on blood pressure, and on heme synthesis), and other effects might not demonstrably affect health.

The bottom of Figure 5-3 presents the most recent information on the distribution of blood lead concentrations in the United States, from NHANES III, phase I, 1988-1991 (JAMA 1994). There has been a remarkable reduction in blood lead concentrations in the United States over the last 15 years. There has been a 78% drop in the average, from 12.8 to 2.8 µg/dL, primarily it is believed, because of the removal of lead from gasoline. But a distribution of blood lead exists in the population, and the data indicate that a small portion of the population has blood lead over 10 µg/dL, as do 9% of children aged 1-5; and 0.2% of the population (over 0.5 million people) have blood lead over 30 µg/dL. Any added lead in the environment might make those people more likely to experience the adverse effects of lead.

The lead emissions of incinerators are highly variable (see Chapter 4, Tables 4-8 and 4-10, and this is reflected in the facts that the mean value of lead emissions from hazardous-waste incinerators is 100 times the median value and that the estimated range of air concentrations due to emissions varies by more than 8 orders of magnitude (from 2.0×10^{-8} to 7 µg/m^3). Although maximal lead air concentrations due to emissions is 7 µg/m^3, which exceeds the ambient-air standards of the EPA, over 95% of the incinerators were estimated to produce ambient concentration increments everywhere less than 0.5 µg/m^3; similarly, maximal lead air concentrations due to emissions from cement kilns was 7 µg/m^3, but 95% would be less than 1.2 µg/m^3. Translating airborne lead to blood lead is complex but has been well studied: for young children and accounting for both the direct route (inhalation) and the indirect route (ingestion of soil, dust, and food contaminated by airborne lead) of exposure, each microgram of airborne lead per cubic meter could increase blood lead by about 4 µg/dL (EPA 1989; CalEPA 1996).

Although the average hazardous-waste incinerator and the average cement kiln would contribute less than 1 µg/dL to the blood lead burden of children around the facilities, there is the potential for the worst-case emitters to add about 20 µg/dL to the lead burden of nearby children. Thus, while the effect of

176

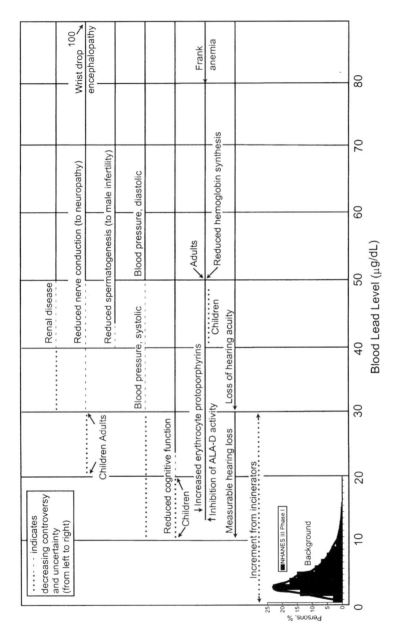

FIGURE 5-3 Blood lead concentrations: background, increment from incineration, and concentrations that have health effects. ‧‧‧‧‧‧‧‧‧‧ indicates decreasing controversy and uncertainty (from left to right). Arrows are used to indicate the blood lead level at which an effect is known to occur or the range in which an effect is known to occur.

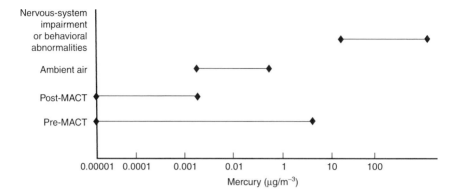

FIGURE 5-4   Mercury concentrations: Background (gas-phase and particle-bound concentrations), increment from incineration, and concentrations at which adverse effects occur for the most-sensitive end points of toxicity.

the average incinerator would be minimal, that of the highest-polluting facilities would be of some concern, and the maximally polluting facilities could add substantially to the lead burden in the local population and raise young children's blood lead to the point where multiple adverse health effects have been reported.

Mercury

Because low concentrations of mercury can have toxic effects, exposure of workers to mercury is of substantial concern to the committee. MACT controls are unlikely to alter the committee's degree of concern, because MACT is not designed to reduce worker exposures. The degree of concern about exposures of the population to mercury is expected to be reduced somewhat under MACT, but, in general, no change is expected regarding the regional level due to the environmental persistence of mercury.

Figure 5-4 compares mercury concentrations that are associated with nervous system impairment and behavioral abnormalities with concentrations found in the environment. Other human health effects associated with exposure to inorganic and organic forms of mercury, as displayed in Table 5-4, were not plotted here, because little human exposure information related to these health effects is available or exposures are uncertain or unknown. However, available data indicate that the major health effect of concern for mercury compounds is nervous system impairment. Other organ-system toxicity produced by mercury is reported to occur only after much-higher exposures. As shown in Figure 5-4, the potential effect of the average incinerator is expected to be minimal; however, a maximally polluting facility could add substantially to the mercury burden in the community. The implementation of MACT technology is expected to reduce exposures to mercury at the local level. Air concentration estimates related to incineration (Pre-MACT and Post-MACT) are based on Tables 4-8 through 4-11 in Chapter 4.

Acidic Gases and Acidic Aerosols

Incinerators directly release both acidic aerosols and gases, as well as acidic aerosol precursors that can be transformed into acid particles in the atmosphere. The acidic gases and vapors released from incinerators are generally of less concern than acids released or formed as aerosols (such as H_2SO_4). Thus, water-soluble acidic gases and vapors (such as SO_2, HCl, and HNO_3), are of low concern because, at ambient concentrations, these are efficiently "scrubbed out" in the trachea before reaching the lung. Particularly strong acidic aerosols, such as those containing H_2SO_4, however, more readily reach into the deepest recesses of the lung and are of greater health concern at ambient concentrations.

Acids released from incinerators therefore warrant a varied degree of concern depending on the form of the acid (particulate or gaseous) and the extent of emission (pre or post compliance with MACT). Acidic gases are of minimal health concern to the local population and of negligible concern at the regional level but represent a moderate concern to workers, given that exposures have the potential to become high. Compliance with MACT regulations further diminishes the concern regarding acidic gases at the local and regional levels, but not in the worker environment.

Acidic aerosols are associated with a somewhat higher degree of concern because of their particulate form and because MACT regulations are not directly aimed at reducing them. However, the acidity concern is reduced after MACT implementation because some MACT controls (such as SO_2 limitations) can be expected indirectly to lower strongly acidic aerosols resulting from such plants.

Carbon Monoxide

Because only about 1% of all CO emissions are attributable to incineration (EPA 1998b,c), the incremental exposure to CO from incinerators is not considered to represent an important increment at either the local or regional level. Although it is possible for workers to be exposed to high levels of CO from incomplete combustion, no data are available to indicate that this has occurred.

CONCLUSIONS AND RESEARCH NEEDS

Conclusions

- Estimates of large increments in ambient concentrations of various pollutants attributable to existing incinerators, particularly heavy metals and dioxins and furans, led to legitimate concerns about potential health effects.
- Pollutants produced and emitted by incinerators that currently appear to have the potential to cause the largest health effects are particulate matter, lead, mercury, and dioxins and furans.
- On the basis of available data, a well-designed and properly operated incineration facility emits relatively small amounts of those pollutants, contributes little to ambient concentrations, and so is not expected to pose a substantial health risk. However, such assessments of risk under normal operating conditions may inadequately characterize the risks or lack of risks because of gaps in and limitations of existing data or techniques used to assess risk, the collective effects of multiple facilities not considered in plant-by-plant risk assessments, potential synergisms in the combined effects of the chemicals to which people are exposed, the possible effect of small increments in exposure on unusually susceptible people, and the potential effects of short-term emission increases due to off-normal operations.
- Reductions in emissions will certainly reduce public health risks from direct and indirect exposure to those emissions. Whether there is a minimal emission rate below which there is no further reduction in health risk has not been established, and the indirect effects of emission reductions (for example, health risks associated with efforts to reduce emissions, as through substitution of other processes or materials, the use of more energy or materials for control equipment, and the manufacture of control equipment) have not yet been evaluated.
- Epidemiologic studies assessing whether adverse effects actually occurred at individual incinerators have been few and were mostly unable to detect any effects. That result is not surprising, given the small populations available to study; the presence of effect modifiers and potentially confounding factors (such as other exposures and risks in the same communities); the long periods that might be necessary for health effects to be manifested; and the low concentrations (and small increments in background concentrations) of the pollutants of concern. Although such results could mean that adverse health effects are not present, they could also mean that the effects may not be detectable using feasible methods and available data sources.
- The potential health effects of particulate matter emitted by incinerators may not have received appropriate attention in traditional risk assess-

ments. In particular, in well-characterized situations (with well-measured emissions) where the contribution of particulate matter to the total ambient particle load is small (around 1%), the acute health effect of emitted particulate matter might be as large as or larger than that of other incinerator-related pollutants. Some past studies have shown the overall urban background of particulate matter already appear to be causing excess mortality and morbidity in the U.S. population, and the particulate-matter increment from all incinerators adds to the existing burden.

- The committee's evaluation was performed based only on emissions under normal operating conditions. Data are not available for levels during off-normal conditions, or the frequency of such conditions. Such information is needed to address whether emissions resulting from off-normal conditions are a concern with respect to possible health effects.

- There is a need to focus health research on the greatest potential for exposure. Based on studies of municipal solid-waste incinerators, workers at these facilities are at much higher risk for adverse health effects from exposure to this technology than local residents. There is evidence that incinerator workers have been exposed to high concentrations of dioxins and toxic metals—particularly lead, cadmium, and mercury—in the past.

- The committee's evaluation of waste incineration and public health has been substantially impaired by the lack of an adequate compilation of the associated ambient concentrations resulting from incinerator emissions. The evaluation was also impaired by the inadequate understanding of the overall contribution of incinerators to pollutants in the total environment, and large variabilities and uncertainties associated with risk-assessment predictions, which, in some cases, limit the ability to define risks posed by incinerators.

- EPA is proceeding to regulate emissions from incinerators by requiring that incinerators reduce emissions to values achieved by the best controlled 12% of the current incinerators, a standard known as maximal achievable control technology, or MACT. Those regulations will affect emissions of the most-important pollutants unevenly; even under MACT, concerns over the widespread effects of persistent pollutants, such as dioxins, lead, and mercury will not be adequately addressed. Other potential effects can be shown to be negligibly small for some facilities on which well-measured emission data are available. However, for some individual facilities with well-measured emissions, health risks are not negligible. Collective potential effects of incinerators on a regional scale and beyond are unknown.

- New or modified facilities that meet the proposed MACT requirements are expected to have substantially lower emissions than previous facilities.

The reduction in emissions will lower the potential exposures and risk to populations surrounding incinerators in the environment in general.

- Based on a consideration of normal operating conditions, implementation of MACT standards is expected to substantially reduce the overall health risks from local impacts of particulate matter, lead, and mercury associated with incineration.
- It is unlikely whether implementation of MACT will substantially reduce the risks at the regional level posed by the persistent environmental pollutants dioxin, lead, and mercury.
- MACT was not designed to protect workers, and MACT regulations are unlikely to reduce worker exposures.

Recommendations

- To increase the power of epidemiologic studies to assess the health effects of incinerators, future multi-site studies should be designed to evaluate combined data from all facilities in a local area as well as multiple localities that contain similar incinerators and incinerator workers, rather than examining health issues site by site.
- In addition to using other exposure assessment techniques, worker exposures should be evaluated comprehensively through biological monitoring, particularly in combination with efforts to reduce exposures of workers during maintenance operations.
- Assessments of health risks that are attributable to waste incineration should pay special attention to the risks that might be posed by particulate matter, lead, mercury, and the dioxin and furans, due to their toxicity and environmental prevalence.
- Health risks attributable to emissions resulting form incinerator upset conditions need to be evaluated. Data are needed on the levels of emissions during process upsets as well as the frequency, severity, and causes of accidents and other off-specification performance to enable adequate risk assessments related to these factors. Such information is needed to address whether or not off-normal emissions are important with respect to possible health effects.
- Database compilers should strive to accumulate data not only on emissions from individual facilities (as in the Hazardous Waste Combustor database), but also the resulting estimates of ambient concentrations. Facilities that have performed emissions testing have also often performed site-specific air dispersion modeling, so that little extra effort would typically be required. Moreover, the overall contribution of incinerators to pollutants in the total environment would be easier to assess if any known site-specific measurements of background concentrations of incinerator-related pollutants were also compiled on a plant-by-plant basis.

6

Regulation Related to Waste Incineration

In the 1960s, state and federal governments began to enact legislation and promulgate regulations calling for increasingly stringent environmental control for municipal solid-waste incinerators. Regulatory activity began with nuisance regulations (related to visible plumes and to odors) and then evolved to emission standards.

Today, waste incinerators must comply with a combination of federal, state, and local regulations that vary from place to place and time to time. In most states, it is also necessary to get permits from state or local governments. For example, a municipal-waste incinerator sited in California must comply with federal laws, as well as get a permit from the California Air Resources Board and one from the local air-pollution control district.

People who live near operating waste-incineration facilities and citizens who are asked to accept such facilities into their local area want assurance that the facilities will be operated safely and in compliance with regulations intended to protect the public health, safety, and the environment. Although the persons most directly affected by a proposed facility might be told much about the minimal hazards associated with new incineration facilities that are normally and efficiently operated, they are likely to be more anxious (perhaps overanxious) about the ability of the facility to be operated over an extended period in compliance with law. They may also be anxious about the risks that arise when equipment breaks down or operations go awry (Davis and Colglazier 1987).

The unintended and uncontrolled release of toxic substances into the environment from waste incineration can occur because of malfunctioning equipment, large changes in the waste feedstream, poor management of the incineration process, or inadequate maintenance or housekeeping. Off-normal operations (e.g., upsets and accidents) at various points in the incineration process might result in explosions; fires; the release of smoke, ash, or noxious odors into the atmosphere; and the spilling or leakage of contaminated or toxic substances. As discussed in Chapter 5, workers at incineration facilities are more at risk than nearby residents due to such occurrences.

Mishaps that are actually experienced by operating facilities that incinerate municipal wastes, hazardous wastes, or medical wastes form a concrete basis for the concerns of nearby residents and other concerned citizens about the safety of waste incineration and the efficacy of regulatory oversight. The fears and worries of residents and concerned citizens are not limited to worst-case scenarios, but extend to events that occur in the normal course of operations at what are otherwise considered properly run and maintained facilities (Curlee 1994). For example, the stream of waste flowing to a hazardous-waste incinerator might be automatically shut off for the purpose of minimizing emissions when operating conditions are outside permitted limits. Automatic waste-feed cutoffs might indicate that an incinerator is not being operated according to good combustion practices. The cutoffs might also affect emissions by leading to a quick shutdown and incomplete combustion. However, if properly managed, the emissions should be minimized. More serious is the use of an emergency bypass or vent stack. Such a stack allows an operator to bypass the air-pollution control equipment following a waste-feed cutoff to prevent the buildup of excessive pressure in an incinerator or to protect the emission control equipment from exceedingly hot flue gases. The frequency of occurrence of such emergency bypass venting by incineration facilities is unknown.

As a result of the possible dangers associated with waste incineration, potentially affected persons expect comprehensive, effective, and responsive regulation that prevents or deters uncontrolled emissions, upsets, and worker injuries, that punishes regulatory infractions, and that promotes decontamination, rectification, and compensation for any harm done.

This chapter examines the structure of waste-incineration regulations with regard to public and occupational health; regulatory oversight; and the policy concerns that are likely to affect future regulatory changes. Particular attention is paid to the different bases on which regulatory standards are formulated and to the extent to which regulations vary with the age and size of a facility. The chapter includes discussions of citizens' concerns regarding the reliability of incineration technology and operations and the effectiveness of regulation.

OVERVIEW OF INCINERATION REGULATIONS RELEVANT TO PUBLIC HEALTH AND THE ENVIRONMENT

Direct federal regulation of facilities and federal oversight of state regulation are primarily the responsibility of the U.S. Environmental Protection Agency (EPA), whose authority arises under the Clean Air Act (CAA) (42 USC §7412), the Clean Water Act (CWA) (42 USC § 1365), the Resource Conservation and Recovery Act (RCRA) (42 USC §§ 6901-6992k), the Toxic Substances Control Act (TSCA) (15 USC §2601), and the Comprehensive Environmental Response, Compensation, and Liability Act (CERCLA), referred to as Superfund (42 USC §9601). To some extent, regulation of facilities that incinerate municipal solid waste, medical waste, or hazardous waste has been effectively delegated to the states, with EPA performing an oversight role. Federal law sets minimal standards for the combustors that the states must implement and enforce, although they are free to impose more-stringent requirements if allowed to do so by state law (Organ 1995). EPA has exclusive jurisdiction, however, over incineration operations that handle polychlorinated biphenyls (PCBs) and hazardous wastes from Superfund cleanup sites.

Incineration regulations generally address emission limitations, good combustion practices, operator training and certification, facility-siting criteria, permit compliance and inspections, and record keeping and reporting requirements. There are wide variations across the country with regard to these subjects. Typically, incineration regulations vary with the type of waste being incinerated, the capacity of the facility, its age, and the overall regulatory environment. The remainder of this overview focuses on relevant requirements of the CAA and RCRA.

The Clean Air Act and Waste Incineration

The CAA requires EPA to establish new source performance standards (NSPS) for new incineration facilities and emission guidelines for existing facilities. Emission guidelines require states to develop plans for controlling emissions from facilities within their jurisdictions. Once EPA has approved the states' plans, they become federally enforceable. Standards and regulations are developed by EPA's Office of Air and Radiation for incinerators that burn municipal solid waste or medical waste. Regulations for hazardous-waste incinerators and cement kilns are developed by EPA's Office of Solid Waste and Emergency Response under the CAA and RCRA (as will be discussed below).

Regulations developed under the CAA are intended to limit atmospheric concentrations of the six criteria pollutants (i.e., carbon monoxide, lead, nitrogen dioxide, ozone, particulate matter, and sulfur dioxide) and control emissions of 188 air toxics (also known as hazardous air pollutants (HAPs)).[1]

Prior to 1990, EPA's efforts to regulate HAPs on the basis of health risk were slowed by conflicts and litigation by interested or affected parties. As a

---

[1] The original list of HAPs included 189 chemicals, but caprolactam was removed from the list in 1996.

result, EPA had developed standards for only seven of the original 189 HAPs. In response to the lack of progress, the 1990 amendments to the CAA shifted the regulatory tool from a risk-based emissions standard to a technology-based standard for the sources of air toxic emissions. The so called "MACT" (Maximum Achievable Control Technology) standards establish control requirements which assure that all major sources of toxic emissions (including waste incineration facilities) have the same level of control already attained by an average of the best performing (top 12%) sources in each pollutant category. The "residual risk" (i.e., the risk remaining) is to be determined in order to assess the risk remaining after the technology-based standard has been met. Section 122(f) stipulates that if an "ample margin of safety" is not reached, then taking into consideration "costs, energy, safety, and other relevant factors," a standard more stringent than the MACT standards alone may be implemented. It is important to clarify that several of the requirements will not be in effect until several years later. In the meantime, for existing municipal solid-waste incinerators, there is a variety of permits of widely varied stringency governing emissions (for example, ranging from uncontrolled for some pollutants to state-of-the-art controls for others).

EPA is charged with measuring the risks that remain after MACT standards are implemented and reporting its measurements to Congress along with data on the methods used to calculate such risks, their health implications, commercially available methods for reducing them, and recommendations as to legislation regarding them (Steverson 1994). EPA has completed a residual risk report to Congress on the methods to be used to assess the residual risk after MACT standards have been promulgated and applied (EPA 1999).[2]

As discussed later in this chapter, EPA has developed NSPS and emission guidelines for large municipal-waste incinerators (units with great than 250 tons per day capacity) of medical-waste incinerators as well as hazardous-waste incinerators, lightweight aggregate kilns, and cement kilns that burn hazardous waste.

National Ambient Air Quality Standards (NAAQS) are in place for the six criteria pollutants mentioned above. The NAAQS set nationwide limits on the atmospheric (ambient) concentrations. If it is determined that an area is not in attainment of any of the NAAQS, that state is expected to develop a State Implementation Plan (SIP) for achieving attainment of the NAAQS through state-selected and enforced controls on emissions. The state must satisfy the EPA that its SIP is adequate. Depending on location, it is possible that some incineration facilities may face additional, more-stringent controls as part of SIP requirements.

---

[2] EPA's (1999) residual risk report was not available until after the committee completed its deliberations.

The Resource Conservation and Recovery Act (RCRA) and Waste Incineration

RCRA gave EPA the authority to control hazardous waste with respect to generation, transportation, treatment, storage, and disposal. RCRA focuses only on active and future facilities and does not address abandoned or historical sites (those are covered by CERCLA). RCRA requires EPA to establish performance, design, and operating standards for all hazardous-waste treatment, storage, and disposal facilities. The regulations developed in response apply to facilities that incinerate hazardous waste. The regulations restrict the emissions of organics, hydrogen chloride, and particulate matter, as well as fugitive emissions.

REGULATIONS APPLICABLE TO MUNICIPAL SOLID-WASTE INCINERATORS

Federal Regulations

Section 111 of the CAA sets national emission standards for municipal solid-waste incinerators. It requires promulgation of performance standards for categories of new and existing stationary sources that might contribute to air pollution reasonably anticipated to endanger public health or welfare (Reitze and Davis 1993).

On December 20, 1989, EPA proposed new source performance standards for new municipal solid-waste incinerators and emission guidelines for existing ones on the basis of best demonstrated technology (BDT)(Subparts Ea and Ca of 40 CFR 60). On February 11, 1991, EPA promulgated those subparts as regulations applicable to municipal-waste incinerators reflecting BDT as determined by the EPA administrator at the time the guidelines were issued. Plants were divided into 3 categories: small (units burning up to 250 tons or Mg/day), large (units burning up to 2,200 Mg/day), and very large (units burning more than 2,200 Mg/day). The regulations included maximum levels that varied with the size of the unit for the following emissions: hydrochloric acid (HCl), oxides of nitrogen (NO_x), opacity for particles, carbon monoxide (CO), sulfur dioxide (SO_2), and dioxins and furans. They also include process parameters, such as load, and flue-gas temperature at the particulate-matter control-device inlet. The regulations also require provisional certification of the chief facility operator and shift supervisors by the American Society of Mechanical Engineers (ASME) or through a state certification program. A site-specific training manual to be used for training other incinerator personnel was required. Reporting is not required for emissions during process upsets, including startup and shutdown. Also, such data are not used to evaluate compliance with standards.

On November 15, 1990, as EPA was preparing final versions of the standards for new and existing municipal solid-waste incinerators, Congress passed the 1990 CAA Amendments; a new provision, section 129 (a)(2), required that emission standards for new incinerators and guidelines for existing ones reflect the maximal achievable degree of emission reduction, taking into consideration the cost and any non-air-quality health and environmental effects and energy requirements of the technology. The level of control was to be based on MACT. Section 129 also effectively added mercury, cadmium, lead, and fly-ash or bottom-ash fugitive emissions to the list of regulated pollutants, expanded the applicability of the standards to some of the smaller plants, and required recalculation of previously promulgated limits for the other pollutants according to a new method.

Municipal solid-waste incinerator "Standards of Performance for New Stationary Sources and Emission Guidelines for Existing Sources," implementing sections 111 and 129 of the CAA, were promulgated on December 19, 1995, Fed. Regist. 60(243):65387-65436.

With regard to new sources, MACT emission standards (the so-called MACT floor) may not be less stringent than the emission control achieved in practice by the best-controlled similar units. As applied to existing incinerators, MACT emission standards may not be less stringent than the average emission limitation achieved in practice by the best-performing 12% of units. EPA has interpreted the former definition to mean the average *performance* level achieved at the uppermost 12th percentile of existing municipal solid-waste incinerators in the United States for which data were available, and the latter to mean the level corresponding to the average *permit level* for the uppermost 12th percentile of existing incinerators for each pollutant.

The data with which MACT floors were to be determined were the subject of some dispute. The MACT floor could have been based on permitted emission levels, levels achieved in practice by currently used technologies, or levels achievable with available technology. Some environmental groups interpreted MACT to mean, in the case of new plants, emission levels equivalent to the best-performing plant's emission levels, and for existing plants, the average performance level achieved at the uppermost 12th percentile of existing municipal solid-waste incinerators both from a worldwide database of facilities. The difference between this interpretation and the permitted-emission-level interpretation is considerable, particularly in the case of existing incinerators, in that permit levels are usually considerably less stringent than the current state-of-the-art performance levels.

Section 129 also requires the setting of numerical emission limits based on MACT. That has been done for all required pollutants except mercury, HCl, and SO_2, for which a dual standard—the less stringent of a numerical limit and a percentage reduction—is proposed. In practice, the percentage reduction usually applies.

Besides the MACT emission limitations, the December 1995 standards and guidelines required that all municipal solid-waste incinerators handling waste at over 35 Mg/day adhere to good combustion practices, which include the following requirements:

- The incinerator load level and the flue-gas temperature at the particulate-matter control-device inlet must be measured and not exceed the levels demonstrated during the most-recent dioxin or furan performance test. EPA did not propose any specific flue-gas temperature requirement for either new or existing plants. Each incinerator is to establish a site-specific maximal flue-gas temperature based on the maximal 4-hr block average temperature measured during the most-recent dioxin and furan compliance test. The incinerator must then be operated in such a way that the flue-gas temperature does not exceed that maximum by more than 17°C (31°F).
- The chief facility operator and shift supervisor must obtain provisional and then full operator certification from ASME, and a provisionally certified control-room operator may "stand in" for the chief facility operator or shift supervisor for an unspecified period.
- All chief facility operators, shift supervisors, and control-room operators are required to complete an EPA or municipal solid-waste incinerator operator-training course. However, uniform course curricula or criteria are not specified in the law.

The rule requires control of flue-gas temperature and load level at the inlet of the particulate-matter control device. Flue-gas temperature at the inlet to the particulate-matter control device, activated-carbon and alkaline-reagent sorbent injection rates, waste-feed rates, and other characteristics are considered surrogates for continuous monitoring of mercury, HCl, and dioxins or furans; and EPA mandates measurement and monitoring of these pollutants under the standards and guidelines.

The CAA amendments of 1990 are being implemented to require the updating of antiquated technologies with more-modern control devices that are not, in the view of EPA, too expensive for both new and large old incinerators. For control of dioxins and furans and mercury, which are the types of the municipal solid-waste incinerator emissions that are most toxic and difficult to remove, and control of acid gases, such as SO_2, NO_x, and HCl, the MACT floors in both the NSPS and the guidelines for large plants are based on use of activated-carbon injection, spray-dryer absorbers with alkaline-reagent injection, fabric-filter particle-control devices, and selective noncatalytic reduction for NO_x control.

Because of concerns about the bioaccumulation of mercury in the environment, EPA considers the incremental costs associated with adding activated-carbon injection to control mercury emission reasonable for new and existing small plants, and it therefore requires the same mercury-emission standards for all municipal solid-waste incinerators—new and old, large and small. EPA considered activated-carbon injection to be the best of three mercury-control technologies but did not evaluate fixed activated-carbon filtration which is a technology that is used at a number of European facilities.

A siting analysis is required for new plants, as is a material-separation plan. The siting analysis is performed to consider "the impact of the affected facility on ambient air quality, visibility, soils and vegetation" and "air pollution control alternatives that minimize, on a site-specific basis, to the maximum extent practicable, potential risks to the public health or the environment" (40 CFR § 60.57b (b)(1), (2)). No other substantive requirements are stated. The requirements for the material-separation plans are largely procedural; EPA has not specified any particular minimum performance levels, separation-system design, or materials to be separated. To ensure proper siting of a landfill or incinerator, it is important to consider current and projected prevention, recycling, and composting levels and the effect of diversion on the character of the resulting waste stream that serves as the incinerator feedstock.

On April 8, 1997, the U.S. Court of Appeals for the District of Columbia Circuit vacated the emission guidelines and new source performance standards as they apply to municipal solid-waste incinerator units with the capacity to combust less than or equal to 250 tons per day of municipal solid waste, and all cement kilns combusting municipal solid waste. As a result the requirements described above apply only to municipal-waste combustor units with the capacity to burn more than 250 tons per day.

On August 25, 1997, EPA amended the emission guidelines and standards that were promulgated on December 19, 1995 (Fed. Regist. 62(164):45116-45127) to make them consistent with the court decision. That amendment document also added supplemental emission guideline limits for four pollutants: hydrogen chloride, sulfur dioxide, nitrogen oxides, and lead. The amendments did not add any additional emission limits to the standards for new facilities. A summary of the emission limits for large municipal solid-waste incinerators is presented in Table 6-1. Proposed emission limits for new and existing small municipal solid waste incinerators are also shown in Table 6-1. Emission limits for carbon monoxide are presented in Table 6-2.

State and Local Regulations

Under the Resource Conservation and Recovery Act (RCRA), states must adopt regulations at least as strict as those required by the new municipal solid-waste incinerator performance standards and guidelines. States where greater stringency is allowed (Organ 1995) might have on the books provisions that are more stringent than those set forth in the new EPA regulations, or they might be writing more-stringent provisions into operating permits. The committee is uncertain as to the extent to which differences exist. There have been a number of calls by environmentalists for moratoriums on new waste incinerators (Ferris

TABLE 6-1 Summary of Emissions Limits for Small and Large Municipal Solid-Waste Incinerators[a,b]

| Pollutant | Small Waste Incinerators (Proposed Rule) | | | | Large Incinerators (Final Rule) | |
| | Existing[c] | | | New | Existing | New |
	Class A	Class B	Class C			
Particulate Matter, mg/dscm	27	34	70	24	27	24
Dioxins and Furans, (total mass basis) ng/dscm	30 60 (with ESP)	123	125	13	30 60 (with ESP)	13
Lead, mg/dscm	0.490	1.6	1.6	0.20	0.49 (by 2000) 0.44 (by 2002)	0.20
Cadmium. mg/dscm	0.040	0.10	0.10	0.020	0.040	0.020
Mercury, mg/dscm	0.080 (or 85% reduction)	.080 (or 85% reduction)	.080 (or 85% reduction)	.080 (or 85% reduction)	.080 (or 85% reduction)	.080 (or 85% reduction)
Hydrogen chloride, ppmv	d	200 (or 50% reduction)	250 (or 50% reduction)	25 (or 95% reduction)	31 (by 2000) 29 (by 2002)	25 (or 95% reduction)
Sulfur dioxide, ppmv	d	55 (or 50% reduction)	80 (or 50% reduction)	30 (or 80% reduction)	31 (by 2000) 29 (by 2002)	30 (or 80% reduction)

Nitrogen oxides, ppmv	e	—	Class I 150 Class II —	240 (by 2000) 180 (by 2002)	150
Carbon Monoxide, ppmv[f]	50-250	—	50-150	50-250	50-150
Opacity	10%	10%	10%	10%	10%
Fugitive Ash	e	Visible emission for no more than 5% of hourly observation period	Visible emission for no more than 5% of hourly observation period	Visible emission for no more than 5% of hourly observation period	Visible emission for no more than 5% of hourly observation period

[a] All emission limits measured at 7% oxygen.

[b] For small incinerators, upper size cutoff is 250 tons/day of combustion capacity; lower size cutoff is 35 tons/day. Large incinerators are defined as greater than 250 tons/day.

[c] Class A units are defined as nonrefactory-type small combustion units located at plants with an aggregate plant capacity greater than 250 tons/day; Class B units are refactory-type small units located at plants with an aggregate plant capacity greater than 250 tons/day; Class C units are small units located at plants with an aggregate plant capacity less than or equal to 250 tons/day.

[d] Not applicable

[e] Not available

[f] See Table 6-2 of this report

Sources: Fed. Regist. 64(Aug. 30):47234-47274; Fed. Regist. 64(Aug. 30):47276-47307; Fed. Regist. 62(Aug. 25):45116-45127; Fed. Regist. 62(Dec. 19):45116-45127.

TABLE 6-2   Summary of Carbon Monoxide Emission Limits for Small and Large Municipal-Waste Incinerators[a]

	Small (Proposed Rule)		Large (Final Rule)	
Type of Combustion Unit	Existing	New	Existing	New
Fluidized bed	100	100	100	100
Fluidized bed, mixed fuel, (wood/refuse-derived fuel)	200	[b]	—	—
Mass-burn rotary refactory	100	100	100	100
Mass-burn rotary waterwall	250	100	250	100
Mass-burn waterwall and refactory	100	100	100	100
Mixed fuel-fired, (pulverized coal/refuse-derived fuel)	150	150	150	100
Modular starved-air and excess air	50	50	50	50
Spreader stoker, mixed fuel-fired (coal/refuse-derived fuel)	200	150	200	150
Stoker, refuse-derived fuel	200	150	200	150

[a] All limits are measured in ppmv at 7% oxygen.
[b] Not applicable

Source: Fed. Regist. 64(Aug. 30):47234-47274; Fed. Regist. 64(Aug. 30):47276-47307; Fed. Regist. 62(Dec. 19):45116-45127.

1995).   A few jurisdictions have enacted moratoriums on construction of new incinerators.   Moratoriums are justified, in the view of their advocates not only because of concern for incineration-related health and environmental risks, but also because of concern that increased incineration capacity might interfere with greater use of waste-reduction strategies, including reuse and recycling, which they believe could reduce the quantity of all emissions and production of toxic byproducts and toxic emissions associated with incineration.

Criticisms of MACT-Standard Regulations

The MACT-standard regulations for municipal solid-waste incinerators might be considered controversial in some cases. Mercury, HCl, and SO_2 have a dual standard: the less stringent of a numerical limit and a percentage reduction. That approach allows for the possibility of higher emissions when waste stream inlet concentrations of a pollutant are high.   EPA's rationale is that low numerical limits might be difficult to achieve when waste-stream inlet concentrations of a pollutant are high.   But the dual standard might effectively reduce the impetus for implementing waste-sorting methods (for example, separation of mercury batteries) to reduce pollutant precursors in the waste stream and reduce inlet pollutant concentrations.

Although research (e.g., Stieglitz and Vogg 1987) in the latter 1980s showed that flue-gas temperatures around 300°C are associated with peak secondary dioxin formation, and research by Environment Canada and others showed that lower emissions of mercury, dioxins, and acid gases occurred when flue-gas temperatures were around 150°C, EPA did not propose any specific flue-gas temperature requirement for either new or existing plants.

The technology on which MACT standards are based might have resulted in greater emissions reductions, but at higher costs if more-advanced technologies—such as activated-carbon filters or static beds for control of dioxin and mercury and selective catalytic reduction for NO_x control—were considered.

The database on which MACT standards were calculated contained important omissions. For example, EPA chose not to include data from any incinerators in Europe in its database for determining the MACT floor.

Finally, environmental activists have called for siting criteria that take into account proximity to sensitive populations (e.g., the old, and the very young), sensitive land uses (e.g., schools and hospitals), facilities with long-term residents (e.g., prisons), and areas with multiple sources of pollution that impose a cumulative burden on residents. The MACT-based regulations do not reflect those concerns.

REGULATIONS APPLICABLE TO HAZARDOUS-WASTE INCINERATORS

Incinerators and other combustors (e.g., light-weight aggregate kilns and cement kilns) that use hazardous wastes as fuels are regulated principally under RCRA and CAA. A hazardous waste is one that either exhibits specific characteristics of ignitability, corrosivity, reactivity, and toxicity or is specifically listed in 40 CFR §§ 261.31 through 261.33. Facilities that treat, store, and dispose of hazardous wastes are comprehensively regulated under RCRA. Operators of hazardous-waste incinerators must obtain an operating permit from either federal or state regulators under standards promulgated by EPA. The permitting process for a new hazardous-waste incinerator generally takes at least 3 years and entails the investment of $5-10 million (Steverson 1994).

Incinerators burning waste contaminated with PCBs, which fall under the Toxic Substances Control Act, must obtain a federal permit. Under Section 112 (d) of the CAA, EPA is required to develop national emission standards for hazardous air pollutants (NESHAP) for major source categories. EPA has determined that industrial/commercial/institutional boilers may be a major source of emissions of one or more HAPs. To the extent that an incinerator discharges pollutants into navigable waters,[3] the operator must also obtain a permit under the Clean Water Act.

---

[3] Traditionally, navigable waters refers to waters that are sufficiently deep and wide for navigation by all, or specified vessels.

The CAA of 1970 authorized a national program of air-pollution prevention and control and required nationwide uniform emission standards for major stationary sources. The act regulated three types of pollutants:

- Criteria pollutants: particulate matter, SO_2, NO_2, hydrocarbons, photochemical oxidants, and CO.
- Hazardous pollutants: under National Emission Standards for Hazardous Air Pollutants, which apply to both new and existing sources.
- Designated pollutants: for pollutants that are neither criteria nor hazardous pollutants, a separate standard is established for existing sources by state agencies under state implementation plans (SIPs).

In the 1970s, the operation of hazardous-waste incinerators was regulated by the states under SIPs case by case. The focus was on emission limits for particles and acid gases (HCl, SO_x, and NO_x).

The RCRA Subpart O incineration regulations, promulgated on January 23, 1981, revolutionized the design and operation of all hazardous-waste incinerators in the United States. The regulation covered the complete operation, including the front-end waste-feed management, waste-feed sampling and analysis, waste-feed rate control and monitoring, combustion-zone operation control and monitoring, air-pollution equipment control and monitoring, and stack-emission testing and monitoring. Emission of particles and HCl, and efficiency of destruction of hazardous organic chemicals were tightly controlled. Stringent reporting and record-keeping requirements were imposed. There are also requirements on personnel training, inspection of equipment, contingency planning, financial responsibility, and closure plans.

The RCRA Subpart H BIF regulation was promulgated on February 21, 1991, and put all remaining hazardous-waste combustion facilities under the RCRA regulatory umbrella. In addition to the Subpart O incinerator requirement just discussed, several new pollutant-emission controls were included, such as products of incomplete combustion, toxic metals, chlorine gas (Cl_2), and dioxin and furan (for facilities with potential to form dioxins and furans in the air-pollution control devices). There were also new and stringent quality-assurance (QA) and quality-control (QC) requirements for continuous flue-gas monitoring.

Since the implementation of Subpart H, EPA, under the RCRA omnibus authority, has imposed the emission standards applicable to BIFs on hazardous-waste incinerators that have high potential to emit the regulated pollutants (Steverson 1994).

As a result of the Subpart O and Subpart H regulations, all hazardous-waste combustion devices in the United States are required to have a detailed compliance-monitoring program. The requirements specify how the equipment can be operated and what levels of emissions are acceptable. Periodic trial-burn testing (for permitted facilities) or certification-of-compliance testing (for BIFs operating under interim status or before a final permit is issued) must be conducted to demonstrate that the stated operating conditions are appropriate and the maximal emission limits are not exceeded. An automatic waste-feed shutoff system is required for all hazardous-waste combustion devices to stop the feeding of hazardous wastes to the combustion chamber immediately if the combustion device does not perform as permitted.

All hazardous-waste incinerators must have four categories of continuous-monitoring devices to ensure that incinerators are operated within the safe operating range as established during the trial burn or certification-of-compliance test. Records of maintenance, calibration, and output of continuous-monitoring instruments must be kept for government regulatory agencies to inspect and ensure compliance.

- The first category is for the waste-feed rate. Limits on ash, heavy metals, and chlorine (and other halogens) also are added to most permits. These must be tracked to ensure that feed rates will not exceed the design capacity of the incinerator and the associated air-pollution control devices.
- The second category deals with the operating characteristics that affect the combustion or destruction of wastes in the incinerator combustion zones, including temperature, oxygen concentration, gas residence time, kiln rotation speed, and liquid-waste atomizing pressure. They must be monitored continuously to ensure that the incinerator is operating properly.
- The third category includes characteristics that affect the control efficiency of the air-pollution control equipment, such as the pressure drop across a scrubber or the power supply to a precipitator. These must be monitored continuously to ensure that the air-pollution control devices are functioning properly.
- The fourth category deals with the actual performance of the incinerator. Flue-gas CO is continuously monitored. CO is known as the most-stable and most-abundant product of incomplete combustion. If most of the CO is destroyed (as indicated by low concentrations in the flue gas), essentially all organic chemicals and products of incomplete combustion will be destroyed as well. Stack continuous emission-monitoring systems are also often required for NO_x, SO_x, HCl, total hydrocarbons, and opacity.

To demonstrate that an incineration facility can indeed perform as well as the designer and operator claim, federal regulations require that all hazardous-waste incinerator owners conduct a trial burn to confirm the incinerator's ability to perform as required. The trial burn is intended to demonstrate the limit of a facility's waste-feed rates under worst-case operating conditions. The demonstrated worst-case operating conditions become the operating limits, for example, the lowest allowable combustion-zone temperature and the maximal allowable waste-feed rate. These critical operating conditions and the associated waste-feed limits are listed in the facility's operating permit. The facility operator is allowed to operate the incinerator at conditions equal to or better than those worst-case conditions. To avoid permit violations and the associated fines, incinerator operators typically push the trial-burn test conditions to as close to the design limit as possible so that there is less likelihood that the operating permit will be violated, even though the facility might never operate under those extreme conditions.

In addition, owners or operators of incinerators must conduct training for people who operate the incinerator, must inspect and maintain the equipment, and must have plans that address emergency procedures. Detailed continuous monitoring, operating, and training records must be kept. Unannounced and periodic inspections are conducted by federal, state, and other agencies to see that incinerator operation is meeting all requirements and that violations are noted if they occur.

On April 19, 1996, EPA (1996b) proposed revised standards and guidelines for hazardous-waste incinerators and for hazardous-waste-burning cement kilns and lightweight-aggregate kilns (Revised Standards for Hazardous Waste Combustors, Part II, Fed. Regist. 61(April 19):17358). Since the proposal for the revised standards, EPA received comments identifying two general types of information that were not considered for the proposed standards: (1) errors in the emissions database used for the proposed standards and (2) new reports on trial burns and certification of compliance. EPA has revised its hazardous-waste combustor database based on those comments and other data collection efforts. EPA published a notice of data availability and request for comments on January 7, 1997 (Fed. Regist. 62(Jan. 7):960-962). In that notice, EPA indicated that changes in the proposed MACT floor levels could result from applying the alternative MACT methodologies discussed in the proposed standards to the updated database. See Table 6-3.

The April 1996 proposed revisions, which were issued under both the CAA and RCRA, would establish MACT standards for dioxins and furans, mercury (Hg), semivolatile metals (cadmium and lead), low-volatility metals (arsenic, beryllium, chromium, and antimony), HCl and Cl_2 combined, particulate matter (PM), CO, and hydrocarbons (HC). Large and small incinerators are held to the same standards, as are wet- and dry-process kilns. In the case of hazardous-waste incinerators, the minimal emission levels or MACT floors from which the MACT standards proceed were determined through analysis of data generated largely during trial burns undertaken to demonstrate compliance with RCRA standards. For kilns, the data came from certifications of compliance obtained under RCRA. MACT floors might not be sufficiently stringent, because data generated for the purpose of achieving flexible permit requirements or during compliance testing reflect worst-case performance and not normal, everyday operating conditions or emissions.

TABLE 6-3 Summary of Emission Limits for Existing and New Hazardous-Waste Incinerators (Final Rule)[a,g]

Pollutant	Existing Combustors			New Combustors		
	Incinerators	Cement Kilns	Lightweight-Aggregate Kilns	Incinerators	Cement Kilns	Lightweight-Aggregate Kilns
Particulate Matter, mg/dscm	34	0.15 kg/Mg[f]	57	34	0.15 kg/Mg[f]	57
Dioxins and Furans, ng TEQ/dscm	.20 .40[b]	.20 .40[b]	.20 .40[b]	.20	.20 .40[b]	.20 .40[b]
Lead and Cadmium (combined emission), µg/dscm	240	240	250	24	180	43
Mercury, µg/dscm	130	120	47	45	56	33
Arsenic, Beryllium, and Chromium, (combined emission) µg/dscm	97	56	110	97	54	110
Carbon monoxide,[e] ppmv	100	100[c,d]	100[c]	100	100[c,d]	100
Hydrocarbons, ppmv as propane	10	10[c] 20[d]	10[c] 20[d]	10	10[c] 20[d]	20

(continued)

TABLE 6-3 Continued

Pollutant	Existing Combustors			New Combustors		
	Incinerators	Cement Kilns	Lightweight-Aggregate Kilns	Incinerators	Cement Kilns	Lightweight-Aggregate Kilns
Hydrochloric acid and Chlorine gas, (combined emission), ppmv	77	130	230	21	86	41

[a] Emission standards and operation requirement apply during startup, shutdown, and malfunction if hazardous waste is in the combustion chamber. Corrected to 7 percent oxygen

[b] Provided that the combustion gas temperature at the inlet to the initial particulate matter control device is 400°F or lower based on the average of the test run average temperatures.

[c] Kilns equipped with a by-pass duct or midkiln gas sampling system

[d] Kilns not equipped with a by-pass duct or midkiln gas sampling system

[e] A source can choose which of these two standards it wishes to continuously monitor for compliance. If a source chooses the carbon monoxide standard, it must also demonstrate compliance with the hydrocarbon emission standard during the performance test. If a source elects to use the hydrocarbon limit for compliance, it must continuously monitor and comply with the hydrocarbon emission standard and need not monitor carbon monoxide emissions.

[f] Also, has an opacity limit of 20%.

[g] A destruction and removal efficiency (DRE) standard is required to ensure MACT control of nondioxin/furan organic hazardous air pollutants. The implementation procedures from the current RCRA requirements for DRE (see §§264.342, 264.343, and 266.104) are adopted. All sources must demonstrate the ability to destroy or remove 99.99 percent of selected principal organic hazardous compounds in the waste feed as a MACT standard. This requirement, commonly referred to as four-nines DRE, is a current RCRA requirement.

Source: Fed. Regist. 64(189):52828-53077, Sept. 30, 1999.

Continuous emission monitoring is required for CO, HC, O_2, PM, and Cl_2. Other requirements pertain to automatic waste-feed cutoffs. Monitoring must continue during cutoffs, and burning of hazardous wastes is not to be resumed until all conditions are within allowable limits. All cutoffs must be documented, and cutoffs in excess of a specified number will trigger increased inspections. Emergency safety-vent openings must be logged and reported to regulatory authorities. The committee was not aware of any operator certification and training requirements like those applicable to municipal solid-waste incinerators and medical-waste incinerators supervisors and operators.

On June 19, 1998, (Fed. Regist. 63(118):33782-33829), EPA finalized some parts of the standards that were proposed in April 1996. The final rule, "Hazardous Waste Combustors: Final Rule—Part I," commonly referred to as the "MACT Fast Track Rule," addresses the following four elements of the proposed standards:

(1) An exclusion from RCRA Subtitle C jurisdiction for hazardous waste-derived fuels that are comparable to fossil fuels

(2) Streamlined procedures to help facility owners and operators comply with their RCRA permits and with forthcoming MACT standards

(3) Affected sources must prepare and submit for public comment a notification identifying the facility's intentions and strategy to comply with the final MACT rule

(4) Waste minimization and pollution prevention criteria when one-year extensions are needed to install waste minimization measures that reduce the amount of hazardous waste entering combustion feedstreams.

The standards were made final in 1999 (see Table 6-3). Phase II of the Hazardous Waste Combustor rule will control emissions form hazardous-waste boilers and halogen-acid furnaces.

INCINERATION IN CONNECTION WITH SUPERFUND CLEANUPS

In 1980, Congress enacted the Comprehensive Environmental Response, Compensation, and Liability Act (CERCLA), which created the Superfund program to clean up the nation's most-contaminated hazardous-waste sites. The most highly contaminated sites are placed on a priority list. Their cleanup has entailed or will entail incineration, including the use of temporarily sited units. Incineration is the nearly exclusive means of destroying PCB- and dioxin-contaminated wastes (GAO 1995a). Under CERCLA, incinerators at Superfund sites must comply with applicable federal regulations. Incineration of PCBs is governed by the Toxic Substances Control Act (TSCA), and incineration of dioxins falls within RCRA.

The Agency for Toxic Substances and Disease Registry (ATSDR) was created under CERCLA in 1980 to assess, among other things, the public-health effects of Superfund sites, including the impact of incineration in the remediation process. ATSDR has undertaken health assessments in communities near incinerators used to burn hazardous substances. (See Chapter 5.)

REGULATIONS APPLICABLE TO MEDICAL-WASTE INCINERATORS

Medical-waste incinerators (MWIs) are used primarily to destroy regulated medical waste that is potentially contaminated with pathogens (also referred to as "hospital/medical/infectious waste"). EPA does not regulate infectious medical waste as hazardous waste (Battle 1994).

EPA has produced final MACT regulations for MWIs under the mandate of the 1990 CAA amendments. On February 27, 1995, EPA proposed new source-performance standards for new MWIs and emission guidelines for existing MWIs to fulfill the requirements of section 129 of the CAA (EPA Proposed Rules of Medical Waste Incinerators, Fed. Regist. 60(38):10653-10691 (proposed Feb. 27, 1995). On June 20, 1996, EPA issued a notice of availability of supplemental information and reopening of public comment period (Fed. Regist. 61(Jun. 20):31736-31779). The notice presented an assessment of the supplemental information submitted following the proposed standards and it solicited public comment on that assessment. Virtually every aspect of the 1995 proposal was changed significantly by the 1996 notice. Final rule-making took place on September 15, 1997 (Fed. Regist. 62(178):48348- 48391).

Medical/infectious waste is defined as any waste generated in the diagnosis, treatment, or immunization of human beings or animals, in research pertaining thereto, or in production or testing of biologicals, including cultures and stocks of infectious agents, human pathological waste, and sharps that have been used in animal or human patient care or treatment. Hospital waste is defined as discards generated at a hospital, except unused items returned to the manufacture. The definition does not include human corpses, remains, and anatomical parts that are intended for internment or cremation. For the purpose of this discussion, all of the above types of waste are referred to as "medical waste."

Incineration facilities that burn medical waste are divided into three source categories based on waste burning capacity: small (less than or equal to 200 lb/hr), medium (greater the 200 to 500 lb/hr), and large (greater than 500 lb/hr). Separate emission standards apply to each subcategory. A summary of the emission limits for new and existing facilities are shown in Table 6-4. Emission limitations have been set for particulate matter, carbon monoxide, dioxins and

TABLE 6-4 Summary of Emission Limits for Hospital/Medical/Infectious Waste Incinerators (Final Rule)[a]

Pollutant (Test Method)	Existing Medical-Waste Incineration Emission Limits				New Medical-Waste Incinerator Emission Limits		
	Small	Small Alternate[b]	Medium	Large	Small	Medium	Large
Particulate matter, mg/dscm	115	197	69	34	69	34	34
Dioxins/furans, ng/dscm total CDD/CDF or ng/dscm TEQ	125 or 2.3	800 or 15	125 or 2.3	125 or 2.3	125 or 2.3	25 or 0.6	25 or 0.6
Lead, mg/dscm	1.2 (or 70% reduction)	10	1.2 (or 70% reduction)	1.2 (or 70% reduction)	1.2 (or 70% reduction)	0.07 (or 98% reduction)	0.07 (or 98% reduction)
Cadmium, mg/dscm	0.16 (or 65% reduction)	4	0.16 (or 65% reduction)	0.16 (or 65% reduction)	0.16 (or 65% reduction)	0.04 (or 90% reduction)	0.04 (or 90% reduction)
Mercury, mg/dscm	0.55 (or 85% reduction)	7.5	0.55 (or 85% reduction)	0.55 (or 85% reduction)	0.55 (or 85% reduction)	0.55 (or 85% reduction)	0.55 (or 85% reduction)
Carbon monoxide, ppmv	40	40	40	40	40	40	40
Hydrogen chloride, ppmv	100 (or 93% reduction)	3100	100 (or 93% reduction)	100 (or 93% reduction)	15 (or 99% reduction)	15 (or 99% reduction)	15 (or 99% reduction)
Sulfur dioxide, ppmv	55	55	55	55	55	55	55
Nitrogen oxides, ppmv	250	250	250	250	250	250	250

[a] "Small" refers to a waste burning capacity equal to or less than 200 lb/hour; "medium" refers to greater than 200 but less than or equal to 500 lb/hour; and "large" refers to greater than 500 lb/hour.
[b] Emission limits for small existing incinerators that meet rural criteria

Source: Fed. Regist. 62(Sept. 15):48348-48391.

furans, hydrogen chloride, sulfur dioxide, nitrogen oxides, lead, cadmium, and mercury. In addition, new or modified large incinerators are subject to a 5% visible emission limit for fugitive emissions generated during ash handling and all existing, new, or modified incinerators are subject to a 10% stack opacity limit. The emission guidelines for existing facilities contain alternative emission limits for small facilities that meet the "rural criteria." Those criteria are based on the MACT levels for small existing medical-waste incinerators that are located at least 50 miles from the nearest Standard Metropolitan Statistical Area and burns no more than 2,000 pounds of hospital waste and medical/infectious waste per week.

In its final rule, EPA estimates that emission limits for existing units are expected to produce reductions of 96-97% for dioxins and furans; 80-87% for lead; 88-92% for particulate matter; 75-82% for CO; 98% for HCl; 75-84% for cadmium; 93-95% for mercury; and 0-30% for SO_2 and NO_x. EPA estimates that in the fifth year after implementation of the standards, there would be nationwide emission reductions of 74-87% for dioxins and furans; 85-92% for particulate matter; 0-52% for CO; 95-98% for HCl; 85-92% for lead; 83-91% for cadmium; 45-74% for mercury; and 0-52% for SO_2 and NO_x. A range of emission reductions is presented to account for the emissions that could occur under a scenario for which no small or medium medical-waste incineration facilities are installed and many of the existing facilities cease operation.

In addition to the medical-waste incineration emission standards and guidelines, EPA includes the requirements listed in Tables 6-5 through 6-9.

On March 2, 1999, a U.S. Circuit Court of Appeals in the District of Columbia (case no. 97-1686) remanded the 1997 rule asking EPA for further justification of its methodology for setting MACT emission limits for new and existing medical-waste incinerators.

CRITICAL COMPARISON OF MACT-BASED REGULATIONS

In its MWC rules, EPA (Fed. Regist. 61(Jun. 20):31736-31779) left NO_x and lead emissions uncontrolled for plants of specified sizes and ages. Also, there are different CO standards, depending on the size, type, and age of plant (see Table 6-2). But that is not the case for hazardous-waste incinerators, which by their nature are at least as diverse in feedstock and combustor design as municipal solid-waste incinerators or medical-waste incinerators. Unlike hazardous-waste combustors under the proposed rules, all municipal solid-waste incinerators and medical-waste incinerators are not regulated within their own categories according to MACT with single numerical emission limits for each pollutant irrespective of plant size, design, age, or feedstock. However, the same types of air-pollution controls (as discussed in Chapter 3) can be applied to large and small facilities. Although there may be other legitimate reasons for doing so, allowing weaker limitations for some designs or sizes provides little incentive for smaller facilities to pursue further achievable emission reductions. Also,

TABLE 6-5 Summary of Additional Requirements Under the Emission Guidelines for Existing Medical-waste Incinerators

Operator Training and Qualification Requirements:
- Complete operator training course.
- Qualify operators.
- Maintain information regarding operating procedures and review annually.

Waste Management Plan:
- Prepare a waste-management plan that identifies the feasibility and approach to separate certain components of a health care waste stream.

Compliance and Performance-Testing Requirements:
- Conduct an initial performance test to determine compliance with the PM, CO, CDD/CDF, HCl, Pb, Cd, and Hg emission limits and opacity limit, and establish operating parameters.
- Conduct annual performance tests to determine compliance with the PM, CO, and HCl emission limits and opacity limit.
- Facilities may conduct performance tests for PM, CO, and HCl every third year if the previous three performance tests demonstrate that the facility is in compliance with the emission limits for PM, CO, and HCl.

Monitoring Requirements:
- Install and maintain equipment to continuously monitor operating parameters including secondary chamber temperature, waste feed rate, bypass stack, and APCD operating parameters as appropriate.
- Obtain monitoring data at all times during operation.

Reporting and Record Keeping Requirements:
- Maintain for 5 years records of results from the initial performance test and all subsequent performance tests, operating parameters, and operator training and qualification.
- Submit the results of the initial performance test and all subsequent performance tests.
- Submit reports on emission rates or operating parameters that have not been recorded or which exceeded applicable limits.

Note: This table depicts the major provisions of the emission guidelines, but does not attempt to show all requirements.

having multiple emission standards for similar devices is inconsistent with minimizing risks of health effects.

SUMMARY OF REGULATIONS RELEVANT TO THE OCCUPATIONAL HEALTH AND SAFETY OF INCINERATION EMPLOYEES

The source and extent of regulation protecting the health and safety of incineration workers depend on the nature of the waste handled by the facility, its

TABLE 6-6   Summary of Additional Requirements Under the Emission Guidelines for Existing Medical-Waste Incinerators that Meet the Rural Criteria

Operator Training and Qualification Requirements:
* Complete operator training course.
* Qualify operators.
* Maintain information regarding operating procedures and review annually.

Inspection Requirements:
* Provide for an annual equipment inspection of the designated facility.

Waste Management Plan:
* Prepare a waste-management plan that identifies the feasibility and approach to separate certain components of a health care waste stream.

Compliance and Performance Testing Requirements:
* Conduct an initial performance test to determine compliance with the PM, CO, CDD/CDF, and Hg emission limits and opacity limit, and establish operating parameters.
* Conduct annual tests to determine compliance with the opacity limit.

Monitoring Requirements:
* Install and maintain equipment to continuously monitor operating parameters including secondary chamber temperature, waste feed rate, bypass stack, and APCD operating parameters as appropriate.
* Obtain monitoring data at all times during operation.

Reporting and Record Keeping Requirements:
* Maintain for 5 years records of results from the initial performance test and all subsequent performance tests, operating parameters, inspections, any maintenance, and operator training and qualification.
* Submit the results of the initial performance test and all subsequent performance tests.
* Submit reports on emission rates or operating parameters that have not been recorded or which exceeded applicable limits.

Note: This table depicts the major provisions of the emission guidelines, but does not attempt to show all requirements.

location, and the nature of the entity operating the facility. Incineration facilities that are operated by government entities and do not handle hazardous waste are exempt from the federal Occupational Safety and Health Act but might be subject to state and local regulation. Thus, municipal solid-waste incinerators operated by government bodies are not subject to the authority of the Occupational Safety and Health Administration (OSHA), whereas such incinerators operated by private companies or contractors might be. The act that created OSHA also enables states to establish their own programs with federal approval if their standards are at least as stringent as OSHA's. Thus, all municipal solid-waste incinerators not operated by government entities must be operated in compliance

TABLE 6-7   Summary of Additional Requirements Under Standards for New Medical-Waste Incinerators

Operator Training and Qualification Requirements:
- Complete operator training course.
- Qualify operators.
- Maintain information regarding operating procedures and review annually.

Siting Requirements:
- Prepare a siting analysis that considers air-pollution control alternatives that minimize, on a site-specific basis and to the maximum extent practicable, potential risks to public health and the environment.

Waste-Management Plan:
- Prepare a waste management plan that identifies the feasibility and approach to separate certain components of a health care waste stream.

Compliance and Performance Testing Requirements:
- Conduct an initial performance test to determine compliance with the PM, CO, CDD/CDF, HCl, Pb, Cd, and Hg emission limits and opacity limit, and establish operating parameters.
- Conduct annual performance tests to determine compliance with the PM, CO, and HCl emission limits and opacity limit.
- Facilities may conduct performance tests for PM, CO, and HCl every third year if the previous three performance tests demonstrate that the facility is in compliance with the emission limits for PM, CO, or HCl.
- Perform annual fugitive testing (large incinerators only).

Monitoring Requirements:
- Install and maintain equipment to continuously monitor operating parameters including secondary chamber temperature, waste feed rate, bypass stack, and air-pollution control device operating parameters as appropriate.
- Obtain monitoring data at all times during operation.

Reporting and Record Keeping Requirements:
- Maintain for 5 years records of results from initial performance test and all subsequent performance tests, operating parameters, any maintenance, the siting analysis, and operator training and qualification.
- Submit the results of the initial performance test and all subsequent performance tests.
- Submit reports on emission rates or operating parameters that have not been recorded or that exceeded applicable limits.
- Provide notification of intent to construct, construction commencement date, planned initial startup date, planned waste type(s) to be combusted, the waste management plan, and documentation resulting from the siting analysis.

Note: This table depicts major provisions of the standards, but does not attempt to show all requirements.

TABLE 6-8   Compliance Times Under the Emission Guidelines for Existing Medical-Waste Incinerators

Requirement	Compliance Time
State plan submittal	Within 1 year after promulgation of EPA emission guidelines
Operator training and qualification requirements	Within 1 year after EPA approval of state plan
Inspection requirements	Within 1 year after EPA approval of state plan
Initial compliance test	Within 1 year after EPA approval of state plan or up to 3 years after EPA approval of state plan if source is granted an extension
Repeat performance test	Within 12 months following initial compliance test and annually thereafter
Parameter monitoring	Continuously, upon completion of initial compliance test
Record keeping	Continuously, upon completion of compliance test
Reporting	Annually, upon completion of initial compliance test; semiannually, if noncompliance

TABLE 6-9   Compliance Times Under the Standard for New Medical-Waste Incinerators

Requirement	Compliance Time
Effective date	6 months after promulgation
Operator training and qualification requirements	On effective date or upon initial startup, whichever is later
Initial compliance test	On effective date or within 180 days of initial startup, whichever is later
Performance test	Within 12 months following initial compliance test and annually thereafter. Facilities may conduct performance tests every third year if pervious three performance tests demonstrate compliance with the emission limits
Operator parameter monitoring	Continuously, upon completion of initial compliance test
Record keeping	Continuously, upon completion of initial compliance test
Reporting	Continuously, upon completion of initial compliance test; semiannually, if noncompliance

with OSHA regulations, although some fall under the jurisdiction of OSHA and some come under approved state plans.

The Superfund Amendments and Reauthorization Act of 1986 requires that OSHA and EPA promulgate regulations to protect workers employed in hazardous-waste operations. OSHA requires that employers develop and implement written programs that cover information and training, personal protective equipment, monitoring, medical surveillance, decontamination, engineering controls and work practices, handling and labeling of drums and containers, exposure and medical-treatment record keeping, and other subjects (29 CFR §1910.120). EPA has extended these requirements to state and local employees who are involved in hazardous-waste operations but do not fall within OSHA's jurisdiction or the jurisdiction of an approved state plan (40 CFR §311).

In November 1990, EPA and OSHA entered into a Memorandum of Understanding on Minimizing Workplace and Environmental Hazards whereby the agencies agreed to coordinate their regulatory efforts on the local, regional, and national levels. The agreement was fleshed out in March 1991. In July 1991, a joint task force formed by the two agencies issued a report summarizing the results of the inspection of 29 hazardous-waste incinerators and made a number of recommendations to EPA and OSHA. On January 25, 1995, the General Accounting Office (GAO) reported that several of the recommendations had not been fully implemented. At that time, EPA had not conducted research on the cause and effects of the use of automatic waste-feed cutoffs and emergency safety vents or vent stacks. Moreover, it had not implemented changes in its inspection protocol to test the effectiveness of worker training, contingency plans, and emergency preparedness (GAO 1995b). OSHA had not improved its inspection expertise, nor had it increased the priority ranking accorded the refuse industry in such a way as to subject a single hazardous-waste incineration to a programmed inspection. Because OSHA considers other industries more dangerous than hazardous-waste incineration, incinerators are inspected on a random basis or in response to complaints, referrals, or accidents. GAO acknowledged that EPA and OSHA were undertaking initiatives not recommended by the task force. For example, EPA and the states targeted combustion facilities for enforcement activities and assessed fines in excess of $9 million. OSHA proposed to require accredited training programs for workers, although it would have no means of ensuring that the accreditation is actually received. GAO suggested that EPA inspections might assist OSHA in assessing the extent of compliance with an accreditation requirement.

MACT-based standards and guidelines applicable to waste combustors required under section 129 of the CAA do not address occupational health and safety directly, although they might affect the well-being of incinerator workers indirectly. For example, it has been estimated that retrofitting incinerators with lime-injection scrubbers will reduce the toxicity of ash. The municipal solid-waste regulations limit visible fugitive emissions from ash-handling and ash-transfer points to no more than 5% of the time, although maintenance and repair activities are exempted from the requirement. Plant operating manuals, operator training, and ASME or equivalent certification of facility operators and shift supervisors might also reduce the likelihood that an incinerator will be run in ways that pose a hazard to its workforce (W. Stevenson, EPA, pers. commun., May 21, 1996).

Certification

In 1988, a committee of the American Society of Mechanical Engineers (ASME) Codes and Standards Division developed a voluntary standard for municipal solid-waste incinerator plant-operator certification. Since then, subcommittees of the Qualification of Resource Recovery Operators Committee (QRO)—some containing members who train operators for firms that design, operate, or construct municipal-waste incinerators—have developed tests for chief facility operators (CFOs) and shift supervisors. In the 1991 New Source Performance Standards, EPA required that CFOs and shift supervisors be certified at the first level of the ASME operator-certification program. EPA's 1991 standards also required that each municipal solid-waste incinerator have a plant-operations manual that each employee was to review. No standards of uniformity were given for preparation of the operations manuals, and no uniform guidelines were proposed to facilitate the review of the manuals by operators.

In the subsequent revision of the municipal solid-waste incinerator emission standards, EPA has proposed requiring that CFOs and shift supervisors be certified at the second, site-specific, level of ASME's operator-certification program. Provisionally certified operators take the site-specific, oral examinations for full certification. In its promulgated municipal solid-waste incinerator standards, EPA requires that a control-room operator who can substitute for a CFO or shift supervisor should also be certified at the first level (on an optional basis). However, there is no stipulated limit on the amount of time that a control-room operator may substitute for a shift supervisor or CFO. With regard to operator training, the standard requires that all CFOs, shift supervisors, and control-room operators complete an municipal solid-waste incinerator operator-training course approved by EPA within 2 years.

The ASME QRO's stipulated minimal qualifications for eligibility to take the first-level operator-certification examinations do not specify any formal academic training, such as college credits in physical science and engineering. Persons with a high-school diploma are permitted to take the examination and receive certification. Furthermore, the requirements proposed by QRO and EPA do not involve demonstration that certified operators have kept up with design changes or are knowledgeable about advances in pollution prevention, combustion efficiency, or emission-control technologies and regulations. The main requirement for renewal of certification is continuation of employment as a CFO or shift supervisor. No periodic testing is required in the operator-certification standard to ensure that the hundreds of operators who received certification in the first few years of credentialing are well versed in the newer technologies or regulations. The ASME created a sister committee to the QRO to develop operator certification for medical-waste incinerator operators.

REGULATORY COMPLIANCE AND ENFORCEMENT

The effectiveness of the panoply of regulations governing waste incineration depends on compliance by incineration facility operators and enforcement of the regulations by federal and state environmental regulators. In general, greater regulatory attention has been paid to securing initial compliance with technologic requirements than with monitoring and ensuring continuing operational compliance (Russell 1990). Audits and assessments of regulatory activity in waste incineration may be rare. Accordingly, it is difficult to assess the full extent of operator noncompliance with regulations or the efficacy of regulatory oversight in policing incineration activity. The absence of data is important. According to Reitze and Davis (1993), "the yet unproven ability of our regulators to effectively control emissions over the life of a facility" is a weakness that undercuts the conclusion that incineration "is a rational option for managing solid waste."

Siting, Startup, and Initial Compliance

Regulations concerning the siting and permitting of new facilities are elaborate. Much regulatory attention is directed toward ensuring that the requisite technology is incorporated and operational in new facilities and that standards are satisfied. Nonetheless, among grassroots environmental activists and mainstream environmentalists there is a high level of dissatisfaction with the regulatory process regarding the siting and permitting of waste incineration. The points that follow do not constitute a complete catalog of the complaints but are among the most-often voiced. Also, it is important to keep in mind that all local residents do not necessarily share these concerns.

Citizens have several kinds of objections to risk assessments. Citizens complain that the data used are hypothetical, that they are not necessarily related to the potentially affected community and the particular facility. To the extent that the facility's risk is assessed in terms of increments to background risk, populations already subject to cumulative risks from other pollution sources are disadvantaged; poor and minority-group communities often feel especially vulnerable in this regard. Moreover, a risk assessment is not a public-health assessment. After a risk assessment has been done and a facility is sited and begins operation, there often is no follow-up to determine whether the assumptions on which the risk assessment was based are true. Community advocates argue that there should be health monitoring in the affected community before a waste-disposal facility opens, during its active operation, and after it closes or is closed. Furthermore, the environmental assessments that precede a siting decision should not be limited to health risks. The risks to the economic and social environment if the facility opens should also be assessed. These concerns are discussed further in Chapter 7 of this report.

Moreover, some citizens doubt that the possibility of upsets, malfunctions, leaks, releases, mishandling, and explosions, whether or not caused by regulatory violations, is taken into account in risk-assessment calculations.

Some citizens believe that they receive notice that their community is a potential site for a waste-disposal facility too late in the siting process. They believe that risk assessments are done to support decisions that have already been made. They do not believe that regulators are impartial decision-makers who will give them a fair chance to present their opposing views and make a fair and impartial decision based on all the evidence. These concerns are addressed in Chapter 7 of this report.

Some citizens believe that stack tests are conducted under ideal or optimally efficient operating conditions when the facilities are in the control of the most-qualified and most-experienced operators and not under ordinary operating conditions, which are likely to be less controlled and therefore more dangerous to the environment and affected populations (Connett and Connett 1994).

Regulators and operators, in contrast, assert that an operator would have little reason to conduct a trial burn at a facility under the best conditions because the limits on emissions in its operating permit are based on conditions determined during that burn, and ordinary operations under less-favorable conditions than the test burn would give rise to permit violations. The operator has an incentive to conduct the trial burn under conditions as close to the design limits of the facility as possible so that there is more latitude in conducting ordinary operations (U.S. Congress 1994). Making it clear to the residents living in the vicinity of a facility being tested, that, contrary to their assumptions, trial burns are conducted under conditions that push (if not exceed) the limits of a facility's design capacity might prompt strong objections from residents.

Continuing Compliance

Regulators engaged in enforcement receive data about ongoing incineration operations through reporting requirements, operator self-monitoring, electronic monitoring, inspections, and citizen complaints.

The inspections of incineration facilities are performed largely by the states with grants from the federal government pursuant to memoranda of agreement that set forth inspection priorities. The priorities are a matter of negotiation between the states and EPA and therefore differ from region to region. Inspectors in the air programs focus on emissions. Inspectors in the waste program

(RCRA) consider combustion and other operations, including the handling of wastes coming into the facility and leaving it. RCRA inspections accordingly will cover personnel training, waste storage, and ash disposal.

A principal goal of an inspection is to determine whether a facility is operating in conformity with the applicable permits or certificates. Logs, strip charts, and data from continuous emission-monitoring systems (CEMS) are important sources of data regarding emissions and cases in which they exceed limits. Data from strip charts and CEMS are particularly reliable because they appear not to be subject to alteration or fraud. CEMS do not exist for all important toxicants though EPA does not require use of all CEMS that are commercially available. The operating conditions set during trial burns are used as surrogate measures of some emissions of substances for which CEMS are not required or for which continuous-monitoring technology has not yet been developed and validated. If the surrogate measures are within compliance ranges, it is assumed that the unmeasured substances are also within compliance ranges. Modems connected to CEMS make it possible for regulators to obtain contemporaneous data on a facility's operations. Such an approach would allow regulators to quickly know about process upsets or bypassing of emission control devices due to emergencies or equipment malfunctions. In Pennsylvania, data from CEMS sent directly and contemporaneously to regulators are used to produce computer-generated records of conditions that exceed limits and the resulting fines that are assessed against a facility, although extenuating circumstances might be the basis for a reduction in fines (Francine Carlini, Air Quality Inspector, Pennsylvania EPA, pers. commun., Nov. 1995). Continuous emission monitoring need not necessarily foreclose the exercise of regulatory discretion. For example, data revealing limit violations that do not appear to be caused by identifiable engineering problems or by technical conditions that can be fixed might not trigger enforcement action (James Topsale, U.S. EPA Region III, pers. commun., Oct. 31, 1995).

Although continuous emission and process monitoring will be required for some pollutants on some incinerators as a result of the CAA, a municipal-waste incinerator will be allowed to exclude data from 25% of its daily operating time and from 10% of the calendar days per quarter when the plant is operating. Furthermore, the proposed standards and guidelines do not indicate which data may or may not be excluded. Separate from the allowable exclusions mentioned above, emissions during startup, shutdown, and upsets are specifically excluded from consideration with respect to compliance. These data exclusions seem to confirm citizen complaints that they do not have access to a full picture of the safety of incinerators.

The timing and frequency of inspections are matters of some criticism by environmentalists. Some jurisdictions conduct inspections without advance notice to the operator; others do not. Environmental activists are concerned that advance notice of an inspection allows an operator to clean up a facility, remove contaminants from the feedstock, hide unfavorable data, and make sure that the most-experienced operators are on duty paying special attention to optimizing combustion and emission controls when the regulators are on site. Regulators respond that announced inspections can be just as effective at turning up violations as unannounced visits. Inspections are undertaken on very short notice; given that the frequency of inspections is mandated by regulations, complete surprise is impossible. Notice facilitates the inspection by ensuring that records and key personnel will be readily available when the inspectors arrive. The inspections depend heavily on data that cannot be concealed or altered (EPA Regulatory Enforcement and Compliance personnel, pers. commun., Feb. 15, 1996).

Incinerator-inspection data are kept at the state, regional, and national levels. Data pertaining to compliance monitoring and enforcement, as well as corrective action, regarding hazardous-waste incineration are supposed to become part of the RCRA Information System (RCRIS). GAO, however, has concluded that RCRIS is difficult to use, does not satisfy the needs of individual states and EPA regions, and fails to serve as a mechanism for maintaining highly reliable data. By and large, RCRIS is not a good basis on which to determine how well incinerators are performing (GAO 1995c).

In the absence of contemporaneous, direct reporting via computerized continuous emission monitoring or the presence of an on-site inspector, regulatory oversight between inspections might depend on reporting by the incinerator operator or complaints from local residents.

The standards and guidelines for municipal solid-waste incinerators require annual testing for dioxin and furan emissions; but to increase the incentive "to optimize performance and achieve emission levels significantly lower" than prescribed limits, less-frequently reporting will be demanded of facilities that meet specified dioxin and furan emission limits for two years. Moreover, reporting requirements have been changed from quarterly to annual to reduce their cost. Such less-frequent testing makes it difficult to determine whether operators have maintained optimal performance.

Violations need not be reported immediately under current EPA proposals. Residents suggest that their complaints are not always accorded the reception that they deserve. Some residents who have lived close to waste combustors complain that their knowledge of the harm that such facilities might produce has been undervalued and ignored. Moreover, although facilities are generating more and more information about their operations, some of it is unavailable to the public and some of it is available but not in a form that is readily usable by laypersons or government officials. It has been suggested that if maintenance and operation records must be kept and made available to the public in a form that is accessible to a lay reader, it would increase the likelihood that incinerators are being operated in compliance with regulations (Reitze and Davis 1993).

Once an environmental protection agency learns of a regulatory infraction or permit violation, a number of options are open to it, including informal responses (site visits or warning letters), administrative remedies (including penalties and compliance orders), and judicial actions (including injunctions and criminal prosecutions). The number of inspections that an agency conducts, the number of facilities shut down or fined, the dollar amount of fines, and the length of time between notice of an infraction and final administrative action with regard to it are all measures of an agency's competence and efficiency in punishing violators and deterring future infractions (Lavelle and Coyle 1992).

Enforcement typically involves the exercise of discretion. Many agencies have enforcement or compliance guidance documents that assist regulators in determining appropriate responses to regulatory infractions (Kuehn 1994). In 1987, EPA revised its Enforcement Response Policy in a way that proved to be problematic. Some industry sources found the policy to be inflexible, insensitive to risk-based concerns, and too harsh in its treatment of minor infractions. But GAO criticized EPA and state regulators operating under EPA oversight for failing to pursue the dictates of the civil-penalty policy, which required penalties stiff enough to deprive violators of the economic benefit of their infractions (GAO 1991). According to GAO, "penalties play a key role in environmental enforcement by acting as a deterrent to violators and by ensuring that regulated entities are treated fairly and consistently with no one gaining a competitive advantage by violating environmental regulations." The size of the penalties was affected by budgetary resources, program targets that favored settlement, regulatory limits on monetary penalties, and the belief of some regulators that working with a violator to achieve voluntary compliance was more effective than imposing heavy fines in deterring future violations.

EPA has issued revised Hazardous Waste Civil Enforcement Response Policy that became effective April 5, 1996. The policy focuses on facilities that pose the greatest risk of exposure to hazardous waste or that are chronic, recalcitrant, or substantial violators of regulatory requirements (EPA 1996c). According to the policy, "an appropriate [enforcement] response will achieve a timely return to compliance and serve as a deterrent to future noncompliance by eliminating any economic advantage received by the violator." The policy indicates that formal proceedings are appropriate in the case of substantial noncompliers; smaller penalties may be assessed against facilities that are unable to pay the full penalty or that discover violations during self-evaluations and audits and promptly disclose and correct them. The policy also sets forth response-time guidelines with allowances for violations under specified circumstances. The policy also specifies the circumstances in which EPA will initiate independent enforcement actions in authorized states.

Persons calling for environmental justice have spawned some reviews of the effectiveness of regulatory activity in areas populated by poor or minority-group citizens. For example, in 1992, the *National Law Journal* published the results of a comprehensive study of completed lawsuits brought by EPA under RCRA and under many other laws, over a seven-year period (Lavelle and Coyle 1992).

The study concluded that the average fines were substantially higher in areas with the highest proportion of white people than in areas with the highest proportion of minority-group people (506% higher in RCRA actions and 306% higher in actions based on violations of multiple laws). The method used by the study for designating white and minority-group communities has been criticized, and later studies, including several undertaken by EPA, have contradicted its findings (Kuehn 1994). EPA recognizes a need for more data with regard to the racial impact of its enforcement activities (Kuehn 1994). Environmental justice issues are discussed further in Chapter 7.

CONCLUSIONS AND RECOMMENDATIONS

Conclusions

- MACT-based regulations vary for incinerators of municipal waste, hazardous waste, and medical waste. There are three to five CO standards, depending on the size, type, and age of municipal solid-waste incinerator. But that is not the case for hazardous-waste incinerators or medical-waste incinerators, which by their nature are at least as diverse in feedstock and combustor design as municipal solid-waste combustors. Within the municipal-waste incinerator rules and medical-waste incinerator rules, several types of emissions (e.g., lead, NO_x, and dioxins and furans) are less-stringently controlled for facilities of specified size (generally smaller) and ages (generally older).
- Workers at incineration facilities tend to be subject to greater risk than other people from exposure to pollutants from normal incinerator operations and from accidents and upset conditions. Ensuring worker safety requires effective coordination of enforcement activities between EPA and OSHA.
- Assessment of both regulatory compliance and the efficacy of regulatory oversight are important to ensure that existing standards are being met in all cases and to satisfy citizens' needs for information concerning incinerator safety.

Recommendations

- In future regulatory decision-making, greater consideration should be given to emission levels achieved in actual performance of incinerators, including process upset conditions. EPA should routinely seek out and use the best and most appropriate data including foreign plant-emission data, and other sources, as well as domestic data, in proposing new standards. In addition, any combustion, emission-control, and continuous emission-monitoring, telemetering and bill boarding technologies and optimum operating practices used in foreign plants should be actively studied and considered for adoption in the United States. In order to give appropriate consideration to combustor and air-pollution control technology and operating techniques used in foreign countries, EPA should develop methods for characterizing the uncertainty of relevant information.

- All regulated medical-waste incinerators and municipal solid-waste incinerators should have uniform limits for each pollutant, irrespective of plant size, design, age, or feedstock, as is the case for hazardous-waste incinerators. The same technology for air-pollution control is applicable to small and large facilities. Although there may be other legitimate reasons for doing so, allowing less-stringent limitations for some designs or sizes is inconsistent with the principle of minimizing risks of health effects.

- Government agencies should encourage research, development, and demonstration of continuous emission monitors (CEMs), telemetering, bill boarding, and computer programs that automatically analyze, summarize, and report CEM data for all types of incineration facilities. In addition to the CEMs already required in the municipal solid-waste incinerator rules, requirement of HCl and particulate-matter CEMs should be considered on all municipal solid-waste incinerators. Also, as soon as a mercury monitor that measures both ionic and metallic forms of mercury emissions has been proven reliable, EPA should consider its use for domestic incinerators. The same approach should be used for other monitors, including those for other heavy metals and the dioxins and furans. EPA should also explore the utility of telemetering and bill boarding of CEM data to regulatory authorities and the public. Providing such data and data summaries on the Internet should be considered.

- In monitoring for compliance, or other purposes, data generated during the intervals in which a facility is in startup, shutdown, and upset conditions should be included in the hourly emissions data recorded and published. It is during those times that the highest emissions are expected to occur, and omitting them systematically from monitoring data records does not allow for a full characterization of the actual emissions from an incineration facility.

- Because operators need to be trained to handle new technologies and follow new requirements, periodic renewal of operator certification for all types of waste incineration should require retesting on new technologies, practices, and regulations. Both provisional and onsite certification should apply to all control room operators, because they can stand in for certified individuals for indeterminate periods of time.

- Government agencies and incineration equipment manufacturers should continue to undertake research to determine how to optimize incineration facility performance for different types of incinerators and how to prevent and mitigate upsets. Detailed guidance should be provided based on the results of this research to state and local regulatory jurisdictions and plant operators, and to the public via the Internet.

- EPA and OSHA should continue striving to improve coordination of enforcement activities between the two agencies to protect the health of incineration workers.

7

Social Issues and Community Interactions

This chapter examines social issues involved in the siting and operation of waste-incineration facilities (such as incinerators and industrial boilers and furnaces), including possible social, economic, and psychological effects of incineration and how these might influence community interactions and estimates of health effects. Issues with respect to perceptions and values of local residents are also considered. In addition, this chapter addresses risk communication issues and approaches for involving the general public to a greater extent in siting and other decisions concerning incineration facilities. The committee recognized at the outset of its study that the social, economic, and psychological effects for a particular waste-incineration facility might be favorable, neutral, or adverse depending on many site-specific conditions and characteristics. However, the current state of understanding for many issues considered in this chapter is such that little or no data specific to waste incineration were available for analysis by the committee. In such cases, the committee identified key issues that should be addressed in the near future.

The social, psychological, and economic impacts of incineration facilities on their locales are even less well documented and understood than the health effects of waste incineration. When environmental-impact assessments are required for proposed federal or state actions, they typically must include socio-economic-impact assessments, but the latter are often sketchy at best. They also might be given short shrift in the decision-making process (Wolf 1980; Freudenburg 1989; Rickson et al. 1990). Furthermore, these socioeconomic assessments attempt to be prospective—that is, they assess the likely effects of proposed actions. Little research has been done to evaluate systematically the socioeconomic impacts of controversial waste-treatment or waste-disposal facilities that have been in place for several years or more (Finsterbusch 1985; Seyfrit 1988; English et al. 1991; Freudenburg and Gramling 1992). Moreover, the committee is not aware of any studies of the effects of removing an established incinerator. One reason for the lack of cumulative, retrospective socioeconomic-impact research is the lack of sufficient data. Although incineration facilities must routinely monitor and record emissions of specified pollutants, health-monitoring studies before or after a facility begins operation are only rarely performed, and periodic studies of the socioeconomic impacts of a facility over time are virtually nonexistent, partly because of methodological problems (Armour 1988) and the absence of regulations that necessitate continued monitoring of socioeconomic impacts.

Whether predictive or retrospective, socioeconomic-impact assessments share the challenge—also faced by health-effects assessments—of confounding factors. Isolating the impacts of a single facility from other contributing conditions is often difficult, especially as those conditions change over time (Greenberg et al. 1995). Furthermore, the demographic composition of the area around the facility can be expected to change as time passes, making it difficult to assess the relationship between the facility and the changing group (Maclaren 1987). Individuals also vary among themselves and over time in their sensitivity to socioeconomic impacts, such as a decline in property values.

The scant information that is available on predicted or observed socioeconomic impacts of various types of controversial waste-treatment or waste-disposal facilities cannot be readily generalized to waste-incineration facilities, nor can the impacts of one waste incinerator be generalized without qualification to other waste incinerators. The host areas and the facilities themselves are, in many instances, too dissimilar to permit drawing inferences from one facility to another without many caveats (Flynn et al. 1983; English et al. 1991). As discussed further below, simply identifying the geographic boundaries of an affected area can present problems.

Much of the following discussion is based on anecdotal evidence related to social issues posed by controversial waste facilities, including waste incinerators. It is clear that much more empirical research is needed on the socioeconomic impacts of waste-incineration facilities on their host areas, but for this research to be feasible on a large scale, detailed socioeconomic data will need to be gathered routinely before and during the operation of such facilities.

It is also clear, however, that citizen concerns about waste incinerators do exist. Newspapers and popular journals report heated disputes about them; the Waste Technologies Industries (WTI) hazardous-waste incinerator in East Liverpool, Ohio, is a prime example. The publications and World Wide Web sites of citizen-group networks, such as the Center for Health, Environment, and Justice (formerly the Citizens Clearinghouse for Hazardous Waste), have routinely reported opposition to incineration of municipal solid waste, medical waste, and hazardous waste. Proposed facilities are often targets of citizen concern; existing facilities appear to receive less attention but are not altogether ignored. Groups opposed to incineration tend to focus on the adverse health and environmental effects of the facility but may also express concerns about socioeconomic impacts. Such groups do not necessarily represent the sentiments of all others living in their vicinity; in fact, it can be expected that a number of community members will be indifferent and that, among those who do care, some will advocate or be willing to consider the startup of the facility while others will be adamantly opposed (Elliott 1984a; Walsh et al. 1993). As is true of identifying the affected area, identifying who should be included as part of the "community" can be difficult.

As a waste-management option, incineration has features that some citizens might find attractive. It can be used to reduce waste volumes, produce electricity and destroy or reduce waste toxicity. Despite those features, the views of other citizens who are inclined to oppose waste incineration need to be heard and understood. If not, conflicts can intensify and they can increase the time and expense of developing waste incinerators that might be socially beneficial. Furthermore, continuing opposition to the facilities can indicate that important concerns are being given short shrift.

This chapter considers four related questions: What defines the affected area? What local concerns, in addition to concerns about direct health effects, can arise in connection with waste incineration? What underlying factors contribute to and help to explain local concerns? How can local concerns best be addressed in interactions with members of the affected area? The chapter points to issues outside the direct health impacts of waste incineration that appear to merit attention in future research and in the siting, licensing, and operation of such facilities.

IDENTIFYING THE AFFECTED AREA

The boundaries of the area potentially affected by a waste-incineration facility are not necessarily the same as the boundaries of the local jurisdiction. The affected area might be a relatively small section of the local jurisdiction; or, as illustrated in Figure 7-1, if an incinerator is at the edge of the jurisdiction, the affected area might extend into one or more other jurisdictions.

Complicating matters is the fact that different impacts (including health effects and socioeconomic effects) have different reaches across space and time. Some impacts, such as those on traffic volume, might occur mainly along narrow corridors; others, such as those on air quality or on property values, might be more diffuse. Some might be relatively transitory, such as those due to a demand for workers during facility construction or to an episode of unusually high emissions due to a process upset or an accident at a facility; others might be cumulative or of long duration, such as those due to chronically high emissions

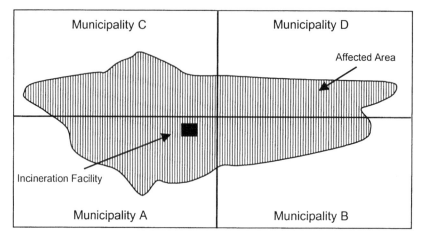

FIGURE 7-1   Hypothetical example of possible transjurisdictional impacts of an incineration facility.

or continued employment opportunities at a facility. Furthermore, health and socioeconomic effects vary in their intensity because of variations at the source, along environmental pathways, and among receptors. Consequently, preliminary mapping of the potentially affected area might necessitate a complex set of overlays for different types of impact (see Figure 7-2), each with its own gradations, boundaries, and time dimensions. Identification of areas for study and

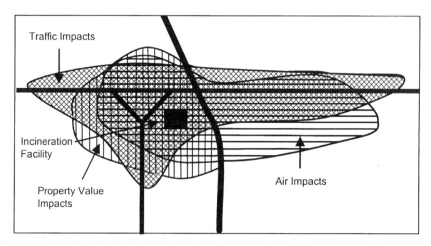

FIGURE 7-2   Hypothetical example of overlay mapping of different types of impacts of an incineration facility.

assessment will necessarily be somewhat arbitrary; what is crucial is that areas that are expected to receive substantial impacts should be included (Flynn et al. 1983).

Identifying an affected area according to where, when, and to what extent impacts occur has at least two important implications for interactions with those who live and work in the vicinity of an incineration facility. First, it highlights considerations for local government decisions concerning waste incineration and other controversial facilities. Second, it describes what constitutes the "community" in less-formal, nongovernment interactions concerning the facility. Both those subjects are introduced below and are addressed in greater detail later in this chapter under the heading Risk Communication.

The Affected Area and Local Decision-Making

Typically, those making decisions or entering into negotiations about an incineration facility's location, size, and so on are the elected or appointed officials of the jurisdiction that the planned facility would be in, such as the mayor and city council or the county executive and county commissioners. When they consider the facility, they are likely to have in mind the interests of the jurisdiction as a whole, not just those of the affected area.

Waste facilities, like other land uses that have potentially undesirable side effects, present the possibility of uneven benefits and costs. A facility might produce substantial benefits, both for the larger region (by providing management capacity for some of its wastes) and for the host jurisdiction as a whole, but might have net adverse impacts on the immediately affected area (Greenberg et al. 1995). For example, as discussed further below, current approaches to the siting of large, controversial facilities sometimes include substantial payments to the host jurisdiction, which are then used as revenue to alleviate taxes or improve local schools, roads, and so forth. Depending on how the extra revenue is allocated, it might or might not benefit the affected area primarily. In some instances, a portion of it must, by prior agreement, be earmarked for improvements in the area immediately surrounding the facility; in other instances, it can be spent at the discretion of the local governing body.

When facilities like hazardous-waste and medical-waste incinerators are proposed, a general rallying of opposition—including opposition by local officials—sometimes occurs, if only because of the fear of the stigma that the facilities may bring. When facilities like municipal solid-waste incinerators are proposed, in contrast, elected officials and most voters in the jurisdiction may favor them, especially to the extent that they can help to meet local waste-management needs. If members of the affected area are only a few among many in local decisions concerning a facility, they run the risk of having their interests and concerns overruled. Decision-making about the facility might then have the appearance, but not the actuality, of fairness and impartiality. To correct for that possibility,

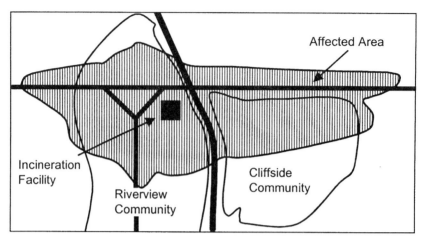

FIGURE 7-3   Hypothetical example of relationship of affected area to local communities.

augmentation of traditional forms of decision-making solely by elected officials or popular referenda is being explored, as discussed further below.

The Affected Area and Community Interactions

Although the term *community* is widely used, what counts as a community is often not clear. Communities are usually thought of as place-based, but the term is also used to refer to groups that, although widely dispersed, share interests (for example, a research community). Even when the term is used in its geographic sense, it is ambiguous: place helps to define a community, but other attributes—particularly those concerning social exchange—are often deemed essential (Catlin 1959; Ladd 1959; Minar and Greer 1969; Poplin 1972). On the basis of such attributes, people living or working in a particular area form their own conceptions of the boundaries and composition of the community. Thus, an affected area does not necessarily constitute a discrete community; instead, it might contain parts or the entirety of several communities (see Figure 7-3), or it might lack the social cohesiveness to have any communities.

If the affected area is not mirrored by a single community, informal interactions by facility proponents and regulators with members of the affected area may be more difficult to conduct. As has been noted elsewhere, "Success for risk communication does not require that every citizen be informed about the risks presented in every regulatory decision, but people need to be confident that some person or group that shares their interests and values is well informed and is representing those positions competently in the political system" (NRC 1989a). However, it is important that every citizen have an opportunity to be informed, whether they become so or not. The opportunity should not entail unnecessary burden. Communication between facility developers, regulators, and members of the affected area is often essential; lacking a single, cohesive community, the conduits for communication with members of the affected area may not be readily apparent. In addition, people who live or work outside or at the far reaches of the affected area may have strong views about a proposed facility but not be part of the community (or communities) in the immediate vicinity of the facility; whether and how to integrate them into informal interactions concerning the facility can be among the most-difficult issues that arise in local interactions.

SOCIOECONOMIC IMPACTS OF INCINERATION FACILITIES

A list of possible socioeconomic effects of an incineration facility is provided in Table 7-1. (It is beyond the scope of this report to discuss the listed effects in detail.) These effects may be favorable or adverse, and they may be economic (such as job creation and decrease in property values), psychological (such as stress and stigma), or social (such as community fractionalization and unity). The effects can occur in individuals, groups in the affected area, or the entire population in the jurisdiction as a whole. In addition, different kinds of impacts

TABLE 7-1 Potential Impacts of Incineration Facilities to Be Considered in Socioeconomic Impact Assessments

- Increase or decrease in population.
- Change in migrational trends.
- Change in population characteristics.
- Disruption of settlement patterns.
- Change in economic patterns.
- Increase or decrease in overall employment or unemployment and change in occupational distribution.
- Increase or decrease in income.
- Change in compliance of land use with land-use plans.
- Increase or decrease in land values.
- Change in taxation resulting from change in land use and income.
- Change in types of housing and in occupancy.
- Change in demand on health and social services.
- Change in demand on educational resources.
- Change in demand on transportation systems.
- Relocation of highways and railroads.
- Change in attitudes and lifestyles.
- Disruption of cohesion.
- Change in tourism and recreational potential.

Source: Adapted from Rau and Wooten 1980.

can interact; for example, impacts that are primarily economic might have psychological and social elements as well.

Different types of incineration facilities will have different effects on their surrounding geographic areas. Some incineration operations are within larger facilities that have other functions, such as manufacturing; others are new, stand-alone facilities. Facilities may be owned and operated under a number of different arrangements, such as by federal, state, or local government; private companies; or joint public-private enterprises. Facilities in residential areas may have greater socioeconomic effects on the surrounding area than facilities in highly industrialized areas.

Economic Impacts

A waste-incineration facility may provide jobs, both directly and by attracting industry to the region because of the services offered by the facility. In addition, such a facility may contribute to the cogeneration of electricity and district heating. The number of jobs will depend on the size and type of incinerator. The number and types of jobs available to *local* residents (whether the jobs are abundant and well-paid or scarce, low-skilled, and low-paid) will depend on the type of facility and its hiring policies, on the policies of local unions and their willingness to accept new members, and on the characteristics of the local population.

A waste facility may have an adverse effect on local economic prospects, however, if businesses leave the affected area or decide not to locate there. Public perceptions may make the risk seem larger (Kasperson et al. 1988) and lead to the stigmatization of affected communities (Edelstein 1988; Slovic et al. 1994). That may be due in part to concern about health and ecological risks, but Gregory et al. (1995) have noted that "stigma goes beyond conceptions of hazard. It refers to something that is to be shunned or avoided not just because it is dangerous but because it overturns or destroys a positive condition; what was or should be something good is now marked as blemished or tainted."

Stigmas can have both direct and indirect economic impacts. Local employment opportunities may be adversely affected, and a stagnation or decline in local retail businesses may necessitate traveling outside the neighborhood to shop for food, clothing and so on. For example Greenberg et al. (1995, p. 259) concluded that the Union County solid-waste incinerator in Rahway City, New Jersey, "will lead to rapid deterioration of the neighborhood because private investors with other choices will choose not to invest in a neighborhood with a prominent technological hazard, except perhaps to site LULUs [locally unwanted land uses]."

A waste-incineration facility may affect local public finances favorably insofar as it adds to local tax revenues or decreases the cost of local-waste disposal. However, such a facility may affect public finances adversely insofar as it increases the need for public services, such as improvements in roads and emergency preparedness, increases the cost of local-waste disposal, or requires large investments of time by local and state officials in permitting and other regulatory activities. In some cases, the net effect on local public finances will depend at least partly on special mitigation and compensation measures. In the case of a proposed municipal solid-waste incinerator to be located in Montgomery County, Pennsylvania, for example, the host municipality, Plymouth Township, was offered a $350,000 annual fee in addition to various other inducements; this helped to build an initial base of support with the township's board members, although they later contested the project because of pressure from a Plymouth citizens' protest group (Walsh et al. 1993).

An incineration facility might also affect property values in its vicinity. Whether it increases or decreases them will depend primarily on what the neighborhood was like before the facility was introduced. Actual property-value differentials near controversial facilities do not consistently reflect anticipated effects (Zeiss and Atwater 1989), but they have been observed in connection with some controversial facilities or contaminated sites partly because of stigma effects associated with those sites (Payne et al. 1987; Smolen et al. 1992); they remain a source of major concern for some people living near an existing waste facility, such as a solid-waste incinerator (Zeiss 1991).

Psychological Impacts

People in the surrounding area may be psychologically affected by the prospect or reality of an incineration facility in their midst. The risk associated with industrial activity is increasingly recognized as including a wide array of adverse and sometimes long-lived psychological impacts, which may be, but are not always, correlated with negative attitudes toward the risk source (Freudenburg and Jones 1991). Concerns about adverse health effects on oneself or one's children, parents, spouse, and so on, as well as fear of adverse economic effects, can contribute to stress or depression, which in turn can produce physical symptoms, such as headaches and sleeplessness (Neutra et al. 1991). Stress or depression may also be experienced if family, work, and social relationships are altered or terminated (through divorce or job loss) because of protracted outlays of time and energy to understand and combat a proposed or existing waste incinerator. In addition, feelings of powerlessness, distrust, and alienation may be fostered if people feel that their neighborhood has been "captured," is not within their control, or lacks protection from the government. Feelings of a lack of community control and of a poor community image can rank with air and water pollution as sources of concern in areas around waste facilities (Zeiss 1991).

Favorable psychological impacts may also be experienced under some circumstances. In particular, people's self-esteem and feeling of social connection may increase as they learn about and interact with others about the facility; they may even become acknowledged regional leaders (Gramling and Freudenburg 1992). Sieber and others have argued that multiple roles help to provide status, security, social prestige, new access to resources, and ego gratification (Sieber 1974; Thoits 1983). However, Menaghan (1989) has cautioned that age and gender expectations will often help to determine whether an individual's role repertoire is personally satisfying. Furthermore, some individuals opposing an incineration facility may feel offended by the notion that such stressful and time consuming efforts may have favorable psychological benefits.

Social Impacts

In addition to having favorable or adverse effects on the economic, physical, and mental well-being of individual people in the affected area, a proposed or existing incineration facility can affect the area's social fabric. Some changes may be precipitated by economic factors, but others may be structural; that is, they may concern the formal and informal relationships of groups and individuals in the area.

Like other potentially controversial land uses, an incineration facility can provoke factionalization in the affected area among those who are opposed to it, those who favor it, and those who do not want the area harmed by heated, widely publicized conflict. Local controversies over major facilities (and sometimes over relatively small-scale ones) can last for years and leave scars and permanently alter formal and informal relationships in the area. As Gramling and Freudenburg (1992) note, "impacts to social systems occur as interest groups form or redirect their energies, promoting or opposing the proposed activity and engaging in attempts to define the activity as involving opportunities or threats." Depending on the outcome and people's perspectives, the altered relationships can be detrimental or beneficial. For example, the "old boy" network may become more-firmly entrenched or, in contrast, community organizations and a more-populist local government may be fostered. An expectation of ongoing factors that affect the quality of life (for example, noise, traffic, odors, and dust) can also be raised as concerns (Greenberg and Schneider 1996). Such concerns will need to be addressed regardless of whether some local residents do not share those concerns.

If individual health and well-being, property values, and the quality of life in a neighborhood are substantially affected by a proposed project (or even if it is *expected* that they will be affected), the neighborhood's character may begin to change. Change may be seen in an increasing ratio of industrial to nonindustrial activities in the area, in the types of homes and businesses, and in the demographic composition of residents. People who can afford to move out may do so, sometimes altering the ethnic mix and age composition of the area. The changes do not happen overnight, but they are likely to be more rapid and destabilizing than the gradual demographic changes that occur in all communities because of births, deaths, and migration.

Similarities and Dissimilarities in the Impacts of Proposed and Existing Facilities

Facilities that are large, controversial, risky, or otherwise out of the range of the host area's ordinary experiences may have observable, measurable social and economic impacts from the time of the earliest announcements or rumors about a project (Freudenburg and Gramling 1992). Years may go by between the proposal of a facility for a particular area and the actual licensing, construction, and operation. During this interval, some of the impacts described above—such as favorable or adverse changes in property values, individual lives, or social structures—may be precipitated by the *likelihood* of the facility and the controversy about whether and how it should be developed. After a facility becomes operational, people in an area are able to use their own experiences (of changing property values, truck traffic, neighborhood composition, and so on) to assess the effects of an existing facility. However, the potential for mobilizing protest and blocking a *proposed* facility can be stronger than those for ensuring that an *existing* facility is being operated properly, with vigilant regulatory oversight and with minimal impacts on the surrounding area.

After a facility has been in operation for a long time, many of the impacts described above, if they occurred at all, will be in the past. The surrounding area may (or may not) have undergone wrenching changes because of the facility when it was in its formative stage, but over the years the area will have altered and adapted to its presence. The facility may continue to have adverse health effects, but it will not be likely to precipitate major new socioeconomic impacts; instead, to the extent that it continues to affect the character of the surrounding area and its residents, it may contribute to feelings of either acceptance or quiet powerlessness and alienation.

If, however, major alterations are proposed for an existing facility (especially an expansion of the volume or types of wastes that it will handle) or if another controversial facility is proposed to be built in or near the already affected area (especially one similar to the existing facility), the existing facility may receive renewed attention, provoking psychological and social impacts similar in many respects to those provoked by the proposal of a new facility. The door is then opened for a revival of broad-scoped interest and concern, even though from a regulatory standpoint only the relatively narrow question of facility expansion or addition of a new facility would typically be at issue. Note, however, that environmental justice concerns about cumulatively disproportionate environmental contaminant burdens on low-income communities or communities of color are beginning to change this regulatory stance in some instances. (See the discussion of environmental justice below.)

PERCEPTIONS AND VALUES

Even economic, psychological, and social impacts considered to be small by scientific and technical experts may be considered unbearable by people living in the affected area. Psychologists have identified a number of characteristics or risk attributes that help to explain the divergence between experts and the lay public—the so-called perception gap. It is well documented (Slovic 1987; Slovic et al. 1982) that members of the public tend to fear most the hazards that they do not impose on themselves voluntarily and that result in severe effects that are delayed (such as cancer). Other attributes, such as blame and distrust, have been singled out as especially important in shaping people's perceptions (Slovic 1993).

The differences between expert and lay perceptions reflect differences in values, however, not merely differences in information and understanding. They therefore cannot be bridged simply by introducing public-education programs and information campaigns. Indeed, efforts to inform and educate the public about controversial technical issues may actually exacerbate public concern and opposition (Johnson 1993). Furthermore, the inherent uncertainties and variability of technical issues make risk communication difficult, and disagreement among experts may itself exacerbate public concern. Members of the public take into consideration intangible attributes of risk in addition to estimates of mortality and morbidity, such as voluntariness, dread, and perceived extent of scientific knowledge; but they also question deeper underlying values. Running beneath the surface of many local, site-specific risk controversies are broader concerns about preservation and protection of the environment, the choice and control of technology, and the structure and function of society in general (Otway 1987; Otway and Wynne 1989; Schwarz and Thompson 1990). Given those concerns, members of the public often raise fundamental questions that are not easily addressed in siting processes or risk-management decision-making. For example, they may ask: Do we as a society really need the technology of incineration? How can we restructure society to produce less waste, rather than building incinerators and landfills?

In a study of conflicts about a variety of technologies, Von Winterfeldt and Edwards (1984) found that the controversies usually revolve around legitimate differences in values among the interested parties. Members of the public often become enraged by technical assessments and public review processes that implicitly or explicitly rule values to be outside the boundaries of discussion (Wynne 1992). Three key issues are whether those responsible for developing, operating, and regulating incineration facilities can be trusted; whether the facilities are needed; and whether fair processes are used to site them.

Social Distrust

The American public has less and less trust in the institutions and people responsible for the siting and management of potentially hazardous facilities, such as incinerators (Weinstein 1988; Renn and Levine 1991; Slovic et al. 1991; Kasperson et al. 1992). In 1983, a survey of attitudes toward the siting of hazardous-waste disposal facilities found that lack of trust in the companies that operate treatment facilities and the government agencies that regulate them was a primary source of concern (Laird 1989). A number of reasons for citizens' distrust or lack of trust are possible (Slovic 1993). Some of the reasons concern the people operating or permitting and regulating a facility; for example, suspicions arise if track records are blemished or if it is sensed that facility proponents and regulators do not share one's own goals for society. Apart from whether facility operators and regulators are regarded as honest, competent, and well intended, scientific uncertainties can contribute to a lack of trust, especially in the case of facilities such as waste incinerators, for which questions remain about environmental transport and fate and about dose-response relationships of various emissions. How uncertainty bounds can be most appropriately communicated and how different tolerances for uncertainty can be reconciled are difficult questions and discussed further in Chapter 8.

Trust is easily lost and difficult to regain (English 1992). Attempts to regain social trust during the period of most risk-communication and facility-siting efforts are naive at best, in that its loss is a broad social phenomenon that affects all social and political institutions. Future risk-communication efforts will be undertaken in the context of continuing, intense social distrust and will have to be designed accordingly.

Need

One of the first questions likely to be asked by people in the prospective host area of a waste-incineration facility (or other waste facility) is, "Is this really needed?" (Chertoff and Buxbaum 1986). If proponents cannot convincingly demonstrate a pressing need for a new or expanded incineration facility, people who are skeptical of the need for the facility, or are otherwise opposed to it, will be disinclined to negotiate on other issues about it.

As grassroots environmental movements began to flourish, pressure was brought to bear to reduce or, if possible, eliminate waste volumes and waste toxicity. With more-stringent regulations and greater corporate attention to release of environmental pollutants, that pressure has led to more-advanced production and waste-management technologies and to reduced demand for hazardous-waste facilities, relative to population size and goods produced. However, there will always be some need for facilities to manage waste byproducts of goods and services, thus society will still have to address how many facilities are needed, of what size, with what technologies, and—the hardest of all—in whose backyards.

Fairness

The second question typically asked by people in the prospective host area is, "Why *here*?" Siting a facility, such as an incinerator, presents an inherent and inescapable need to address equity: Whatever site is chosen, potential health risks and other adverse impacts are necessarily borne by a relatively small group, but the benefit (waste treatment or disposal) can accrue to a larger population. Using more and smaller, local facilities may avoid some of the larger-scale inequities (Morrell 1987), but it cannot escape the dilemma and may create other problems, such as concerns about an overall increase in emissions. Although building new facilities may create new jobs and provide substantial tax revenues and other benefits (such as property-value guarantees), there is a well-documented tendency (known as loss aversion) for people to focus on adverse impacts more than on possible benefits (Tversky and Kahneman 1991). In a survey of five communities in Canada, Zeiss (1991) found that residents were far more likely to be concerned about the possible adverse impacts of a waste facility than to be attracted by its benefits.

Aside from actual or perceived outcomes (benefits and costs), the process for choosing a particular site is often perceived to be unfair (English 1992; Kunreuther et al. 1993). In many cases, residents living near a proposed facility will feel that they have been unfairly singled out and that they have had little or no substantive input into the decision. Possible attendant risks will be considered an imposition, and the degree of involuntariness will exacerbate perceived risk.

Competing Distributional Principles

To answer the question "Why here?" appeals may be made to various distributional principles. For example, appeals can be made to the principle of "the greatest good for the greatest number," which suggests that communities must be prepared to host a waste facility if they are physically and locationally well suited to do so and if society will thereby reap substantial benefits. Laws that provide for state preemption of local control over facility siting decisions (Tarlock 1987) tacitly invoke this utilitarian principle. In contrast, appeals can be made to the principle of equality, which suggests that each community has equal rights and deserves equal treatment; none should be singled out and compelled to take a waste facility if it will benefit outsiders primarily, but each should bear an appropriate share of responsibility for waste management. This principle is tacitly invoked in siting approaches that provide for "simultaneous siting of numerous new facilities in accordance with regional needs and with local patterns of equity" (Morrell 1987, p. 118).

To mediate between those opposing views, some have argued for a market-driven distributive principle, whereby communities would voluntarily host waste facilities because of handsome incentive packages (Kunreuther et al. 1993). In contrast, it may be argued—using a "justice as fairness" concept (Rawls 1971)—that if the procedures for selecting a host area are open, inclusive, and scrupulously fair, their outcomes will by definition be fair. However, neither market-driven nor process-driven approaches have as yet had widespread success in overcoming various concerns—particularly concerns about adverse health effects, about who will derive monetary and other benefits from the facility, and about who can consent on behalf of the affected area (or, if a public referendum is used, who should be able to vote, how large a majority is needed, and whether the referendum will be binding on elected officials).

Environmental Justice

When an incineration facility is placed in a disadvantaged community, concerns about fairness are likely to become more pressing. Minority groups, low-income groups, and urban dwellers probably suffer disproportionately from exposure to air pollutants (Berry 1977; Wernette and Nieves 1992). In addition, African-American children have the highest blood-lead concentrations among all groups, and blood-lead concentrations among Hispanics, urban dwellers, and low-income groups are higher than the national average (Montgomery and Carter-Pokras 1993; Schwartz and Levin 1992). Results of several studies indicate that hazardous-waste treatment, storage, and disposal facilities are more likely to be located within or adjacent to low-income and minority-group communities (United Church of Christ Commission for Racial Justice 1987; Bullard 1990; Bryant and Mohai 1992; Goldman and Fitton 1994). Some researchers have disputed those findings, however (Anderton et al. 1994), by citing differences in study design that make it impossible to compare results and reach a conclusive answer. On the basis of a review of the available scientific literature and the information obtained from various site visits, IOM (1999) concluded that there are identifiable communities of concern that experience higher levels of exposure to environmental contaminants. In addition, such communities are less able to deal with these exposures as a result of limited knowledge and disenfranchisement from the political process. Moreover, factors directly related to their socioeconomic status, such as poor nutrition and stress, can make people in those communities more susceptible to the adverse health effects of environmental hazards and less able to manage them by obtaining adequate health care.

Often low-income communities and communities of color speculate that some sources of environmental degradation were placed in their communities because land was cheap and the citizens lacked economic and political power—including access to technical and legal expertise—to keep it out. Other sources of environmental degradation may have been in place for a long time, with disadvantaged communities growing up around them because of low incomes, low property values, and discriminatory exclusion from other areas.

Results of some studies indicate that additional factors worsen the environmental burden on low-income and minority-group people. Environmental pollution in the areas where they live and work may be worsened by lax and irregular enforcement of regulations and by inadequate public services, such as water-treatment and sewage systems. And the effects on local residents may be graver for a number of reasons. For example, members of sensitive populations (children, the elderly, pregnant women and their fetuses, people with impaired immune systems, and people with chronic diseases) who are poor often cannot afford to move elsewhere. In addition, because of necessity or tradition, their diets rely heavily on locally caught fish and home-grown food and to recreation on local lands and in local water. Also, those living in the affected areas often are exposed to occupational health hazards.

A growing awareness that health problems, as well as socioeconomic problems, are, at least partly, environmental in their origins has led a number of disadvantaged communities to mobilize against further assaults on the air, water, and land in their locales. They face an uphill battle, however. Many of the problems are slow to surface, so cause-effect arguments are difficult to make. And many problems result from factors that are cumulative and interactive, whereas regulatory standards have tended to focus on single sources of pollution, on single chemicals in a single transport medium, and on health and environmental effects exclusively, with little or no attention to related socioeconomic effects (EPA 1992e). However, even if it were determined that facilities like waste incinerators contribute only marginally to increased environmental degradation, they may be regarded by their host areas as presenting unacceptable additional increases in risk.

Numerous initiatives have been undertaken over the last 3 years to address environmental justice. In 1992, the U.S. Environmental Protection Agency (EPA) created the Office of Environmental Justice (now in EPA's Office of Enforcement and Compliance Assurance). On February 11, 1994, President Clinton issued executive order 12898, directing all federal agencies to develop strategies designed to promote enforcement of health and environmental statutes in areas with minority-group and low-income populations, ensure public participation, improve research and data collection on the health and environment of these populations, and identify differential patterns of natural-resource consumption among these populations. More recently, Title VI of the 1964 Civil Rights Act has been used as a basis for prohibiting federal agencies and recipients of federal financial assistance from issuing permits that—because they were expected to result in disproportionate environmental burdens—were determined to have the effect of discriminating based on race, color, or national origin. Moreover, a number of states have enacted environmental-justice legislation and have considered environmental-justice bills (National Conference of State Legislatures Environmental Justice Group 1995). Finally, industry initiatives have recently begun to come to terms with these problems. Much of the attention, however, is still at the stage of "understanding the problem."

RISK COMMUNICATION

Background

Risk communication has been broadly defined as any "purposeful exchange of information about health or environmental risks between interested parties" (Covello et al. 1986, p. 112) and by the National Research Council Committee on Risk Perception and Communication as "an interactive process of exchange of information and opinion among individuals, groups, and institutions" (NRC 1989a). A wide variety of activities can be considered risk communication, including individual exchanges (such as phone conversations) among members of the public, agency officials, and industry representatives; informational campaigns by government agencies and industry; and elaborate processes of stakeholder involvement and decision-making.

Good risk communication may ease community concerns, but there are no guarantees. Poor risk communication will almost certainly exacerbate public concerns. As noted by the National Research Council Committee on Risk Perception and Communication, "even though good risk communication cannot always be expected to improve a situation, poor risk communication will nearly always make it worse" (NRC 1989a). In spite of those caveats and the relative newness of the field, guidelines for conducting risk communication are emerging. Those are reviewed below, although the purpose here is not to provide a manual for communicating about incineration issues.

To know whether a risk-communication process is good or bad, one needs to know what the *goals* of the process are and what constitutes *success*. Different participants in the process may have strikingly dissimilar goals and criteria for success. Generically, risk communication can have several goals (Covello et al. 1986):

- To inform and educate.
- To encourage behavioral changes and protective actions.
- To notify in case of emergencies.
- To encourage joint problem-solving and conflict resolution.

Communicating about the health and environmental risks of incineration facilities is likely to involve elements of the 1st and 3rd goals, but most commonly will focus on the 4th. Given the extent of public distrust of industry and government roles in siting processes in general, the public often views risk communication with severe skepticism (Kasperson et al. 1992). Members of the public sometimes view risk-communication efforts as a thinly veiled attempt to foist unwanted facilities on unwilling communities rather than as sincere efforts in joint problem-solving and conflict resolution. Even well-meaning efforts to inform and educate the public may be viewed as, at best, attempts to bring public perceptions into line with expert assessments of risk and, at worst, attempts to obfuscate the issues and belittle public concerns. Some government and industry representatives see risk communication merely as a means to a particular end (in this context, a siting decision).

In keeping with this skepticism and disagreement over goals, there may also be no consensus on what constitutes a successful risk-communication process. The facility owners or operators may see siting the facility as a success, whereas members of the public may see preventing the facility from being sited or operated as success. Members of the public may see the siting of a new incineration facility as a threat to the successful development of recycling and source-reduction programs. Even if a majority of local residents agree to a siting decision after an elaborate process of public discussion and negotiation, many members of the public could claim that the siting process was unfair, undemocratic, and invalid. To avoid that situation, a measure of success has been developed in terms of the *process* of risk communication, independent of the *outcome*. Risk communication is construed to be successful "to the extent that it raises the level of understanding of relevant issues or action for those involved and satisfies them that they are adequately informed within the limits of available knowledge" (NRC 1989a).

If demand for incineration diminishes, siting controversies may become less common, although they are unlikely to disappear soon. Some underserved regions could move to site new facilities, and some old facilities could be replaced (by the building of new facilities at the same locations or at new locations). But risk communication should remain as a continuing process at existing facilities or at future facilities after siting. Overt expressions of public concern may diminish after siting, but they could come to the fore again if accidents occur; if changes in facility design, ownership, and operation (such as a decision to accept medical or other hazardous wastes) are planned; if cancer clusters or other health effects that may be attributable to a facility are discovered; or if a facility is shut down and needs to be cleaned up.

Possible Approaches

Given the nature of the problem (differing perceptions and values, procedural and outcome inequities, and public distrust), what is the best way to proceed in communicating about the risks associated with incineration facilities? There are three possible strategies: ignore public perceptions and concerns (including equity), try to change them, or work with them (Hadden 1991). Ignoring public perceptions and merely "informing" the public about events—the decide-announce-defend (DAD) approach—is now considered undemocratic, undesirable, and ineffective. Trying to change public perceptions, attitudes, and concerns through education to bring them more into line with expert views of the issues is doomed to disappoint; the public has a rich, multidimensional view of risk that is extremely resistant to change (Kahneman and Tversky 1982; Hadden 1991), and risk controversies often result from deeper debates about the relationship between technology and society (Ruckelshaus 1985; Otway 1987; Wynne 1992). Working with public perceptions and concerns—accepting them as legitimate and involving the public in consultative and participatory processes—makes the most sense, but how should one proceed?

General Principles

The field of risk communication generally has evolved from early efforts to transmit technical information to elaborate efforts involving public participation and empowerment (Fischoff 1995; NRC 1996; The Presidential/Congressional Commission on Risk Assessment and Risk Management 1997). That progression mirrors Sharon Arnstein's ladder of citizen participation (Arnstein 1969). When the field of risk communication began, the emphasis was on the duty to inform members of the public about risky activities. With the recognition that public and expert perceptions of risks differ greatly, the emphasis shifted to trying to educate the public and thereby narrow the perception gap; during this stage of development, researchers and practitioners looked for ways to improve the content and design of risk messages (Covello et al. 1988). It soon became evident that public perceptions, attitudes, and beliefs are extremely resistant to change, in part because they reflect deeply rooted values, not merely a lack of knowledge and understanding. It also became apparent that the members of various nontechnical communities had knowledge and expertise relevant to the resolution of risk controversies (Wynne 1992). Thus, there was a shift to consultative, two-way risk communication. Many researchers and practitioners in the field are now beginning to see that risk communication often must move beyond notions of two-way risk communication to more elaborate models of citizen participation and empowerment (Chess et al. 1995). The National Research Council report *Understanding Risk: Informing Decisions in a Democratic Society* (1996) emphasizes the need for a recursive analytical and deliberative process that involves all interested and affected parties from the earliest stage in the risk decision-making process.

Fiorino (1990) identifies three reasons (normative, substantive, and instrumental) why a good participatory approach is appropriate:

- People have a democratic right to be involved in decisions that affect them.
- Affected and interested parties could have important information that is relevant to the analytical and decision-making process.
- Such involvement could in some cases be the only way to decrease conflict and reach a sustainable decision.

During the practical and theoretical development of the field, several authors have proposed general principles and designed excellent guidelines for risk communication (for example, Sandman 1985; Covello and Allen 1988; Covello et al. 1988; Hance et al. 1988, 1990). The guidelines often advocate involvement by interested and affected parties, although paradoxically they tend to have a unidirectional focus in that they assume that a company or government agency is "driving" the process (Otway and Wynne 1989). Risk communication is usually initiated by the proponent of a particular technology or by those responsible for its regulation and oversight, so it is not surprising that the guidelines place the burden of responsibility on those in control, who are admonished to do "better." The field of risk communication has come a long way in the last 15 years, and there is a need for continued research. For example, many valuable guidelines have been derived from common-sense observations, but these sometimes lack well-developed theoretical bases and strict empirical validation (Kasperson and Palmlund 1989; Morgan et al. 1992). Sandman (1985) developed one of the earliest sets of guidelines which remains pertinent (see Table 7-2). Similar guidelines were developed by Covello and Allen (1988). More-elaborate guidance can be found in Chess et al. (1988) and Hance et al. (1990). Much of the literature on risk communication overlaps with other social-science research on siting issues in general, and we see a similar development from the traditional DAD approach that was common through the 1970s to more-participatory processes (Kunreuther et al. 1993, p. 302). DAD failed because it alienat-

TABLE 7-2   Some Typical Guidelines for Risk Communication According to Sandman (1985)

- Acknowledge community power to stop siting process.
- Avoid implying that community opposition is irrational or selfish.
- Instead of asking for trust, help community to rely on its own resources.
- Adapt communication strategy to known dynamics of risk perception.
- Do not ignore issues other than health and safety.
- Make all planning provisional so that community consultation is required.
- Involve community in direct negotiations.
- Establish open information policy, but accept community need for independent information.
- Consider new communication methods.

TABLE 7-3   The Facility-Siting Credo

Procedural Steps:
- Institute a broad-based participatory process.
- Seek consensus.
- Work to develop trust.
- Seek acceptable sites through a volunteer process.
- Consider competitive siting processes.
- Set realistic timetables.
- Keep multiple options open at all times.

Desired Outcomes:
- Agreement that the status quo is unacceptable.
- Solution that best addresses the problem.
- Guarantee that stringent safety standards will be met.
- All negative aspects of the facility are fully addressed.
- Host community is better off.
- Contingent agreements are used.
- Geographic fairness.

Source: Adapted from Kunreuther et al. 1993.

ed many of the interested parties, especially the public, which ultimately recognized its ability to stymie the siting process (Morell and Magorian 1982; Kunreuther and Linnerooth 1983; O'Hare et al. 1983; Susskind and Cruikshank 1987; Portney 1991).

In an attempt to move beyond those limited approaches, a national facility-siting workshop in 1990 developed the Facility-Siting Credo, which is a summation of much that has been learned over the last two decades of research on facility siting (Table 7-3). The credo includes a set of guidelines intended to achieve a siting process that is fairer to all parties, but it is not intended as a how-to manual. The guidelines include procedural steps and desired outcomes. The following discussion of the credo is adapted from Kunreuther et al. (1993, p. 304). The credo was empirically tested in a survey of 104 persons involved in 29 waste-facility siting cases by Kunreuther et al. (1993). Of the 29 facilities, 24 incinerated some type of waste. Assuming that success is defined by the siting of a facility, they found that "the siting process is most likely to be successful when the community perceives the facility design to be appropriate and to satisfy the community's needs. Public participation also is seen to be an important process variable, particularly if it encourages a view that the facility best meets community needs" (Kunreuther et al. 1993). They concluded that "participatory siting procedures may stand a far better chance of success than the Decide Announce Defend (DAD) approach or legislated siting procedures." They also found that "a siting process that encourages public participation and contributes to the formation of a view that the facility best meets community needs, explains

siting outcomes, whether or not the interested parties trust the facility support-
ers." In other words, public participation could increase the likelihood of success in siting without necessarily increasing trust, although such participation
will usually result in increased trust.

Elements for Consideration in Public-Involvement Programs

Given the general public opposition to the siting of waste-incineration facilities and the countless possible effects of such a facility, relationships between
facility operators, developers, regulators, and members of the affected community are often strained at best. Fundamental differences in interests and values
need to be acknowledged, but the relationships can be improved by carefully
crafted methods of public participation. We briefly outline some of the major
elements to be considered in crafting this process. The reader should consult
other publications on risk decisions and public involvement for more-detailed
consideration of the pros and cons of various approaches (for example, NRC
1996).

Sustained Discussion

Sustained discussion requires mechanisms that provide the opportunity not
only for basic information exchange but also for all interested and affected parties to talk with each other in concrete terms over an extended period. The
process of public involvement should be open and substantive, not merely window dressing. Members of the community should be involved early and often,
and the DAD approach should be avoided. All possible approaches should be
provisional and up for discussion; otherwise, the community will soon recognize
that it has little useful input into the decision-making process. Once members of
the community and the developers of a proposed facility acknowledge that the
community has the power to stop the siting process, the discussion can proceed
on a more equal footing. Because public empowerment—access to power and
control over events—is often one of the underlying issues in incineration controversies, such discussion could evolve into a negotiating process, and the mechanisms adopted need to be designed accordingly.

A community advisory committee (CAC) is often the mechanism used for
sustained discussions. CACs typically have 10-20 members, with a mixture of
citizens from the community and local officials and with facility developers,
operators, and regulators present in an *ex officio* capacity (Lynn and Busenberg
1995). Such groups may meet over several months or years and may discuss
various facility-related topics. For example, they may review and provide reactions to proposed facility plans at various stages. Task forces (comprising members of the community and technical experts) could be used to address particular
issues (such as alternatives for ash disposal and health effects of particular emis-

sions). Commitment of sponsors to the CAC may be a key factor in determining the effectiveness of the CAC.

Whatever the small-group mechanism used, the following questions (and many others) will have to be addressed explicitly for all members at the outset when the group is being established: Who is to participate, and how will they be chosen? Will members serve for set terms? If so, how will their replacements be chosen? Will members represent their own views, or will they be responsible to particular constituencies? How will decisions be made—by vote or consensus? Who sets the agendas? What are the goals of the group, and what issues will it tackle? How much (if any) decision-making power will the group have, or will it be merely advisory? What resources will be available to the group and how much discretion does the group have in the use of these resources?

Assessment of Needs and Concerns

One of the likely first tasks of any advisory group will be to assess the needs and concerns of the community. There is enormous diversity of opinion among members of the public regarding human-health and environmental-risk controversies; the public is not monolithic. Typically, members of the public are concerned about much more than direct health and safety impacts. As noted above, members of the public may also be concerned about quality of life (such as increased noise and traffic), socioeconomic effects (such as employment opportunities and property values), and other less tangible issues, such as stigma and equity. Because these effects are more difficult to characterize and measure than health effects, they are often omitted from traditional risk assessments; but their omission could be perceived by members of the public as an effort to ignore or belittle their concerns and is likely to exacerbate mistrust. In addition to concerns about the broad range of effects, members of the public are often concerned about oversight and control.

As mentioned previously, members of the public are likely to ask why an incinerator is needed and why wastes can't be reduced by recycling, source reduction, and so on. Even if need can be overwhelmingly demonstrated, the targeted community will likely ask, Why us? The developer, regulators, and advisory committee will have to address those questions, not as irrational or selfish responses to a proposed incinerator, but as valid "coherent positions that deserve respectful responses" (Sandman 1985, p. 446). Demonstrating need is the essential element of the siting process (Kunreuther et al. 1993, p. 303).

In developing this information, the advisory group could decide to form separate task forces from among its members or to recruiting additional expertise. Because autonomy is an important part of empowerment, the participatory process may encourage and facilitate the use of indigenous and independent expertise. It is reasonable for experts on behalf of the operators and regulators to expect to be trusted, but it is also reasonable for local citizens to withhold trust and insist on relying on their own judgment and that of independent experts. Also, although it has been amply demonstrated that citizens are capable of mastering technical detail when sufficiently motivated (Lynn 1987), this may not be the most-efficient way of gathering information. Hiring independent consultants that members of the community trust may be best (Hadden 1991). Ways to make funds available for that purpose may need to be explored.

Ideally, information from facility developers and regulators should be timely, substantive, and honest. They should be forthright about who will make decisions and about what opportunities community members will have to influence the decisions. They should also make clear how much is known about issues relevant to the facility; for example, if dioxins are released from the facility, the debate within scientific and regulatory communities about the carcinogenic effects of dioxins should be stated.

Information from the members of the public and their representatives should also be timely, substantive, and honest. This information should, if possible, be provided to facility developers, operators, and regulators well in advance of key decisions so that they can take it into account. Information provided by members of the public should straightforwardly articulate the values of those speaking or writing. Information about the community and about facility effects that have been noticed (such as truck traffic, smoke, and lower property values) should be as well documented as possible, and the extent to which this information is current and reliable should be explained.

In sum, the goal is to develop an open process that maximizes information flow. Neither community nor facility proponents should hide any information, because "failure to disclose a relevant fact can poison the entire process once the information has wormed its way out—as it invariably does. . . . Any information that would be embarrassing if disclosed later, should be disclosed now" (Sandman 1985). Failure to provide information will exacerbate distrust on both sides of the debate.

Negotiation

Sustained dialogue provides the means for developers, operators, and regulators to listen to and exchange views with community members through detailed, fairly informal discussion. Dialogue may or may not entail negotiation that leads to formal contractual agreements about specific facility-related issues. If a small-group process is intended to lead to negotiations or if there is a possibility that it will, the ground rules for such negotiations need to be laid out explicitly at the beginning of the process so that all participants and members of the community observing the process are clear about their respective roles and responsibilities and about overall expectations.

Apart from the up-or-down question of whether a new facility will be sited or an existing facility expanded, a number of issues are appropriate for negotiation with the affected community. These may include caps on the facility size and restrictions on waste types and imports; provisions to ensure that the facility's layout and appearance are as compatible as possible with the surrounding neighborhood; mitigation of potential nuisances, hazards, and burdens on local services through such measures as road improvements; property-value guarantees provided by the facility owner to property owners within a specified distance from the facility; guarantees from the facility owner about acceptable uses of the land when the facility closes; guarantees from the facility owner about local hiring and purchasing policies; and special compensations, both monetary (such as percentage of profits or a flat fee) and nonmonetary (such as parks and school improvements and free waste-disposal arrangements).

Two particularly important items for negotiation are issues of oversight and control and compensation. Psychometric studies of risk have found controllability to be one of the most-important attributes that shape risk perception (Slovic 1987). In siting and operating incineration facilities, measures of public oversight and control can be incorporated in various ways from the initial planning and design of a facility through its operation and eventual decommissioning. These may include citizen oversight boards and monitoring teams with the authority to close the facility. Results of several surveys indicate that the extent of control is a major influence on public attitudes and perceptions. For example, Elliott (1984b) found that a proposal to allow public officials and citizens to conduct safety inspections had the greatest effect on public attitudes toward a hazardous-waste facility. Kunreuther et al. (1990) found that respondents in Nevada would be more likely to accept a high-level nuclear-waste facility if a local panel had the authority to shut it down. Lynn (1987) found citizen review of the design and operation of a facility a critical element in the decision to accept its siting. "Citizens have even accepted incinerators when a group of citizens and local officials were designated to be present during any burns they choose in order to insure that optimal operating temperatures are achieved and maintained" (Hadden 1991). Similarly, public concerns about an incinerator in Japan declined when large neon signs were used to indicate the operating conditions of the facility (Hershkowitz and Salerni 1987). Providing that kind of information enhances public perceptions of control and reduces anxiety.

Compensation has been a common response to the problem of finding sites for noxious and hazardous facilities. Some compensation programs have achieved their goals of getting a facility sited, but other programs have run into enormous difficulties and extreme public opposition. Much has been written about the design and use of compensation programs, and we present here only the highlights.

There is a public distaste for trading health for money, and any compensation scheme needs to avoid the appearance of bribing a community. Because it is unseemly to compensate for risk itself, the goal should be to reduce risks first and compensate only for the non-health-related effects that cannot be reduced or avoided. Compensation needs to be closely targeted to possible economic losses (such as loss in property values), and the form and content of any compensation package must be carefully negotiated with the affected parties. As noted above, siting an incineration facility necessarily creates inequities, and compensation may be the only way to try to redress them. There are two types of equity. One is procedural (who is involved and how) and the other is outcome (the distribution of harms and benefits). A compensation package needs to address both forms. Thus, negotiating the compensation package in an open, inclusive, participatory process that attempts to minimize any harms (for instance, with property-value guarantees) and maximize benefits (for instance, via preferential employment and purchasing) is more likely to be successful.

It should be noted that compensation is not a cure-all. In the absence of a good participatory decision-making process, no amount of compensation can ensure public acceptance. A Wyoming survey found respondents less inclined to consider accepting a facility deemed risky if they were offered compensation than if they were given good information and the opportunity to participate (Davis 1986).

Resolution of issues like these can help to reduce concerns about adverse health, environmental, and socioeconomic effects of a facility and can increase the benefits that the facility provides to the community. Resolution of the issues will not, however, make a suspect facility palatable; they should be tackled only after such issues as the suitability of the facility's site and its technology and the adequacy of the operator's record have been resolved.

Procedural questions become all the more crucial in sustained communication, including negotiation. What are the boundaries of the "community"? On which issues should the community have the final say? Who should speak for the community, future community residents, and the environment itself? How should division in community opinion be resolved? Should the interests and opinions of those who are most immediately and substantially affected carry the greatest weight even if they are outnumbered by others? Those are daunting questions, not easily resolved, but they should be tackled to ensure that interactions among the facility developer, the operator, regulators, the local government, and the affected community are substantive.

CONCLUSIONS

Assessing Socioeconomic Impacts

During and after the siting and building of a waste incineration facility, it may have various effects on members of the surrounding area in addition to physical health effects. The effects may be favorable or adverse, and they may be economic (such as job creation or decrease in property values), psychological (such as stress or stigma), or social (such as community factionalization or unity). They can affect individuals, groups, or the entire population in the surrounding area.

There is little reliable information on the socioeconomic impacts of waste-incineration facilities on their host areas. This chapter has identified issues that appear to merit attention, but these issues will not necessarily arise in the case of every incineration facility. Much more empirical research is needed, including longitudinal research on effects during the siting of the facility, as well as during its operation.

When research is conducted on the socioeconomic, health, and environmental impacts of a facility, the boundaries of the potentially affected area should not be predetermined; instead, they should be defined as a function of where, when, and to what extent various impacts may occur. That approach permits a more-accurate and more-comprehensive analysis of the nature of the impacts. It also permits a better understanding of problems that may arise in connection with local interactions and decision-making concerning the facility.

Understanding Citizen Concerns

Even though a large body of research is not yet in hand on the possible health, environmental, and socioeconomic impacts of different types of waste-incineration facilities in different settings, it is clear that citizen concerns about these facilities do exist. The concerns need to be heard and understood, if only because escalated conflicts over waste incinerators may result and increase the time and expense of developing facilities that are potentially beneficial to society. Opposition to facilities also can indicate that important concerns are being given short shrift. Differences between expert and lay perceptions are not due merely to differences in information and understanding; they are also due to differences in values—particularly values concerning trust, need, and equity.

One of the first questions often asked by members of the prospective host area is likely to be, "Is this facility really needed?" If facility proponents cannot convincingly demonstrate a pressing need for a new or expanded waste facility, people who are skeptical of or opposed to the facility will be disinclined to negotiate on other issues regarding the facility.

Siting a facility like a waste incinerator presents an inherent and inescapable need to address equity. Whatever site is chosen, the associated health risks, if any, and other effects are necessarily borne by relatively small groups, whereas the benefits of waste treatment or disposal (for example, jobs and substantial tax revenues) can accrue to a larger population. More and smaller local facilities may alleviate, but cannot eliminate, this dilemma and may create other problems, such as an increase in total emissions. Equity issues are exacerbated when a facility is placed in a low-income or otherwise disadvantaged community, where it raises broader concerns about disproportionate health, environmental, and socioeconomic burdens already being borne.

Public Involvement

A good risk-communication program is not a panacea, but poor risk communication will nearly always make matters worse. Good risk communication is a continuing process at existing facilities or future facilities after siting.

People's perceptions are extraordinarily resistant to change, in part because they reflect underlying values. Efforts that ignore or try to change these perceptions radically are likely to fail. Given fundamental value differences, concern over procedural and outcome inequities, and distrust, there is growing consensus among academics and practitioners that effective risk communication should accept as legitimate the perceptions and concerns of various members of the public and involve them in consultative, participatory processes. Not only do members of the public have a democratic right and responsibility to be involved in the assessment and management of hazards in their communities, but such involvement may result in improved assessments and management strategies.

Developing effective participatory programs is extraordinarily difficult, but some general principles are beginning to emerge. The process of public involvement should be open, inclusive, and substantive, and members of the affected area should be involved early and often. Two of the first tasks in any process should be to solicit the broad array of people's concerns and to address the question of need for the proposed facility. Other major concerns are likely to include issues of safety, compensation, and local oversight and control. Risk communication can be considered successful, according to one measure, to the extent that it raises the level of understanding for those involved and satisfies them that they are adequately informed.

RECOMMENDATIONS

- The social, psychological, environmental, and economic effects of proposed and existing waste incineration facilities should be assessed, and mitigation of or compensation for such effects should be considered where appropriate.
- To enable large-scale empirical research on the socioeconomic impacts of waste-incineration facilities on their host areas, detailed socioeconomic data should be gathered routinely before and during the operation of such facilities.
- The boundaries of an area potentially affected by a waste incinerator should not be defined at the outset by a particular community's political boundaries or jurisdiction. Instead, the assessment area should be based on the geographic extent over which various effects could reasonably occur.
- Citizens, as well as all other parties involved, should ensure that their communications are timely, substantive, and honest. In addition, their

written or spoken statements should clearly indicate the extent to which they represent the views of others.

- Risk-communication efforts concerning waste-incineration facilities should be designed and conducted with an understanding that citizens and experts may have different values, not simply different levels of knowledge and understanding, and that the phenomenon of social distrust is broad, intense, and likely to continue.

- Continued communication concerning a facility—during its development and operation, as well as in the proposal stage—may best be conducted with a citizens advisory group. However, small-group exchanges should supplement, not supplant, participatory opportunities for the general public.

- Proponents of an incineration facility should assume, in their interactions with local communities, that they (the proponents) should make the case for the new or expanded facility, especially if a waste combustor is not used solely within a manufacturing facility to incinerate waste on site.

- If a new or expanded facility is contemplated, local citizens might consider conducting their own assessments of the proposed facility and its effects through various approaches, including, for example, hiring independent consultants that members of the community trust, seeking technical-assistance grants from the government, or finding technical advisors who are acceptable to both sides.

- Particular attention should be paid to equity issues when a facility is to be placed in a community that is already experiencing disproportionate health, environmental, and socioeconomic burdens.

- Participatory programs should be evaluated by the participants and external researchers to identify elements that can be used with benefit elsewhere and elements that should be avoided.

8

Uncertainty and Variability

Estimating potential human exposures to environmental concentrations resulting from emissions of waste-incineration facilities and estimating the risk of possible health effects of the exposures is a complex task. Such tasks involve the use of computational models coupled with large amounts of data to predict institutional performance, individual human behavior, engineered-system performance, contaminant transport in the environment, human contact with contaminants, and dose-response relationships. Comprehensive assessments also involve treatment of the variability and uncertainty associated with those data.[1] This chapter addresses how, in the context of waste incineration, uncertainty and variability are defined, characterized, and treated in the risk-assessment and risk-communication process. This process includes consideration of hazard identification, dose-response characterization, emission-source characterization, exposure assessment, risk characterization, and risk communication. Additional information on these issues can be found in NRC (1983, 1993, 1994, 1996), Morgan and Henrion (1990), Cullen (1995), and EPA (1999).

---

[1] Variability refers to the individual-to-individual differences in quantities associated with predicted risk. For example, actual human exposures vary according to individual differences in location, breathing rates, food consumption, activity patterns, and so forth. Uncertainty refers to the lack of precise knowledge as to what the truth is, whether qualitative or quantitative. There can be uncertainty in the magnitude of an individual quantity that can be measured (e.g., the imprecision of a stack emission rate measurement). Other uncertainties pertain to gaps in the scientific theory that is required to make predictions on the basis of causal inferences. For example, the appropriate model to use for predicting the relationship between the dose of a toxic substance and the health response.

The committee identified aspects of uncertainty and variability likely to have important scientific and policy implications for the potential health effects attributable to waste incineration. One overarching issue is how uncertainty and variability can influence the utility of estimates of health effects of waste incineration or of alternative technologies for waste management. As uncertainty and variability become larger, it becomes more difficult for interested or affected parties to decide how to interpret results and assign relevance to the magnitude of estimated risk. If the range is too large, different people might base their interpretation of the results on their prior opinion of waste incineration. That is, those who favor use of incineration technology might tend to focus on results in the middle range (for example, the median or mean of either variability or uncertainty distributions) of postulated effects. Those who oppose the technology might tend to focus on any results that suggest harmful effects (for example, the upper 5-10% of the possible range of outcomes). When the uncertainties or variabilities are large, there can be a large difference between those two parts of the range of possible outcomes.

To give some perspective on how uncertain and variable information can influence the characterization of health impacts, Figure 8-1 provides a schematic of the major components that must be characterized to assess possible health effects. Listed next to each component are the major types of information or models needed to map the output from one logical stage of this system into the next. Two important issues are evident. First, there are large variations in the precision and accuracy[2] of the information needed to characterize the sequence of steps as listed on the left of the figure. Second, this process is open, so that each component is not solely influenced by the previous one. For example, the concentration of dioxin congeners in the atmosphere near an incinerator is not linked solely to emissions from the incinerator, but may also be attributable to other sources in and out of the region. And a health effect might be connected to the facility, but understanding the etiology of any disease requires consideration of a variety of potential factors. As has been pointed out by Oreskes et al. (1994) such open-ended systems models—which are common in earth sciences, economics, engineering, and policy-making—cannot be fully verified or validated, because the operative processes are always incomplete. Nevertheless, such models can be confirmed and can be used to put bounds on the likely range of outcomes; in this sense, they offer something of value to the policy-making process.

The following five factors determine the reliability of a health-risk assessment: specification of the problem (scenario development), formulation of the

---

[2] Precision refers to the agreement among individual measurements of the same property of the sample. Accuracy refers to the agreement of a measurement (or an average of measurements of the same property) with an accepted reference or true value.

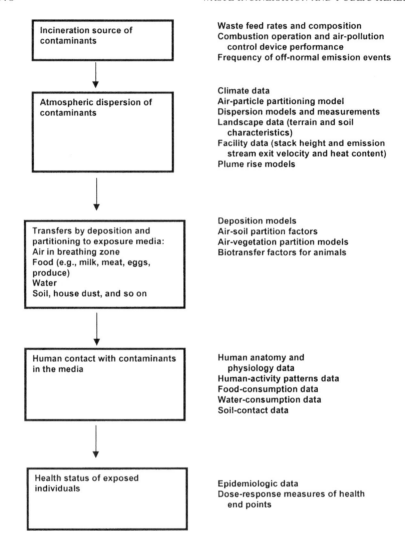

FIGURE 8-1   Schematic of major components included in assessment of possible health effects of waste incineration. Other important considerations such as social and economic impacts (see Chapter 7) are not included in this figure.

conceptual model (the influence diagram), formulation of the computational model, estimation of input values, and calculation and interpretation of results, including uncertainties. Uncertainty analysis should be an iterative process, moving from the identification of generic uncertainties to more refined analyses for chemical-specific or facility-specific uncertainties (NRC 1994). The use of uncertainty analysis in health risk assessment for exposure to chemical contaminants became widespread in the 1980s (Bogen and Spear 1987).

Uncertainty analysis must confront the distinction between variability and true uncertainty characterizing possible outcomes. Variability occurs in such factors as location (affecting such local properties as rainfall, soil characteristics, weather patterns), and human characteristics. Those factors are inherently variable (from person to person) and cannot be represented by single values. In contrast, true uncertainty refers to a factor that is not known beyond a certain degree of precision because of measurement or estimation error.

CONFRONTING VARIABILITY AND UNCERTAINTY

Risk-based management strategies often operate on the premise that, with sufficient funding, science and technology will ultimately provide an obvious and cost-effective solution to the problems of protecting human health and the environment. However, there are many sources of uncertainty and variability in assessing possible impacts on human-health (and ecological) risk assessment, and many of these uncertainties and variabilities are not reducible in a practical sense. Effective policies are possible under conditions of limited knowledge, but they must take the uncertainty and variability into account.

For such technologies as waste incineration, it is rare to *measure* the magnitude of human exposure and the resulting health risks. Such aspects are often estimated by models that vary in complexity. Regardless of the model complexity, there are two approaches by which one can assess how model predictions are influenced by model reliability and data precision: uncertainty analysis and sensitivity analysis. To address sensitivity and uncertainty, one can think of a model as producing an output, that is a function of several inputs.

For example, the output could be the dioxin concentration in human tissue, and the inputs could refer to dioxin emission source strengths, wind speed and direction, exposure factors, and uptake rates. Uncertainty analysis involves the determination of the variation or imprecision in the output based on the collective variation of the model inputs, whereas sensitivity analysis involves the determination of the size of the changes in model output as a result of changes of known size in individual model inputs.

Data and Modeling Adequacy

Uncertainty in model predictions arises from a number of sources, including specification of the problem, formulation of the conceptual model, formulation of the computational model, estimation of input values, and interpretation of the results. Of those, only uncertainties due to estimation of input values can be quantified in a straightforward manner using the usual methods of uncertainty propagation. Some of the uncertainties that arise from misspecification of the problem, model-formulation errors, and interpretation, can be assessed with more-complex processes, such as decision trees and event trees based on expert opinions. An additional uncertainty that can plague modeling efforts is straight-forward quality assurance and quality control (QA/QC) problems in model implementation (so the results calculated are not the results of the intended computations)—these are in principle easy to correct, but may dominate other uncertainties.

Influence of Uncertainty on Perception of Risk

The decision to spend money to identify, estimate, and manage risk carries with it an implicit valuation of the risk being controlled. Because of the uncertainty inherent in risk characterization and risk management, it is important to consider how individuals and societies value uncertainty in knowledge of adverse consequences. One expects such valuations to be expressed in terms of relative preferences, economic preferences, or ethical constraints.

In managing health risks, the results of a risk characterization are integrated with social, economic, and political considerations to provide input to the risk-management process. A variety of techniques have been proposed and used to apply the values held by different stakeholders to the evaluation of risks. Some of the commonly used techniques of risk valuation are the elicitation of individual and societal preferences, decision analysis, and application of theories of science policy, social-welfare economics, and ethics. Use of those valuation strategies can yield an important input to the risk-management decision.

Exposure and dose-response estimates are important for understanding risks, but they fail to provide all of the tools used by individuals and societies to manage risk. As discussed in Chapter 7, it is also important to understand the process by which people perceive health risks and then decide how acceptable these risks are. Risk management decisions involving costs and benefits face the economic problem of individual and societal valuation of life—a problem which has been considered by Raiffa et al. (1977) among many others. For example, although it is difficult to set a value on a statistical year of life lost, some data have been compiled so as to make consideration of years of life lost more feasible (Murray and Lopez 1996).

Because risk assessments and risk management decisions must be made in the absence of complete information, they implicitly involve judgments made with uncertainty. Psychologists have observed that when people are asked to make judgments involving uncertainty, they appear to adopt a number of heuristics, or rules of thumb, for decision-making. In particular, it appears that belief about the likelihood or severity of a given event is related to the ease with which previous occurrences of the event or a similar event can be recalled. Kahneman et al. (1982) proposed a number of interesting rules of thumb regarding the acceptability or trading of risks, among them that people prefer to reduce one risk to zero instead of lowering multiple risks, that gains and losses tend not to be valued the same, that people usually are not willing to trade gains against losses, and that people are usually not willing to make risk-risk tradeoffs.

The limitations of risk-estimation methods make it clear that risk managers should be aware of the uncertainty in risk estimates and include this awareness in their decisions and in their communication of risk to the public. It should be recognized that as uncertainties become large and the range of possible outcomes (that is exposure or risk) becomes difficult to characterize precisely, the beliefs of the interested or affected parties regarding risk acceptability are likely to be as important, or even more important, than results of a risk assessment in a decision-making process (Kahneman et al. 1982; NRC 1982, 1989a).

Performance Characteristics and Contaminant-Source Terms

Characterizations of contaminant source terms for a waste-incineration facility are often derived from emission test-data of the facility itself or by extrapolating from similar ones, and from assumptions regarding the waste feed and incineration system performance. System performance includes both the operation of the combustor and the operation of the pollution-control equipment. Similar facilities may vary in emissions as a result of emission limitation requirements, control technologies, and operating practices, as well as more obvious differences. For example, older permits can be more lenient than those written more recently. In addition, some emission standards are different for different sizes of incinerators (See Chapter 6).

To the extent that there is uncertainty regarding the performance of an incineration facility, there will be lack of precision in information on the magnitude and composition of emissions (for example, see Frey et al. 1999). Moreover, emissions from a single facility will vary with time, for example due to changes in operating conditions. Thus, assessment of health risk for waste-incineration facilities should include consideration of such variations, including emissions resulting from off-normal activities, in addition to routine stack and fugitive emissions. Because they involve unusual events on which there is little advance information, assessing the frequency of occurrence and progression of off-normal emissions is likely to be a highly uncertain process.

Hazard Identification

Hazard identification involves the determination that a health hazard is or might be associated with a chemical exposure or physical factor. It is a *sine qua non* in the risk-assessment process and can be based on simple screening methods, short- and long-term assays of living cells or multicellular organisms, or preliminary human health surveys. Such approaches generally cannot yield a yes or no answer, but a probability that approaches yes or no only as the probability approaches 1 or zero. For example, one assay used to determine whether a chemical might be a human carcinogen is the Ames bacterial-revertant assay. A principal uncertainty associated with this assay is whether a positive response (or negative response) in the assay means that the chemical is capable (incapable) of producing cancer in humans. Other examples include epidemiologic studies of factors associated with the onset of respiratory or reproductive effects.

Table 8-1 summarizes the utility of different types of information provided by assays used in hazard identification. Such assays are followed by qualitative extrapolations of various types from the assay to the human-exposure situations associated with incinerators. The extrapolations are smallest for the last assay listed in Table 8-1, and largest for the first. Even using the results of epidemiologic surveys generally requires qualitative (as well as quantitative) extrapolations, because various aspects of the exposure situations are usually different. From in vivo assays in other organisms, an additional interspecies extrapolation is required that may easily result in misclassifications of agents as hazardous or not hazardous, and the qualitative jump is even larger for the first two assays. It is important to also keep in mind that different approaches could be inconsistent (e.g., positive animal studies and negative epidemiologic studies).

Environmental Transport and Human Exposure

Some form of exposure assessment is required in a number of health-related assessments, including risk assessments, status and trends analyses, and epidemiologic studies. In assessing exposures to environmental concentrations resulting from the emission of contaminants from incineration facilities, a multimedia environmental approach is needed. In such an approach, all contaminant releases to the environment are traced through all environmental media—air, water, soil, sediment, vegetation, food, etc.—taking into account any changes in the form of the contaminants.

Estimating the effects of human exposures resulting from incineration emissions generally require information on the following:

- The quantities of contaminants released to air or the concentrations measured or estimated in air, soil, plants, and water in the vicinity of the source.
- The transfer rates of contaminants onto (and out of) environmental media to which humans may be exposed. Such transfer rates must take account of contaminant degradation, partitioning, bioconcentration, dilution, and other physical, chemical, and biologic processes.
- The frequency, magnitude, and duration of human contact with the contaminated exposure medium.

TABLE 8-1 Comparison of Uncertainties in and Reliability of Various Strategies for Hazard Identification

Strategy	Comments
Predictions based on description of the chemical substituents, molecular weight, energy and molecular orbital calculations, structure-activity relationships or other factors that might predict toxicological characteristics	These prediction methods are generally inexpensive and constitute a rapid screen. They can be applied to a large number of chemical agents. However, assessing the likely effects of mixtures is difficult. Validation is important for the overall reliability of the hazard measure. Predictions should be viewed as hypotheses that require more-detailed evaluation.
In vitro assays, short-term assays, effects on cells in culture, analysis of effects on specific cellular functions	Less expensive and more rapid than in vivo assays. These assays may be used to characterize a site and mechanism of action. They provide useful adjuncts to in vivo assays. The measure of effect is often sensitive to both the material used in the assay and the protocol used to assess damage.
In vivo assays, toxic response to the agent, development of disease in exposed animals or humans	These are generally more-expensive and time-consuming assays. They are thought to represent the most complete and biologically integrated assays for characterizing toxicity of a chemical. Despite their integrative nature, there can be considerable disagreement on the validity of animal data for use in assessing human hazards.
Epidemiologic studies of health effects in response to chemical exposures, including cancer and reproductive, developmental, and immune disease	In all epidemiologic studies, the focus of attention is on health effects in human populations. Uncertainties in these studies can refer to characterization of the exposure, the etiologic agent, confounding factors, and understanding causation in heterogeneous human populations.

- The dose-response relationships for the particular contaminants. Descriptions of such relationships must include the effect of time and intensity of exposure.

The uncertainty and variability typical of these aspects have been assessed in a number of papers. McKone and Ryan (1989) considered the overall uncertainty in estimating the link between atmospheric concentrations and food concentrations for 2,3,7,8-tetrachlorodibenzo-p-dioxins (TCDD) and for metals. McKone (1993) compiled a list of the cross-media transfer coefficients typically incorporated in a multimedia exposure assessment and estimated the uncertainty (measured as a coefficient of variation[3]) in these estimates. Table 8-2 lists McKone's (1993) estimates of uncertainties in correlation models for partition coefficients of organic chemicals between soil and plants, between air and plants, and between animal intake and animal food products. These early estimates of uncertainty include components that might be related to variability (for example, between soils or between locations) in a more modern assessment.

Cullen (1995) has considered the degree to which the uncertainty of the results of a probabilistic risk assessment for waste incinerators is contingent on certain model assumptions. She found that the risk-assessment results are very sensitive to the selection of models for representing the fate and transport of incinerator contaminants, especially to their assumptions about gas-particle partitioning in the stack and downwind atmosphere.

For lipophilic contaminants—such as dioxins, furans, and polychlorinated biphenyls—and for such metals as lead and mercury, exposures through food have been demonstrated to be major contributors to total dose in non-occupationally exposed populations (Travis and Hester 1991). Overall uncertainties in estimating potential doses through food chains are much larger than uncertainties associated with direct exposure pathways (McKone and Ryan 1989; McKone and Daniels 1991). Intake of substances in food varies widely among individuals, among age groups, among regions of the country, and among seasons of the year (NRC 1993). It is possible that exposures via one environmental medium dominates the health concerns related to an emission source. Therefore, depending on the magnitude of error considered acceptable for a particular investigation, a focus on the medium of greatest exposure might be appropriate.

Dose-Response Characterization

Dose-response characterization is the process of defining the site of action in the body, the mechanism of action, and the dose-effect relationship, for a material causing adverse effects. In this process, a series of models usually are relied on. The models may be of various types, including statistical models and biologically based models (e.g., physiologically based pharmacokinetic or biologically based dose-response models). Each has limitations in representing the actual toxic or human-hazard effects and as a result each has associated various degrees of uncertainty (see Table 8-3).

Model uncertainty (that is, being unsure about the true nature of the relationship between dose and response) is likely to be highly important in dose-response characterization. Despite the admitted large uncertainty, simple dose-response models are the most commonly used for predicting human health effects

---

[3] The coefficient of variation is the ratio of the standard deviation to the estimated arithmetic mean value of a parameter.

TABLE 8-2   Summary of Methods for Estimating Intermedia Transfer Factors and Coefficient of Variation Associated with Estimation Errors

Factor Description and Symbol	Estimation Formula or Value	Coefficient of Variation[a]	Units	Key References
Octanol-water partition coefficient, K_{OW}	Chemical specific	n/a	kg (water)/kg (octanol)	Lyman et al. 1982; Verschueren 1983; Howard 1990a,b, 1991
Organic-carbon partition coefficient, K_{OC}	$0.41\ K_{ow}$	1	kg(water)/kg (carbon)	Karickhoff (1981)
Soil/soil-water partition coefficient, K_D	$f_{oc} \times K_{oc}$	1	kg(water)/kg (soil solids)	Karickhoff (1981)
Plant-soil partition coefficient for surface soil due to rainsplash, rain K_{ps}	0.0034	1	kg(soil)/kg (plant FM)[b]	Dreicer et al. (1984)
Plant-soil partition coefficient from root-zone soil to above-ground plant parts, K_{ps}	$7.7\ K^{-0.58}_{ow}$	4	kg(soil)/kg (plant FM)	Travis and Arms (1988)
Plant-soil partition coefficient from root-zone soil to roots (used for protected produce), K_{ps} (roots)	$270\ K^{-0.58}_{ow}$	4	kg(soil)/kg (plant FM)	Topp et al. (1986)
Plant-air partition coefficient for gas-phase contaminant, K^{gs}_{ap} Riederer (1990)	$[0.5 + (0.4 + 0.01 \times K_{ow}) \times (RT/H)] \times 10^{-3}$	14	m^3(air)/kg (plant FM)	Bacci et al. (1990)
Plant-air partition and coefficient for particle-bound contaminant, K^{pt}_{ap}	3300	1.5	m^3(air)/kg (plant FM)	McKone and Ryan (1989)
Biotransfer factor for meat concentration versus intake for beef cattle, B_t	$2.5 \times 10^{-8} K_{ow}$	11	day/kg(meat)	Travis and Arms (1988)
Biotransfer factor for milk concentration versus intake for dairy cattle, B_k	$7.9 \times 10^{-9} K_{ow}$	6	day/kg(milk)	Travis and Arms (1988)
Biotransfer factor for egg concentration versus intake for chickens, B_e	$1.6 \times 10^{-6} K_{ow}$	14	day/kg(eggs)	McKone (1993)

[a]A high coefficient of variation implies that the intermedia transfer factor is not well understood.
[b]FM refers to fresh mass.

Source: McKone 1993. Copyright 1993, Overseas Publishers Association N.V.; permission received from Gordon and Breach Publishers.

TABLE 8-3 Aspects That Contribute to Uncertainty in Models Used for
Dose-Response Characterization

Model Description	Aspects that Contribute to Uncertainty
Statistical dose-effect models	
Dose-response characterizations are statistical descriptions of the relationship between dose and disease. The models contain little or no biologic insight.	Typically, large doses are given to experimental animals for statistical and predictive reasons. Therefore, it is necessary to scale the dose-response relationship across species by weight or surface-area adjustments.
If the dose-effect models are derived from epidemiologic data, lack of biologic insight is of little concern. If they are derived from animal data, relevance might be a concern.	Toxicokinetic and toxicodynamic characteristics determine dose-response relationships. Does the observed dose-response relationship define the severity of disease, proportion of the population with the disease, or both?
	How does one correct for species differences in the dose-response relationship, surface area, or body weight?
Biologic models	
Biologic models provide qualitative or quantitative descriptions of the site and mechanism of action. They attempt to define how and where an agent acts to produce disease, how sites and mechanisms differ across species, and what effect the differences have in the prediction of human disease.	Is the parent compound or a metabolite responsible for the disease?
	What are the differences across species in toxicokinetics and toxicodynamics?
	Are there species differences in the physiologic factors that influence the response of different species to the agent (chemical or metabolite) associated with disease?
	What physiologic factors influence the chemical-disease relationship?
	What are the quantitative and qualitative descriptors of uncertainty and variability in dose-response characterization?
	Is the biologic model reasonably correct?

in dose or exposure regions (usually low doses of exposures) that are not exper-
imentally accessible. Such models have often been used in establishing policy.
As use of risk assessment grows, the need for sophistication of the models and
the accuracy or completeness of their representation of biologic processes is also
likely to grow. For example, pharmacokinetic (PK) models provide tools to
estimate the distribution of a chemical throughout the body and thus assists in

evaluating the amount delivered to the target site. It is believed that such models can improve the procedures used to extrapolate to low doses or between species.

UNCERTAINTY AND SENSITIVITY ANALYSES

A Tiered Approach to Uncertainty and Variability Analyses

An important, and often ignored, step in the risk-characterization process is the characterization of variability and uncertainty. This process has often been passed over in practice. To adequately confront variability and uncertainty in risk assessments, it is necessary to incorporate the treatment of both from the very beginning. One approach is to take a tiered approach to such analyses. Three tiers are involved, and may be applied separately to variability and uncertainty.

First, the variation, and if possible the co-variation of all input values should be clearly stated, with clear separation between variation due to variability and that due to uncertainty. One approach might be to evaluate variances and covariances on suitable scales of measurement, or the components of such variances and covariances attributable to variability and uncertainty. More generally, the most-complete specifications available of joint variability and uncertainty distributions are needed. A clear assignment of the variation in input values to variability or uncertainty requires a careful and clear summary and justification of the assumptions used for each aspect of the modeling in which these input variables are used.

Second, a sensitivity analysis may be used to assess how model predictions are affected by variation in input values. The goal of the sensitivity analysis is to rank input parameters on the basis of their contributions to variance in the output. The ranking should take account separately of the input-output sensitivity of the models, and the variability and uncertainty variances of the inputs (so that a highly variable or highly uncertain input might have high rank even if the input-output sensitivity is not high).

Third, some form of variance-propagation methods (such as Monte-Carlo methods) should be used to map how the overall variability and uncertainty of risk estimates is tied to the variability and uncertainty associated with the models, inputs, and exposure scenarios. The sensitivity analysis may be used to limit the number of inputs for which full variability or uncertainty information is required, for example by retaining or demanding full distributional information only for the high rank inputs. Distributional information for lower-ranked inputs might be represented by approximations (e.g., using just constant estimates, or first and second moment estimates). Modern methods have now eliminated computer-program constraints and available computational power as limitations on retaining distributional information—if the information is available, it can now readily be used. The major limitation is now on gathering distributional information about the inputs, and a sensitivity analysis allows finding the inputs on which most resources should be expended.

Methods for Representing and Propagating Parameter Variance

Describing the uncertainty, variability, or both in a risk estimate obtained as an output from a model involves quantification of various statistics of that output. Such statistics might include its range, its arithmetic mean, its arithmetic or geometric standard deviation, and various percentiles, like the 5% and 95% percentile. Such information is encoded in, and may be presented by using the probability-density function or the cumulative-distribution function. Such functions of risk can be obtained only with usable estimates of the probability distributions of the input variables. What is usable depends on factors including the amount and quality of information available, the understanding of the appropriate biologic models, careful consideration of individual variations in susceptibility, population-wide confounders, and, of course, the needs of a particular analysis.

Five main steps in an input uncertainty analysis (IAEA 1989) are described below, and very similar comments apply to analysis of variability:

- Identify the inputs that could contribute substantially to the uncertainty in the predictions of outcome by a model. Care should be exercised not to discard potentially important uncertainties without good cause.
- Construct for each input a probability-density function both to define the range of values that an input parameter can have and to reflect the belief that the parameter will take on the various values within that range.
- Account for dependencies (correlations) among the input data and how they affect uncertainty.
- Propagate the uncertainties through the model to generate a probability-density function of predicted outcome values.
- From the probability-density function of predicted values of the outcome variable, derive confidence limits and intervals to provide a quantitative statement about the effect of input uncertainty on the model predictions.

The value of information derived from an input uncertainty or variability analysis depends heavily on the care given to the process of constructing input-parameter probability-density functions. One begins the process of constructing a probability-density function for a given input by assembling values from measurements. The values should be consistent with the model and its particular application. The values will vary as a result of spatial and temporal variability, measurement error, extrapolation of data from one situation to another, extent of knowledge, and so on. The process of constructing a probability-density function from limited and imprecise data can be highly subjective. The analyst must often apply judgment to the process, and so requires expertise and wisdom. The process is likely to become more objective as the amount of data on a given input increases. However, a large set of data does not necessarily imply easy construction of a suitable probability-density function for any particular application.

UNCERTAINTIES IN THE COMMUNICATION OF RISK INFORMATION

Most risk assessment documents for waste-incineration facilities have been assembled mainly to comply with regulatory requirements and guidelines. There is even now an EPA document that attempts to formalize some of the requirements for risk assessments for hazardous-waste incinerators (EPA 1998a). However, if regulatory requirements or guidelines are the only motivation for investing time and energy in these a risk assessment document, the result is usually badly presented. Such documents are difficult to use as a basis for interacting with affected communities, and it is hard to argue that the documents provide much in the way of public service. If an important goal of a risk assessment is to address the effect of a process or facility on public health and to address the concerns of the affected community, there is a need to provide information that can be used to support informed debate on community health issues. However, in many risk-assessment documents, much of the information related to those issues is not readily accessible in the text, equations, tables, and appendixes. The sense of incomprehensibility and exclusion that is experienced by affected communities can result in unnecessary polarization. One way to avoid that is to include early in the risk-assessment process a summary of the local-community concerns and a brief description of how they are addressed in the risk assessment. Failure to express community concerns in a risk assessment leaves one with the impression that the concerns of the regulators are all that matter.

CONCLUSIONS

There are many reasons for uncertainty and variability in the information used to assess possible health effects of waste incineration. There are large variations from facility to facility with regard to types of waste combusted, operating practices, allowable emission levels, emission-control technologies, types of chemicals emitted, environmental characteristics, proximity to other sources of contaminant exposure, frequency of off-normal emissions, and the biologic and behavioral characteristics of the people who might be exposed to the contaminants in the environment. Some uncertainties are peculiar to incineration, and some are inherent in any activity that releases contaminants into the environment.

Some of the uncertainties and variabilities can be reduced or better accounted for; others, by their nature, remain unchanged. Nonetheless, there is a need to make decisions concerning the siting, design, operation, and regulation of incineration facilities. The most effective decisions are the ones that take uncertainty and variability fully into account.

When key uncertainties become very large, quantitative estimates of risk may do little to change the previously held beliefs of interested or affected parties. Those who favor the use of incineration technology tend to focus on results in the middle range, and those who are opposed to incineration tend to focus on scenarios associated with the high exposures.

RECOMMENDATIONS

- Decisionmakers should coordinate with risk assessors in identifying the uncertainties and variabilities associated with estimating the health risks of waste incineration that are likely to have the greatest impact on the specific decision to be made.
- Decisionmakers should consider individual and societal values regarding uncertain adverse consequences by eliciting individual or societal preferences, using decision analysis, and applying theories of science policy, social-welfare economics, and ethics.
- Assessments of public-health risk posed by waste incineration should consider, through the use of sensitivity analyses or otherwise, the importance of emissions resulting from off-normal activities in addition to routine stack emissions or fugitive emissions.
- Incinerator risk assessments should include the following components of uncertainty and variability analyses:
  — An estimate of the variability and uncertainty distributions of all input values and their effect on final estimates.
  — A sensitivity analysis to assess how model predictions are related to variations in input data.
  — Variance-propagation models that show how the variability and uncertainty of final results are tied to the uncertainties and variabilities associated with various models, their inputs, and assumptions used throughout the risk assessment.
- Risk assessments should provide information that can be used to support informed debate on the various issues of concern regarding the health of community members. Assessments should include a summary of local community concerns and a description of how they have been or will be addressed in the risk-assessment process.

References

Abbey, D.E., P.K. Mills, F.F. Petersen, and W.L. Beeson. 1991. Long-term ambient concentrations of total suspended particulates and oxidants as related to incidence of chronic disease in California Seventh-Day Adventists. Environ. Health Perspect. 94:43-50.

Abbey, D.E., M.D. Lebowitz, P.K. Mills, F.F. Petersen, W.L. Beeson, and R.J. Burchette. 1995. Long-term ambient concentrations of particulates and oxidants and development of chronic disease in a cohort of nonsmoking California residents. Inhalation Toxicol. 7:19-34.

Abbey, D.E., N. Nishino, W.F. McDonnell, R.J. Burchette, S.F. Knutsen, W. Lawrence Beeson, J.X. Yang. 1999. Long-term inhalable particles and other air pollutants related to mortality in nonsmokers. Am. J. Respir. Crit. Care Med. 159(2):373-382 .

Adams, C.R., D.K. Ziegler, and J.T. Lin. 1983. Mercury intoxication simulating amyotrophic lateral sclerosis. JAMA 250:642-643.

Adams, M.A., P.M. Bolger, and E.L. Gunderson. 1994. Dietary intake and hazards of arsenic. Pp. 41-49 in Arsenic: Exposure and Health, W.R. Chappell, C.O. Abernathy and C.R. Cothern, eds. Northwood, U.K.: Science and Technology Letters.

Afonso, J.F., and R.R. de Alvarez. 1960. Effects of mercury on human gestation. Am. J. Obstet. Gynecol. 80(1):145-154.

AHA (American Hospital Association). 1993. An Ounce of Prevention: Waste Reduction Strategies for Health Care Facilities. American Society for Healthcare Environmental Services of the American Hospital Association. Chicago, IL: The Society.

Al-Saleem, T. 1976. Levels of mercury and pathological changes in patients with organomercury poisoning. Bull. WHO 53(Suppl):99-104.

Allred, E.N., E.R. Bleecker, B.R. Chaitman, T.E. Dahms, S.O. Gottlieb, J.D. Hackney, M. Pagano, R.H. Selvester, S.M. Walden, and J. Warren. 1989. Short-term effects of carbon monoxide exposure on the exercise performance of subjects with coronary artery disease. N. Engl. J. Med. 321(21):1426-32. [published erratum appears in N. Engl. J. Med. 1990 322(14):1019]

Amdur, M.O., J. Doull, and C.D. Klaassen. 1991. Casarett and Doull's Toxicology: The Basic Science of Poisons. New York: Pergamon Press.

Amin-Zaki, L., S. Elhassani, M.A. Majeed, T.W. Clarkson, R.A. Doherty, M.R. Greenwood, and T. Giovanoli-Jakubczak. 1976. Perinatal methylmercury poisoning in Iraq. Am. J. Dis. Child. 130(10):1070-1076.

Anderton, D.L., A.B. Anderson, J.M. Oakes, and M.R. Fraser. 1994. Environmental equity: The demographics of dumping. Demography 31(2):229-248.

Angerer, J., B. Heinzow, D.O. Reimann, W. Knorz, and G. Lehnert. 1992. Internal exposure to organic substances in a municipal waste incinerator. Int. Arch. Occup. Environ. Health 64(4):265-273.

Armour, A. 1988. Methodological problems in social impact monitoring. Environ. Impact Assess. Rev. 8:249-265.

Arnstein, S. 1969. A ladder of citizen participation. J. Am. Inst. Planners 35:216-224.

Aronow, R., C. Cubbage, R. Wiener, B. Johnson, J. Hesse, and J. Bedford. 1990. Mercury exposure from interior latex paint—Michigan. MMWR 39(8):125-126.

ASME (The American Society of Mechanical Engineers). 1995. The Relationship Between Chlorine in Waste Streams and Dioxin Emissions from Waste Combustor Stacks, CRTD-Vol. 36. Fairfield, N.J.: ASME Press.

Assennato, G., C. Paci, M.E. Baser, R. Molinini, R.G. Candela, B.M. Altamura, and R. Giorgino. 1987. Sperm count suppression without endocrine dysfunction in lead-exposed men. Arch. Environ. Health 42(2):124-127.

Astle, J.W., F.A.P.C. Gobas, W.J. Shiu, and D. Mackay. 1987. Lake sedimentation in historic records of atmospheric contaminants by organic chemicals. Pp. 57-77 in Sources and Fates of Aquatic Pollutants. R.A. Hites, and S.J. Eisenreich, eds. Washington, D.C.: American Chemical Society.

ATSDR (Agency for Toxic Substances and Disease Registry). 1990. Toxicological Profile for Benzo(a)Pyrene (Update). PB/90/258245. Agency for Toxic Substances and Disease Registry, Atlanta, Ga.

ATSDR (Agency for Toxic Substances and Disease Registry). 1993a. Study of Symptom and Disease Prevalence, Caldwell Systems, Inc. Hazardous Waste Incinerator, Caldwell County, North Carolina. Final Report. ATSDR/HS-93/29. U.S. Department of Health and Human Services, Agency for Toxic Substances and Disease Registry, Atlanta, Ga. 144 pp.

ATSDR (Agency for Toxic Substances and Disease Registry). 1993b. Toxicological Profile for Chromium. ATSDR/TP-92/08. Prepared by Syracuse Research Corp. under subcontract to Clement International Corp. U.S. Department of Health and Human Services, Agency for Toxic Substances and Disease Registry. Atlanta, Ga. 227 pp.

ATSDR (Agency for Toxic Substances and Disease Registry). 1993c. Toxicological Profile for Beryllium (Update). ATSDR/TP-92/04. U.S. Department of Health and Human Services, Agency for Toxic Substances and Disease Registry. Atlanta, Ga. 129 pp.

ATSDR (Agency for Toxic Substances and Disease Registry). 1994. Toxicological Profile for Mercury (Update). ATSDR/TP-93/10. Prepared by Clement International Corp. under contract 205-88-0608 for U.S. Department of Health and Human Services, Agency for Toxic Substances and Disease Registry, Atlanta, Ga.

ATSDR (Agency for Toxic Substances and Disease Registry). 1995. Toxicological Profile for Polycyclic Aromatic Hydrocarbons. Prepared by Research Triangle Institute, under contract 205-93-0606 for U.S. Department of Health and Human Services, Agency for Toxic Substances and Disease Registry, Atlanta, Ga.

ATSDR (Agency for Toxic Substances and Disease Registry). 1997a. Toxicological Profile for Cadmium. Draft. Prepared by Research Triangle Institute, under contract 205-93-0606 for U.S. Department of Health and Human Services, Agency for Toxic Substances and Disease Registry, Atlanta, Ga.

ATSDR (Agency for Toxic Substances and Disease Registry). 1997b. Toxicological Profile for Lead. (Draft). Prepared by Research Triangle Institute, under contract 205-93-0606 for U.S. Department of Health and Human Services, Agency for Toxic Substances and Disease Registry, Atlanta, Ga.

ATSDR (Agency for Toxic Substances and Disease Registry). 1998a. Toxicological Profile for Chlorinated Dibenzo-p-Dioxins. Prepared by Research Triangle Institute, under contract 205-93-0606 for U.S. Department of Health and Human Services, Public Health Service, Agency for Toxic Substances and Disease Registry, Atlanta, GA.

ATSDR (Agency for Toxic Substances and Disease Registry). 1998b. Toxicological Profile for Arsenic (Update). Prepared by Research Triangle Institute, under contract 205-93-0606 for U.S. Department of Health and Human Services, Public Health Service, Agency for Toxic Substances and Disease Registry, Atlanta, GA.

ATSDR (Agency for Toxic Substances and Disease Registry). 1998c. Toxicological Profile for Polychlorinated Biphenyls (Update). Prepared by Research Triangle Institute, under contract 205-93-0606 for U.S. Department of Health and Human Services, Agency for Toxic Substances and Disease Registry, Atlanta, Ga.

ATSDR (Agency for Toxic Substances and Disease Registry). 1999. Toxicological Profile for Mercury (Update). Prepared by Research Triangle Institute, under contract 205-93-0606 for U.S. Department of Health and Human Services, Agency for Toxic Substances and Disease Registry, Atlanta, Ga.

Avol, E.L., W.S. Linn, D.A. Shamoo, K.R. Anderson, R.C. Peng, and J.D. Hackney. 1990. Respiratory responses of young asthmatic volunteers in controlled exposures to sulfuric acid aerosol. Am. Rev. Respir. Dis. 142(2):343-348.

AWMA (Air & Waste Management Association). 1994. Medical Waste Disposal. Medical Waste Committee (WT-3), Technical Council, Air & Waste Management Association. J. Air Waste Manage. Assoc. 44:1176.

Axelson, O., E. Dahlgren, C.-D. Jansson, and O. Rehnlund. 1978. Arsenic exposure and mortality: A case referent study from a Swedish copper smelter. Br. J. Ind. Med. 35:8-15.

Bacci, E., D. Calamari, C. Gaggi, and M. Vighi. 1990. Bioconcentration of organic chemical vapors in plant leaves: Experimental measurements and correlation. Environ. Sci. Technol. 24:885-889.

Bacci, E., M.J. Cerejeira, C. Gaggi, G. Chemello, D. Calamari, and M. Vighi. 1992. Chlorinated dioxins: Volatilization from soils and bioconcentration in plant leaves. Bull. Environ. Contam. Toxicol. 48(3):401-408.

Bache, C.A., W.H. Gutenmann, M. Rutzke, G. Chu, D.C. Elfving, and D.J. Lisk. 1991. Concentrations of metals in grasses in the vicinity of a municipal refuse incinerator. Arch. Environ. Contam. Toxicol. 20(4):538-542.

Baes, C.F., III, R.D. Sharp, A.L. Sjoreen, and R.W. Shor. 1984a. A Review and Analysis of Parameters for Assessing Transport of Environmentally Released Radionuclides Through Agriculture. ORNL-5786. Oak Ridge, Tenn.: Oak Ridge National Laboratory. Available from NTIS, Springfield, Va., Doc. No. DE85-000287.

Baes, C.F., III, R.D. Sharp, A.L. Sjoreen, and O.W. Hermann. 1984b. TERRA: A Computer Code for Simulating the Transport of Environmentally Released Radionuclides Through Agriculture. ORNL-5785. Health and Safety Research Division, Oak Ridge, Tenn.: Oak Ridge National Laboratory.

Baghurst, P.A., E.F. Robertson, A.J. McMichael, G.V. Vimpani, N.R. Wigg, and R.R. Roberts. 1987. The Port Pirie cohort study: Lead effects on pregnancy outcome and early childhood development. Neurotoxicology (3):395-401.

Baker, E.L., Jr., P.J. Landrigan, A.G. Barbour, D.H. Cox, D.S. Folland, R.N. Ligo, and J. Throckmorton. 1979. Occupational lead poisoning in the United States: Clinical and biochemical findings related to blood lead levels. Br. J. Ind. Med. 36(4):314-322.

Baker, E.L., R.G. Feldman, R.F. White, and J.P. Harley. 1983. The role of occupational lead exposure in the genesis of psychiatric and behavioral disturbances. Acta Psychiatry Scand. Suppl. 303:38-48.

Bakir, F., S.F. Damluji, L. Amin-Zaki, M. Murtadha, A. Khalidi, N.Y. al-Rawi, S. Tikriti, H.I. Dahahir, T.W. Clarkson, J.C. Smith, and R.A. Doherty. 1973. Methylmercury poisoning in Iraq. Science 181(96):230-241.

Barisotti, D.A., L.J. Abrahamson, and J.R. Allen. 1979. Hormonal alterations in female rhesus monkeys fed a diet containing 2,3,7,8-tetrachlorodibenzo-*p*-dioxin. Bull. Environ. Contam. Toxicol. 21(4/5):463-469.

Barregård, L., G. Sällsten, and B. Järvholm. 1990. Mortality and cancer incidence in chloralkali workers exposed to inorganic mercury. Br. J. Ind. Med. 47(2):99-104.

Barrow, C.S., H. Lucia, and Y.C. Alarie. 1979. A comparison of the acute inhalation toxicity of hydrogen chloride versus the thermal decomposition products of polyvinylchloride. J. Combust. Toxicol. 6:3-12.

Barton, R.G., P.M. Maly, W.D. Clark, W.R. Seeker, and W.S. Lanier. 1987. Prediction of the Fate of Toxic Metals in Hazardous Waste Incinerators, Final Report. Prepared by Energy and Environmental Research Corporation, Irvine, CA., EPA Contract 68-03-3365.

Barton, R.G., W.D. Clark, and W.R. Seeker. 1990. Fate of metals in combustion systems. Combust. Sci. Technol. 74:327-342.

Barton, R.G., A.R. Trenholm, S. Shoraka-Blair, G.A. Jungclaus, and G.D. Hinshaw. 1996. Estimation of Unmeasured Hazardous Emissions from Hazardous Waste Incinerators. Presented at International Conference on Incineration and Thermal Treatment Technologies, 1996.

Bascom, R., P.A. Bromberg, D.A. Costa, R. Devlin, D.W. Dockery, M.W. Frampton, W. Lambert, J.M. Samet, F.E. Speizer, and M. Utell. 1996. Health effects of outdoor pollution. Am. J. Respir. Crit. Care Med. 153(1):3-50.

Battle, L.C. 1994. Regulation of medical waste in the United States. Pace Environ. L. Rev. 11(2):517-585.

Bellinger, D.C., H.L. Needleman, A. Leviton, C. Waternaux, M.B. Rabinowitz, and M.L. Nichols. 1984. Early sensory-motor development and prenatal exposure to lead. Neurobehav. Toxicol. Teratol. 6(5):387-402.

Bernard, A.M., and R. Lauwerys. 1989. Cadmium, NAG activity, and 2-microglobulin in the urine of cadmium pigment workers. [Letter] Br. J. Ind. Med. 46:679-680.

Berry, B.J.L. 1977. The Social Burdens of Environmental Pollution: A Comparative Metropolitan Data Source. Cambridge, Mass.: Ballinger Publishing.

Bidstrup, P.L., and R.A.M. Case. 1956. Carcinoma of the lung in workmen in the bichromates-producing industry in Great Britain. Br. J. Ind. Med. 13:260-264.

Biggeri, A., F. Barbone, C. Lagazio, M. Bovenzi, and G. Stanta. 1996. Air pollution and lung cancer in Trieste, Italy: Spatial analysis of risk as a function of distance from sources. Environ. Health Perspect. 104(7):750-4.

Bluhm, R.E., J.A. Breyer, R.G. Bobbitt, L.W. Welch, A.I. Wood, and R.A. Branch. 1992. Elemental mercury vapour toxicity, treatment, and prognosis after acute, intensive exposure in chloralkali plant workers. Part I: History, neuropsychological findings and chelator effects. Hum. Exp. Toxicol. 11(3):201-210.

Bobak, M., and D.A. Leon. 1992. Air pollution and infant mortality in the Czech Republic,1986-88. Lancet 340(8826):1010-1014.

Bodek, I., W.J. Lyman, W.F. Reehl, and D.H. Rosenblatt. 1988. Environmental Inorganic Chemistry: Properties, Processes, and Estimation Methods. New York: Pergamon.

Bogen, K.T., and R.C. Spear. 1987. Integrating uncertainty and interindividual variability in environmental risk assessments. Risk Anal. 7(4):427-436.

Borgmann, U., and D.M. Whittle. 1991. Contaminant concentration trends in Lake Ontario lake trout (*Salvelinus namaycush*): 1977 to 1988. J. Great Lakes Res. 17(3):368-381.

Bornschein, R.L., J. Grote, T. Mitchell, P.A. Succop, K.N. Dietrich, K.M. Krafft, and P.B. Hammond. 1989. Effects of Prenatal Lead Exposure on Infant Size at Birth in Lead Exposure and Child Development: An International Assessment, M.A. Smith, L.D. Grant, and A.I. Sors, eds. Boston: Kluwer Academic.

Bovet, P., M. Lob, and M. Grandjean. 1977. Spirometric alterations in workers in the chromium electroplating industry. Int. Arch. Occup. Environ. Health 40:25-32.

Bowen, H.J.M. 1979. Environmental Chemistry of the Elements. New York: Academic.

Braver, E.R., P. Infante, and K. Chu. 1985. An analysis of lung cancer risk from exposure to hexavalent chromium. Teratog. Carcinog. Mutagen. 5(5):365-378.

Bresnitz, E.A., J. Roseman, D. Becker, and E. Gracely. 1992. Morbidity among municipal waste incinerator workers. Am. J. Ind. Med. 22:363-378.

Briggs, G.G., R.H. Bromilow, A.A. Evans, and M. Williams. 1983. Relationships between lipophilicity and the distribution of non-ionized chemicals in barley shoots following uptake by the roots. Pestic. Sci. 14(5):492-500.

Brower, R., J. Gerritsen, K. Zankel, A. Huggins, N. Peters, S. Campbell, and R. Nilsson. 1990. Risk Assessment Study of the Dickerson Site. Prepared for Power Plant and Environmental Review Division, Maryland Department of Natural Resources. PPSE-SH-4. 3 Vols.

Brown, B., and K. Felsvang. 1991. Control of Mercury and Dioxin Emissions from United States and European Municipal Solid Waste Incinerators by Spray Dryer Absorption Systems. Conference papers and abstracts from the Second Annual International Specialty Conference, Tampa, Fla. April 15-19, 1991.

Brown, I.A. 1954. Chronic mercurialism: A cause of the clinical syndrome of amyotrophic lateral sclerosis. Arch. Neurol. Psychiatry 72:674-681.

Bryant, B., and P. Mohai. 1992. Race and the Incidence of Environmental Hazards. Boulder, Colo.: Westview.

Brzuzy, L.P., and R.A. Hites. 1996. Global balance for chlorinated dibenzo-p-dioxins and dibenzofurans. Environ. Sci. Technol. 30(6):1797-1804.

Buchet, J.P., H. Roels, A. Bernard, and R. Lauwerys. 1980. Assessment of renal function of workers exposed to inorganic lead, cadmium, or mercury vapor. J. Occup. Med. 22(11):741-750.

Buchet, J.P., R. Lauwerys, H. Roels, A. Bernard, P. Bruaux, F. Claeys, G. Ducoffre, P. de Plaen, J. Staessen, Amery A., et al. 1990. Renal effects of cadmium body burden of the general population. Lancet 336(8717):699-702.

Bullard, R.D. 1990. Dumping in Dixie: Race, Class, and Environmental Quality. Boulder, Colo.: Westview.

Burnett, R.T., R.E. Dales, M.E. Raizenne, D. Krewski, P.W. Summers, G.R. Roberts, M. Raad-Young, T. Dann, and J. Brook. 1994. Effects of low ambient levels of ozone and sulfates on the frequency of respiratory admissions to Ontario hospitals. Environ. Res. 65(2):172-194.

CalEPA (California Environmental Protection Agency) Air Resources Board. 1996. Proposed Identification of Inorganic Lead as Toxic Air Contaminant, Part B Health Assessment, Draft. Sacramento, Calif. January.

Campbell, B.C., P.A. Meredith, and J.J.C. Scott. 1985. Lead exposure and changes in the renin-angiotensin-aldosterone system in man. Toxicol. Lett. 25:25-32.

Campbell, J.S. 1948. Acute mercurial poisoning by inhalation of metallic vapor in an infant. Can. Med. Assoc. J. 58:72-75.

Cantox, Inc. 1988. Health Hazard Evaluation of Specific PCDD, PCDF, and PAH in Emissions from the Proposed Petrosun/SNC Resource Recovery Incinerator and from Ambient Background Sources. Petrosun/SNC Operations Ltd., Brampton, Ontario. (February).

Caporaso, N., R.B. Hayes, M. Dosemeci, R. Hoover, R. Ayesh, M. Hetzel, and J. Idle. 1989. Lung cancer risk, occupational exposure, and the debrisoquine metabolic phenotype. Cancer Res. 49(13):3675-3679.

Carpi, A., L.H. Weinstein, and D.W. Ditz. 1994. Bioaccumulation of mercury by sphagnum moss near a municipal solid waste incinerator. J. Air Waste Manage. Assoc. 44:669-672.

Carroll, G.J., R.C. Thurnau, and D.J. Fournier, Jr. 1995. Mercury Emissions from a Hazardous Waste Incinerator Equipped with a State-of-the-Art Wet Scrubber. J. Air Waste Manage. Assoc. 45:730-736.

Catlin, G.E.G. 1959. The meaning of community. Pp. 114-134 in Community, NOMOS II, C. J. Friedrich, ed. New York: Liberal Arts Press.

Chambers, A., M. Knecht, N. Soelbert, D. Eaton, D. Roberts, and T. Broderick. 1998. Mercury Emissions Control Technologies for Mixed Waste Thermal Treatment. Presented at International Conference on Incineration and Thermal Treatment Technologies, May 11-15, 1998, Salt Lake City, Utah.

Chaney, R.L., J.A. Ryan, Y-M. Li, and S.L. Brown. 1999. Soil cadmium as a threat to human health. In Cadmium in Soils and Plants, M.J. McLaughlin, and B.R. Singh, eds. Dordrecht.: Kluwer Academic.

Chen, C.-J., Y.-C. Chuang, T.-M. Lin, and H.-Y. Wu. 1985. Malignant neoplasms among residents of a blackfoot disease-endemic area in Taiwan: High-arsenic artesian well water and cancers. Cancer Res. 45:5895-5899.

Chen, C.J., Y.C. Chuang, S.L. You, T.M. Lin, and H.Y. Wu. 1986. A retrospective study on malignant neoplasms of bladder, lung and liver in blackfoot disease-endemic area in Taiwan. Br. J. Cancer 53(3):399-405.

Chertoff, L., and D. Buxbaum. 1986. Public perceptions and community relations. Pp. 31-42 in The Solid Waste Handbook: A Practical Guide, W.D. Robinson, ed. New York: John Wiley & Sons.

Chess, C., B.J. Hance, and P.M. Sandman. 1988. Improving Dialogue with Communities: A Short Guide for Government Risk Communication. Trenton, N.J.: New Jersey Department of Environmental Protection.

Chess, C., K.L. Salomone, B.J. Hance, and A. Saville. 1995. Results of a national symposium on risk communication: Next steps for government agencies. Risk Anal. 15:115-125.

Chisolm, J.J., Jr., D.J. Thomas, and T.G. Hamill. 1985. Erythrocyte porphobilinogen synthase activity as an indicator of lead exposure in children. Clin. Chem. 31(4):601-605.

Chmiel, K.M., and R.M. Harrison. 1981. Lead content of small mammals at a roadside site in relation to the pathways of exposure. Sci. Total. Environ. 17:145-154.

Cinca, I., I. Dumitrescu, P. Onaca, A. Serbanescu, and B. Nestorescu. 1980. Accidental ethyl mercury poisoning with nervous system, skeletal muscle, and myocardium injury. J. Neurol. Neurosurg. Psychiatry 43(2):143-149.

CKRC (Cement Kiln Recycling Coalition). 1995. An Analysis of Technical Issues Pertaining to the Determination of MACT Standards for the Waste Recycling Segment of the Cement Industry. Prepared by Environmental Risk Sciences, Inc., Environomics, and Rigo & Rigo Associates, Inc., for Cement Kiln Recycling Coalition, Washington, D.C.

Clayton, C.A., R.L. Perritt, E.D. Pellizzari, K.W. Thomas, R.W. Whitmore, L.A. Wallace, H. Ozkaynak, and J.D. Spengler. 1993. Particle Total Exposure Assessment Methodology (PTEAM) study: Distributions of aerosol and elemental concentrations in personal, indoor, and outdoor air samples in a southern California community. J. Expo. Anal. Environ. Epidemiol. 3(2):227-250.

Coate, D., and R. Fowles. 1989. Is there statistical evidence for a blood lead-blood pressure relationship? J. Econ. 8:173-184.

Collett, R.S., K. Oduyemi, and D.E. Lill. 1998. An investigation of environmental levels of cadmium and lead in airborne matter and surface soils within the locality of a municipal waste incinerator. Sci. Total Environ. 209(2-3):157-167.

Connett, P., and E. Connett. 1994. Municipal waste incineration: Wrong question, wrong answer. Ecologist 24(1):14-20.

Cooney, G.H., A. Bell, W. McBride, and C. Carter. 1989a. Low-level exposures to lead: The Sydney lead study. Dev. Med. Child Neurol. 31(5):640-649.

Cooney, G.H., A. Bell, W. McBride, and C. Carter. 1989b. Neurobehavioural consequences of prenatal low-level exposures to lead. Neurotoxicol. Teratol. 11(2):95-104.

Cordier, S., F. Deplan, L. Mandereau, and D. Hemon. 1991. Paternal exposure to mercury and spontaneous abortions. Br. J. Ind. Med. 48(6):375-381.

Covello, V.T., and F.W. Allen. 1988. Seven Cardinal Rules of Risk Communication. OPA-87-020. U.S. Environmental Protection Agency, Washington, D.C.

Covello, V.T., D. von Winterfeldt, and P. Slovic. 1986. Communicating scientific information about health and environmental risks: Problems and opportunities from a social and behavioral perspective. Pp. 109-134 in Risk Communication: Proceedings of the National Conference on Risk Communication, J.C. Davies, V.T. Covello, and F.W. Allen, eds. Washington, D.C.: The Conservation Foundation.

Covello, V.T., P.M. Sandman and P. Slovic. 1988. Risk Communication, Risk Statistics and Risk Comparisons: A Manual for Plant Managers. Washington D.C.: Chemical Manufacturers Association.

Cramer, G.M. 1994. Exposure of U.S. Consumers to Methylmercury from Fish. Pp. 103-118 in DOE/FDA/EPA Workshop on Methylmercury and Human Health, P.D. Moskowitz, L. Saroff, M. Bolger, J. Circmanec, and S. Durkee, eds. Brookhaven National Laboratory, Upton, N.Y.

Cullen, A.C. 1995. The sensitivity of probabilistic risk assessment results to alternative model structures: A case study of municipal waste incineration. J. Air Waste Manage. Assoc. 45(7):538-546.

Curlee, T.R. 1994. Waste-to-Energy in the United States: A Social and Economic Assessment. Westport, Conn.: Quorum Books.

Czuczwa, J.M., and R.A. Hites. 1986. Airborne dioxins and dibenzofurans: Sources and fates. Environ. Sci. Technol. 20(2):195-204.

Dabeka, R.W., A.D. McKenzie, G.M. Lacroix, C. Cleroux, S. Bowe, R.A. Graham, H.B. Conacher, and P. Verdier. 1993. Survey of arsenic in total diet food composites and estimation of the dietary intake of arsenic by Canadian adults and children. J. AOAC Int. 76(1):14-25.

Damokosh, A.I., J.D. Spengler, D.W. Dockery, J.H. Ware, and F.E. Speizer. 1993. Effects of acidic particles on respiratory symptoms in 7 United States communities. Am. Rev. Respir. Dis. 147(4):A632.

Darmer, K.I., Jr., E.R. Kinkead, and L.C. DiPasquale. 1974. Acute toxicity in rats and mice exposed to hydrogen chloride gas and aerosols. Am. Ind. Hyg. Assoc. J. 35(10):623-631.

Davies, J.M., D.F. Easton, and P.L. Bidstrup. 1991. Mortality from respiratory cancer and other causes in United Kingdom chromate production workers. Br. J. Ind. Med. 48:299-313.

Davies, K. 1989. Concentration and dietary intake of selected organochlorines, including PCBs, PCDFs in fresh food composites grown in Ontario, Canada. Chemosphere 17:263-276.

Davis, C. 1986. Public involvement in hazardous-waste siting decisions. Polity 19(2):296-304.

Davis, G., and W. Colglazier. 1987. Siting hazardous waste facilities: Asking the right questions. Pp. 167-182 in America's Future in Toxic Waste Management: Lessons from Europe, B.W. Piasecki, and G.A. Davis, eds. New York: Quorum Books.

Della Porta, G, T.A. Dragani, and G. Sozzi. 1987. Carcinogenic effects of infantile and long-term 2,3,7,8-tetrachlorodibenzo-p-dioxin treatment in the mouse. Tumori 73(2):99-107.

Dellinger, B., J.L. Torres, W.A. Rubey, D.L. Hall, and J.L. Graham. 1984. Determination of the Thermal Decomposition Properties of 20 Selected Hazardous Organic Compounds. University of Dayton Research Institute. EPA-600/2-84-138.

Deml, E., I. Mangelsdorf, and H. Greim. 1996. Chlorinated dibenzodioxins and dibenzofurans (PCDD/F) in blood and human milk of non occupationally exposed persons living in the vicinity of a municipal waste incinerator. Chemosphere 33(10):1941-1950.

Dempsey, C.R., and E.T. Oppelt. 1993. Incineration of hazardous waste: A critical review update. J. Air Waste Manage. Assoc. 43:25-73.

Dockery, D.W., and C.A. Pope, III. 1994. Acute respiratory effects of particulate air pollution. Annu. Rev. Public Health 15:107-132.

Dockery, D.W., J. Schwartz, and J.D. Spengler. 1992. Air pollution and daily mortality: Associations with particulates and acid aerosols. Environ. Res. 59(2):362-373.

Dockery, D.W., C.A. Pope III, X. Xu, J.D. Spengler, J.H. Ware, M.E. Fay, B.G. Ferris, Jr., and F.E. Speizer. 1993. An association between air pollution and mortality in six U.S. cities. N. Engl. J. Med. 329(24):1753-1759.

Dockery, D.W., J. Cunningham, A.I. Damokosh, L.M. Neas, J.D. Spengler, P. Koutrakis, J.H. Ware, M. Raizenne, and F.E. Speizer. 1996. Health effects of acid aerosols on North American children: Respiratory symptoms. Environ. Health Perspect. 104(5):500-505.

DOS (New York City Department of Sanitation). 1992. A Comprehensive Solid Waste Management Plan for New York City and Final Generic Environmental Impact Statement, Appendix Volume 1.1, Waste Stream Data. New York City Department of Sanitation.

Dreicer, M., T.E. Hakonson, G.C. White, and F.W. Whicker. 1984. Rainsplash as a mechanism for soil contamination of plant surfaces. Health Phys. 46(1):177-187.

DTSC (Department of Toxic Substances Control). 1992a. Documentation of assumptions used in the decision to include and exclude exposure pathways. Ch. 2 in Guidance for Site Characterization and Multimedia Risk Assessment for Hazardous Substances Release Sites, Vol. 2. UCRL-CR-103460. A report prepared for the State of California, Department of Toxic Substances Control, by Lawrence Livermore National Laboratory, Livermore, Calif.

DTSC (Department of Toxic Substances Control). 1992b. Guidelines for the documentation of methodologies, justification, input, assumptions, limitations, and output for exposure models. Ch. 3 in Guidance for Site Characterization and Multimedia Risk Assessment for Hazardous Substances Release Sites, Vol. 2. UCRL-CR-103460. A report prepared for the State of California, Department of Toxic Substances Control, by Lawrence Livermore National Laboratory, Livermore, Calif.

Duvall, D.S., and W.A. Rubey. 1977. Laboratory Evaluation of High-Temperature Destruction of Polychlorinated Biphenyls and Related Compounds. EPA-600/2-77-228. U.S. Environmental Protection Agency, Cincinnati, Ohio.

Ecologistics. 1993a. 1991 Field Year Biomonitoring Program Proposed Rotary Kiln Project, Laidlaw Environmental Services, Ltd. Prepared for Laidlaw Environmental Services, Burlington, Ont. Feb. 26.

Ecologistics. 1993b. 1992 Field Year Biomonitoring Program Proposed Rotary Kiln Project, Laidlaw Environmental Services, Ltd. Prepared for Laidlaw Environmental Services, Burlington, Ont. Aug. 4.

Ecologistics. 1994. Final Report, 1992 Field Year Biomonitoring Program. Prepared for Laidlaw Environmental Services, Ltd. Burlington, Ontario.

Edelstein, M.R. 1988. Contaminated Communities: The Social and Psychological Impacts of Residential Exposure to Toxic Exposure. Boulder, Colo.: Westview.

Eitzer, B.D. 1995. Polychlorinated dibenzo-p-dioxins and dibenzofurans in raw milk samples from farms located near a new resource recovery incinerator. Chemosphere 30(7):1237-1248.

Elias, R.W., Y. Hirao, and C.C. Patterson. 1982. The circumvention of the natural biopurification of calcium along nutrient pathways by atmospheric inputs of industrial lead. Geochim. Cosmochim. Acta 46:2561-2580.

Elinder, C.G. 1985. Cadmium: Uses, occurrence and intake. Pp. 23-64 in Cadmium and Health: A Toxicological and Epidemiological Appraisal. Vol. I. Exposure, Dose and Metabolism, Effects and Response. Boca Raton, Fla.: CRC Press.

Ellenhorn, M.J., and D.G. Barceloux. 1988. Medical Toxicology: Diagnosis and Treatment of Human Poisoning. New York: Elsevier.

Elliott, M.L.P. 1984a. Coping with conflicting perceptions of risk in hazardous waste facility siting disputes. Ph.D. dissertation. Massachusetts Institute of Technology, Department of Urban Studies and Planning, Cambridge, Mass.

Elliott, M.L.P. 1984b. Improving community acceptance of hazardous waste facilities through alternative systems for mitigating and managing risk. Hazard. Waste Hazard. Mater. 1(3):397-410.

Elliott, P., M. Hills, J. Beresford, I. Kleinschmidt, D. Jolley, S. Pattenden, L. Rodrigues, A. Westlake, and G. Rose. 1992. Incidence of cancers of the larynx and lung near incinerators of waste solvents and oils in Great Britain. Lancet 339(8797):854-858.

Elliott, P., G. Shaddick, I. Kleinschmidt, D. Jolley, P. Walls, J. Beresford, and C. Grundy. 1996. Cancer incidence near municpal solid waste incinerators in Great Britain. Br. J. Cancer 73(5):702-710.

Engleson, G., and T. Hermer. 1952. Alkyl mercury poisoning. Acta Paediat. Scand. 41:289-294.

English, M.R. 1992. Siting Low-Level Radioactive Waste Disposal Facilities: The Public Policy Dilemma. New York: Quorum Books.

English, M.R., M.N. Murray, L.R. Blank, and D.J. Delffs, Jr. 1991. The Economic and Social Impacts of Chem-Nuclear's Low-level Radioactive Waste Disposal Facility on Barnwell County, South Carolina. Knoxville, Tenn.: University of Tennessee, Energy, Environment, and Resources Center.

Enterline, P.E., and G.M. Marsh. 1982. Cancer among workers exposed to arsenic and other substances in a copper smelter. Am. J. Epidemiol. 116(6):895-911.

Environment Canada. 1985. National Incinerator Testing and Evaluation Program: Two-Stage Combustion (Prince Edward Island). Environment Canada Report No. EPS 3/UP/1, Vol. I-IV. ISBN 0-662-14311-6.

EPA (U.S. Environmental Protection Agency). 1985. Cadmium Contamination of the Environment: An Assessment of Nationwide Risk. EPA-440/4-85-023. U.S. Environmental Protection Agency, Office of Water Regulations and Standards. Washington, D.C.

EPA (U.S. Environmental Protection Agency). 1986a. Air Quality Criteria for Lead. 4 vols. EPA-600/8-83/028aF-dF. U.S. Environmental Protection Agency, Environmental Criteria and Assessment Office, Research Triangle Park, N.C. Available from NTIS, Springfield, Va., Doc. No. PB87-142378.

EPA (U.S. Environmental Protection Agency). 1986b. Second Addendum to Air Quality Criteria for Particulate Matter and Sulfur Oxides (1982): Assessment of Newly Available Health Effects Information. EPA-600/8-86-020F. U.S. Environmental Protection Agency, Office of Health and Environmental Assessment, Washington, D.C. Available from NTIS, Springfield, Va., Doc. No. PB87-176574.

EPA (U.S. Environmental Protection Agency). 1988a. Drinking water regulations maximum contaminant level goals: National primary drinking water regulations for lead and copper. Federal Register 53:31515-31578.

EPA (U.S. Environmental Protection Agency). 1988b. Special Report on Ingested Inorganic Arsenic. Skin Cancer; Nutritional Essentiality. EPA/625/3-87/013. Risk Assessment Forum, U.S. Environmental Protection Agency, Washington, D.C.

EPA (U.S. Environmental Protection Agency). 1989. Evaluation of the Potential Carcinogenicity of Lead and Lead Compounds: In Support of Reportable Quantity Adjustments Pursuant to CERCLA (Comprehensive Environmental Response, Compensation and Liability Act) Section 102. EPA/600/8-89-045A. U.S. Environmental Protection Agency, Office of Health and Environmental Assessment, Washington, D.C. Available from NTIS, Springfield, Va., Doc. No. PB89-181366.

EPA (U.S. Environmental Protection Agency). 1990. Methodology for Assessing Health Risks Associated with Indirect Exposure to Combustor Emissions, Interim Final. EPA/600/6-90-003. Office of Health and Environmental Assessment, U.S. Environmental Protection Agency, Washington, D.C.

EPA (U.S. Environmental Protection Agency). 1991a. Maximum contaminant level goals and national primary drinking water regulations for lead and copper. Federal Register 56:26461-26564.

EPA (U.S. Environmental Protection Agency). 1991b. Feasibility of Environmental Monitoring and Exposure Assessment for a Municipal Waste Combustor: Rutland, Vermont Pilot Study. EPA/ 600/8-91/007. U.S. Environmental Protection Agency, Office of Health and Environmental Assessment, Cincinnati, Ohio.

EPA (U.S. Environmental Protection Agency). 1992a. Screening Procedures for Estimating the Air Quality Impact of Stationary Sources, Revised. EPA/454/R-92/019. U.S. Environmental Protection Agency, Office of Air Quality Planning and Standards, Research Triangle Park, N.C.

EPA (U.S. Environmental Protection Agency). 1992b. Integrated Risk Information System (IRIS). [Online]. Available: HtmlResAnchor http://www.epa.gov/ngispgm3/iris/index.html. U.S. Environmental Protection Agency, Office of Health and Environmental Assessment, Environmental Criteria and Assessment Office, Cincinnati, Ohio.

EPA (U.S. Environmental Protection Agency). 1992c. Guidelines for Exposure Assessment. Fed. Regist. 57(May 29):22888-22938.

EPA (U.S. Environmental Protection Agency). 1992d. Dermal Exposure Assessment: Principles and Applications. Interim Report. EPA/600/8-91/011B. U.S. Environmental Protection Agency, Office of Health and Environmental Assessment, Washington, D.C.

EPA (U.S. Environmental Protection Agency). 1992e. Workgroup Report to the Administrator, Vol. 1. R.M. Wolcott and W.A. Banks, eds. Report No. EPA/230/P-92/008. U.S. Environmental Protection Agency, Washington, D.C..

EPA (U.S. Environmental Protection Agency). 1993. Air Quality Criteria for Oxides of Nitrogen. Vol. 1 of 3. EPA/600/8-91/049AF. U.S. Environmental Protection Agency, Office of Health and Environmental Assessment, Office of Research and Development, Research Triangle Park, N.C.

EPA (U.S. Environmental Protection Agency). 1994a. Characterization of Municipal Solid Waste in the United States: 1994 Update. EPA/530/S-94/042. Prepared by Franklin Associates, Ltd., Prairie Village, KS., for the U.S. Environmental Protection Agency, Office of Solid Waste and Emergency Response, Washington, D.C.

EPA (U.S. Environmental Protection Agency). 1994b. Estimating Exposure to Dioxin-Like Compounds, Vol. 2: Properties, Sources, Occurrence, and Background Exposures. EPA/600/6-88/ 005Cb. External Review Draft. U.S. Environmental Protection Agency, Office of Health and Environmental Assessment, Washington, D.C.

EPA (U.S. Environmental Protection Agency). 1994c. National Incinerator Testing and Evaluation Program: The Environmental Characterization of Refuse-derived Fuel (RDF) Combustion Technology, Mid-Connecticut Facility, Hartford, Connecticut. EPA-600/R-94-140.

EPA (U.S. Environmental Protection Agency). 1994d. Health Assessment for 2,3,7,8-Tetrachlorodibenzo-*p*-dioxin (TCDD) and Related Compounds, Vol. 3, p. 9-3. EPA/600/BP- 92/001c. Draft. U.S. Environmental Protection Agency, Office of Health and Environmental Assessment, Washington, D.C.

EPA (U.S. Environmental Protection Agency). 1995a. SCREEN3 User's Guide. EPA/454/B-95/ 004. U.S. Environmental Protection Agency, Office of Air Quality Planning and Standards, Research Triangle Park, N.C.

EPA (U.S. Environmental Protection Agency). 1995b. Screening Procedures for Estimating the Air Quality Impact of Stationary Sources, Revised. EPA-450/R-92-019. U.S. Environmental Protection Agency. Research Triangle Park, N.C.

EPA (U.S. Environmental Protection Agency). 1996a. National Dioxin Emissions from Medical Waste Incinerators. Emissions Standards Division, Office of Air Quality Planning and Standards, U.S. Environmental Protection Agency, Research Triangle Park, N.C.

EPA (U.S. Environmental Protection Agency). 1996b. Revised Standards for Hazardous Waste Combustors. Proposed Rule. Fed. Regist. 61(April 19):17358-17536.

EPA (U.S. Environmental Protection Agency). 1996c. New EPA RCRA enforcement policy offers compliance flexibility. Inside EPA Weekly Report 17(13):1, 6. March 29.

EPA (U.S. Environmental Protection Agency). 1997a. RCRA, Superfund & EPCRA Hotline Training Module. Introduction to: Boilers and Industrial Furnaces (40 CFR Part 266, Subpart H). EPA530-R-97-047. Office of Solid Waste and Emergency Response, U.S. Environmental Protection Agency.

EPA (U.S. Environmental Protection Agency). 1997b. Risk Assessment for the Waste Technologies Industries (WTI) Hazardous Waste Incineration Facility (East Liverpool, Ohio). EPA/905-R97-002, 002a-002h.

EPA (U.S. Environmental Protection Agency). 1997c. Mercury Study Report to Congress, Volume 1: Executive Summary. EPA-452/R-97-003. Office of Air Quality Standards and Office of Research and Development, U.S. Environmental Protection Agency.

EPA (Environmental Protection Agency). 1998a. Human Health Risk Assessment Protocol for Hazardous Waste Combustion Facilities, Volumes One-Three. EPA530-D-98-001A-C. Peer Review Draft. U.S. EPA, Office of Solid Waste, Center for Combustion Science and Engineering.

EPA (U.S. Environmental Protection Agency). 1998b. National Air Pollutant Emission Trends Update:1970-1997. EPA 454/E-98-007. (December).

EPA (U.S. Environmental Protection Agency). 1998c. National Air Quality and Emissions Trends Report, 1997 . Office of Air Quality Planning and Standards. Emissions Monitoring and Analysis Division. EPA 454/R-98-016 (December)

EPA (U.S. Environmental Protection Agency). 1999. Residual Risk. Report to Congress. EPA-453/R-99-001. Office of Air Quality Planning and Standards. Environmental Protection Agency. (March).

Facchetti, S., and F. Geiss. 1982. Isotopic Lead Experiment: Status Report. Publ. No. EUR 8352 EN. Luxembourg: Commission of the European Communities.

Fagala, G.E., and C.L. Wigg. 1992. Psychiatric manifestations of mercury poisoning. J. Am. Acad. Child Adolesc. Psychiatry 31(2):306-311.

Fangmark, I., B. van Bavel, S. Marklund, B. Strömberg, N. Berge, and C. Rappe. 1993. Influence of combustion parameters on the formation of polychlorinated dibenzo-p-dioxins, dibenzofurans, benzenes, and biphenyls and polyaromatic hydrocarbons in a pilot incinerator. Environ. Sci. Technol. 27(8):1602-1610.

Felsvang, K.S., and O. Helvind. 1991. Results of Full Scale Dry Injection Tests at MSW-Incinerators Using a New Active Absorbent. Pp. 542 in Municipal Waste Combustion: Conference Papers and Abstracts from the Second Annual International Specialty Conference, April 15-19, 1991, Tampa, Fla.

Fernandez-Salguero, P.M., D.M. Hilbert, S. Rudikoff, J.M. Ward, and F.J. Gonzalez. 1996. Arylhydrocarbon receptor-deficient mice are resistant to 2,3,7,8-tetrachlorodibenzo-*p*-dioxin-induced toxicity. Toxicol. Appl. Pharmacol. 140:173-179.

Ferris, D. 1995. Burning injustice: An overview of federal incinerator policy in the United States. Clearinghouse Rev. 29(4):424-435.

Fiedler, D.A., K.W. Brown, J.C. Thomas, and K.C. Donnelly. 1991. Mutagenic potential of plants grown on municipal sewage sludge-amended soil. Arch. Environ. Contam. Toxicol. 20(3):385-390.

Fingerhut, M.A., W.E. Halperin, D.A. Marlow, L.A. Piacitelli, P.A. Honchar, M.H. Sweeney, A.L. Greife, P.A. Dill, K. Steenland, and A.J. Suruda. 1991. Cancer mortality in workers exposed to 2,3,7,8-tetrachlorodibenzo-*p*-dioxin. N. Engl. J. Med. 324(4):212-218.

Finkelstein, A., R. Klicius, and D. Hay. 1987. The National Incinerator Testing and Evaluation Program (NITEP). Presented at the International Workshop on Municipal Waste Incineration. Montreal, Quebec, October 1-2, 1987.

Finsterbusch, K. 1985. State of the art in social impact assessment. Environ. Behav. 17(2):193-221.

Fiorino, D.J. 1990. Citizen participation and environmental risk: A survey of institutional mechanisms. Sci. Technol. Hum. Value 15(2):226-243.

Fischoff, B. 1995. Risk perception and communication unplugged: Twenty years of process. Risk Anal. 15(2):137-145.

Fitzgerald, W.F., and T.W. Clarkson. 1991. Mercury and monomethylmercury: Present and future concerns. Environ. Health Perspect. 96:159-166.

Fitzgerald, W.F., R.P. Mason, and G.M. Vandal. 1991. Atmospheric cycling and air-water exchange of mercury over mid-continental lacustrine regions. Water Air Soil Pollut. 56:745-768.

Flynn, C.B., J.H. Flynn, J.A. Chalmers, D. Pijawka, and K. Branch. 1983. An integrated methodology for large-scale development projects. In Social Impact Assessment Methods, K. Finsterbusch, L.G. Llewellyn, and C.P. Wolf, eds. Beverly Hills, Calif.: Sage Publications.

Foulkes, E.C. 1978. Renal tubular transport of cadmium-metallothionein. Toxicol. Appl. Pharmacol. 45(2):505-512.

Fournier, D.J., Jr., and L.R. Waterland. 1989. The Fate of Trace Metals in a Rotary Kiln Incinerator with a Single-Stage Ionizing Wet Scrubber. EPA Contract 68-03-0038.

Fournier, D.J., Jr., W.E. Whitworth, Jr., J.W. Lee, and L.R. Waterland. 1988. The Fate of Trace Metals in a Rotary Kiln Incinerator with a Venturi/Packed Column Scrubber. EPA Contract 68-C9-3267.

Frampton, M.W., P.E. Morrow, C. Cox, P.C. Levy, J.J. Condemi, D. Speers, F.R. Gibb, and M.J. Utell. 1995. Sulfuric acid aerosol followed by ozone exposure in healthy and asthmatic subjects. Environ. Res. 69(1):1-14.

Franchini, I., and A. Mutti. 1988. Selected toxicological aspects of chromium (VI) compounds. Sci. Total Environ. 71:379-387.

Franklin Associates (Franklin Associates, Ltd.). 1997. Solid Waste Management at the Crossroads. Franklin Associates, Ltd., Prairie Village, KS.

Franklin Associates (Franklin Associates, Ltd.). 1998. Characterization of Municipal Solid Waste in the United States: 1997 Update. Prepared for U.S. Environmental Protection Agency, Municipal and Industrial Solid Waste Division, Office of Solid Waste, Report No. EPA530-R-98-007. Franklin Associates, Ltd., Prairie Village, KS.

Franzblau, A., and R. Lilis. 1989. Acute arsenic intoxication from environmental arsenic exposure. Arch. Environ. Health 44:385-390.

Freudenburg, W.R. 1989. Social scientists' contributions to environmental management. J. Social Issues 45(1):133-152.

Freudenburg, W.R., and R. Gramling. 1992. Community impacts of technological change: Toward a longitudinal perspective. Social Forces 70(4):937-955.

Freudenberg, W.R., and R.E. Jones. 1991. Criminal behavior and rapid community growth: Examining the evidence. Rural Sociology 56(4)619-645.

Frey, H.C., R. Bharvirkar, and J. Zheng. 1999. Quantification of Variability and Uncertainty in Emission Factors. Paper No. 99-267, Proceedings of the 92nd Annual Meeting, June 20-24, 1999, St. Louis, MO., Air and Waste Management Association, Pittsburgh, PA.

Friberg, L. 1950. Health hazards in the manufacture of alkaline accumulators with special reference to chronic cadmium poisoning. Acta Med. Scand. 138(Suppl. 240):1-124.

Friberg, L., S. Hammarstrom, and A. Nystrom. 1953. Kidney injury after chronic exposure to inorganic mercury. Arch. Ind. Hyg. Occup. Med. 8:149-153.

Friberg, L., M. Piscator, G. Nordberg, and T. Kjellstrom. 1974. Cadmium in the Environment, 2d Ed. Cleveland: CRC.

Friberg, L., G.F. Nordberg, and V.B. Vouk. 1979. Pg. 505 in Handbook on the Toxicology of Metals. New York: Elsevier.

Fruin, S.A., M. Elnabarawy, and D.P. Duffy. 1994. A Multi-media Environmental Impact Study for a Manufacturing Facility Incinerator. Paper 94-TA260.13P. 87th Annual Meeting of the American Waste Management Association, Pittsburgh, PA.

Fulton, M., G. Raab, G. Thomson, D. Laxen, R. Hunter, and W. Hepburn. 1987. Influence of blood lead on the ability and attainment of children in Edinburgh. Lancet 1(8544):1221-1226.

Furst, P., C. Furst, and W. Groebel. 1990. Levels of PCDDs and PCDFs in food-stuffs from the Federal Republic of Germany. Chemosphere 20(7-9):787-792.

Galloway, J.N., J.D. Thornton, S.A. Norton, H.L. Volchok, and R.A. McLean. 1982. Trace metals in atmospheric deposition: A review and assessment. Atmos. Environ. 16:1677-1700.

GAO (U.S. General Accounting Office). 1991. Hazardous Waste: Incinerator Operating Regulations and Related Air Emissions Standards. October. U.S. General Accounting Office, Washington, D.C..

GAO (U.S. General Accounting Office). 1995a. SUPERFUND: EPA Has Identified Limited Alternatives to Incineration for Cleaning up PCB and Dioxin Contamination. GAO/RCED-96-13. U.S. General Accounting Office, Washington, D.C.

GAO (U.S. General Accounting Office). 1995b. Hazardous Waste Incinerators: EPA's and OSHA's Actions to Better Protect Health and Safety Not Complete. GAO/RCED-95-17. Gaithersburg, MD., U.S. General Accounting Office.

GAO (U.S. General Accounting Office). 1995c. Hazardous Waste: Benefits of EPA's Information System Are Limited. GAO/AMID-95-167. August. U.S. General Accounting Office, Washington, D.C.

Gartrell, M.J., J.C. Craun, D.S. Podrebarac, and E.L.Gunderson. 1986. Pesticides, selected elements, and other chemicals in adult total diet samples, October 1980-March 1982. J. Assoc. Off. Anal. Chem. 69(1):146-159.

Garvey, D.J., and L.D. Longo. 1978. Chronic low level maternal carbon monoxide exposure and fetal growth and development. Biol. Reprod. 19(1):8-14.

Gaspar, J.A. 1998. New Processes for the Removal of Mercury from Incinerator Flue Gases. Presented at International Conference on Incineration and Thermal Treatment Technologies, May 11-15, 1998, Salt Lake City, UT.

Gaspar, J.A., M.C. Widmer, J.A. Cole, and W.R. Seeker. 1997. Study of Mercury Speciation in a Simulated Municipal Waste Incinerator Flue Gas. Presented at International Conference on Incineration and Thermal Treatment Technologies, May 12-16, 1997, San Francisco, CA.

Gilman, A., R. Newhook, and B. Birmingham. 1991. An updated assessment of the exposure of Canadians to dioxins and furans. Chemosphere 23(11-12):1661-1668.

Glass, G.E., J.A. Sorensen, K.W. Scmidt, G.R. Rapp, Jr., D. Yap, and D. Fraser. 1991. Mercury deposition and sources for the upper Great Lakes region. Water Air Soil Pollut. 56:235-250.

Gleason, M.N., R.E. Gosselin, and H.C. Hodge. 1957. Clinical Toxicology of Commercial Products. Baltimore, MD.: Williams & Wilkins. 154 pp.

Goebel, H.H., P.F. Schmidt, J. Bohl, B. Tettenborn, G. Kramer, and L. Gutmann. 1990. Polyneuropathy due to acute arsenic intoxication: Biopsy studies. J. Neuropathol. Exp. Neurol. 49(2):137-149.

Goldman, B.A., and L.J. Fitton. 1994. Toxic Wastes and Race Revisited. Washington, D.C.: Center for Policy Alternatives, National Association for the Advancement of Colored People, and United Church of Christ Commission for Racial Justice.

Graham, D.L., and S.M. Kalman. 1974. Lead in forage grass from a suburban area in northern California. Environ. Pollut. 7:209-215.

Gramling, R., and W.R. Freudenburg. 1992. Opportunity-threat, development, and adaptation: Toward a comprehensive framework for social impact assessment. Rural Sociol. 57(2):216-234.

Gray, E.J., J.K. Peat, C.M. Mellis, J. Harrington, and A.J. Woolcock. 1994. Asthma severity and morbidity in a population sample of Sydney school children: Part I - Prevalence and effect of air pollutants in coastal regions. Aust. N.Z. J. Med. 24(2):168-175.

Greenberg, M., D. Schneider, and J. Parry. 1995. Brown fields, a regional incinerator and resident perception of neighborhood quality. Risk Health Safety Environ. 6(3):241-259.

Greenberg, M.R., and D. Schneider. 1996. Environmentally Devastated Neighborhoods: Perceptions, Policies and Realities. New Brunswick, N.J.: Rutgers University Press.

Greene, T., and C.B. Ernhart. 1991. Prenatal and preschool age lead exposure: Relationship with size. Neurotoxicol. Teratol. 13(4):417-427.

Greenspan, B.J., P.E. Morrow, and J. Ferin. 1988. Effects of aerosol exposures to cadmium chloride on the clearance of titanium dioxide from the lungs of rats. Exp. Lung Res. 14:491-500.

Gregory, R., J. Flynn, and P. Slovic. 1995. Technological stigma. Am. Sci. 83(3):220-223.

Grose, E.C., J.H. Richards, R.H. Jaskot, M.G. Menache, J.A. Graham, and W.C. Dauterman. 1987. A comparative study of the effects of inhaled cadmium chloride and cadmium oxide: Pulmonary response. J. Toxicol. Environ. Health 21(1-2):219-232.

Gundel, L.A.,V.C. Lee, K.R.R. Mahanama, R.K. Stevens, and J.M. Daisey. 1995. Direct determination of the phase distributions of semi-volatile polycyclic aromatic hydrocarbons using annular denuders. Atmos. Environ. 29(14):1719-1733.

Gunderson, E.L. 1995. FDA total diet study, July 1986-April 1991, dietary intakes of pesticides, selected elements, and other chemicals. J. AOAC Int. 78 (6):1353-1363.

Gustavsson, P. 1989. Mortality among workers at a municipal waste incinerator. Am. J. Ind. Med. 15:245-253.

Hadden, S.G. 1991. Public perception of hazardous waste. Risk Anal. 11(1):47-57.

Hafez, E.S.E. 1974. Reproduction in Farm Animals, 3d ed. Philadelphia: Lea & Febiger.

Hall, B.D., R.A. Bodaly, R.J.P. Fudge, J.W.M. Rudd, and D.M. Rosenberg. 1997. Food as the dominant pathway of methylmercury uptake by fish. Water Air Soil Pollut. 100 (1-2):13-24.

Hance, B.J., C. Chess, and P.M. Sandman. 1988. Improving Dialogue with Communities: A Risk Communication Manual for Government. New Jersey Dept. of Environmental Protection. Trenton, N.J.: Division of Science and Research.

Hance, B.J., C. Chess, and P.M. Sandman. 1990. Industry Risk Communication Manual: Improving Dialogue with Communities. Boca Raton, Fla.: Lewis.

Hanley, Q.S., J.Q. Koenig, T.V. Larson, T.L. Anderson, G. Van Belle, V. Rebolledo, D.S. Covert, and W.E. Pierson. 1992. Response of young asthmatic patients to inhaled sulfuric acid. Am. Rev. Respir. Dis. 145(2 Part 1):326-331.

Hanna, S.R., G.A. Briggs, R.P. Hosker, Jr., and J.S. Smith. 1982. Handbook on Atmospheric Diffusion. QC880.4.D44.H36. Oak Ridge, Tenn.: Technical Information Center, Oak Ridge National Laboratory, U.S. Department of Energy.

Hänninen, H., P. Mantere, S. Hernberg, A.M. Seppäläinen, and B. Kock. 1979. Subjective symptoms in low-level exposure to lead. Neurotoxicology 1(2):333-348.

Harada, M. 1976. Intrauterine poisoning, clinical and epidemiological studies and significance of the problem. Bull. Inst. Const. Med. Kumamoto Univ. 25(Suppl.) 1-60.

Hardy, H.L., and I.R. Tabershaw. 1946. Delayed chemical pneumonitis occurring in workers exposed to beryllium compounds. J. Ind. Hyg. Toxicol. 28(5):197-211.

Harrington, J.M., J.P. Middaugh, D.L. Morse, and J. Housworth. 1978. A survey of a population exposed to high concentrations of arsenic in well water in Fairbanks, Alaska. Am. J. Epidemiol. 108(5):377-385.

Hart, R.P., C.S. Rose, and R.M. Hamer. 1989. Neuropsychological effects of occupational exposure to cadmium. J. Clin. Exp. Neuropsychol. 11(6):933-943.

Hartzell, G.E., H.W. Stacy, W.G. Switzer, D.N. Priest, and S.C. Packham. 1985. Modeling of toxicological effects of fire gases. IV. Intoxication of rats by carbon monoxide in the presence of an irritant. J. Fire Sci. 3:263-279.

Harvey, P.G., M.W. Hamlin, R. Kumar, G. Morgan, et al. 1988. Relationships between blood lead, behaviour, psychometric and neuropsychological test performance in young children. Br. J. Dev. Psychol. 6(2):145-156.

Hattemer-Frey, H.A., and C.C. Travis. 1989. Comparison of human exposure to dioxin from municipal waste incineration and background environmental contamination. Chemosphere 18(1-6):643-650.

Hayes, R.B., A.M. Lilienfeld, and L.M. Snell. 1979. Mortality in chromium chemical production workers: A prospective study. Int. J. Epidemiol. 8(4):365-374.

Hershkowitz, A., and E. Salerni. 1987. Garbage Management in Japan: Leading the Way. New York, NY : INFORM.

Hill, W.H. 1943. A report on two deaths from exposure to the fumes of di-ethyl mercury. Can. J. Public Health 34:158-160.

Hill, W.J., and W.S. Ferguson. 1979. Statistical analysis of epidemiological data from a chromium chemical manufacturing plant. J. Occup. Med. 21:103-106.

Hirano, S., N. Tsukamoto, S. Higo, and K.T. Suzuki. 1989a. Toxicity of cadmium oxide instilled into the rat lung. II. Inflammatory responses in broncho-alveolar lavage fluid. Toxicology 55(1/2):25-35.

Hirano, S., N. Tsukamoto, E. Kobayashi, and K.T. Suzuki. 1989b. Toxicity of cadmium oxide instilled into the rat lung. I. Metabolism of cadmium oxide in the lung and its effects on essential elements. Toxicology 55(1/2):15-24.

Hoffman, F.O., C.F. Baes, III, D.E. Dunning, Jr., D.E. Fields, C.A. Little, C.W. Miller, T.H. Orton, E.M. Rupp, D.L. Shaeffer, and R.W. Shor. 1979. A Statistical Analysis of Selected Parameters for Predicting Food Chain Transport and Internal Dose of Radionuclides. NURRG/CR-1004, ORNL, NUREG/TM-282. Prepared for the U.S. Nuclear Regulatory Commission by Oak Ridge National Laboratory, Oak Ridge, Tenn.

Howard, P.H. 1990a. Handbook of Environmental Fate and Exposure Data for Organic Chemicals, Vol. I. Chelsea, Mich.: Lewis Publishers.

Howard, P.H. 1990b. Handbook of Environmental Fate and Exposure Data for Organic Chemicals, Vol. II. Chelsea, Mich.: Lewis Publishers.

Howard, P.H. 1991. Handbook of Environmental Fate and Exposure Data for Organic Chemicals, Vol. III. Chelsea, Mich.: Lewis Publishers.

Hsu, C.-C., M.-L.M. Yu, Y.-C.J. Chen, et al. 1994. The Yu-Cheng rice oil poisoning incident. Pp. 661-684 in Dioxins and Health. A. Schecter, ed. New York: Plenum Press.

Hu, H. 1991. Knowledge of diagnosis and reproductive history among survivors of childhood plumbism. Am. J. Public Health 81(8):1070-1072.

Huang, J.X., F.S. He, Y.G. Wu, and S.C. Zhang. 1988. Observations on renal function in workers exposed to lead. Sci. Total Environ. 71(3):535-537.

Hudson, R.J.M., S.A. Gherini, C.J. Watras, and D.B. Porcella. 1994. Modeling the biogeochemical cycle of mercury in lakes: The mercury cycling model (MCM) and application to the MTL study lakes. Pp. 473-523 in Mercury Pollution Integration and Synthesis, C.J. Watras, and J.W. Huckabee, eds. Boca Raton, Fla.: Lewis Publishers.

Hülster, A., and H. Marschner. 1993. Transfer of PCDD/PCDF from contaminated soils to food and fodder crop plants. Chemosphere 27(1-3):439-446.

Hunt, G., B. Maisel, and M. Hoyt. 1991. Post-Operational Monitoring for PCDDs/PCDFs in the Vicinity of Resource Recover Facilities in the State of Connecticut. Prepared by ENSR for State of Connecticut Department of Environmental Protection.

IAEA (International Atomic Energy Agency). 1989. Evaluating the Reliability of Predictions Made Using Environmental Transfer Models, Safety Series 100. Vienna: International Atomic Energy Agency.

IARC (International Agency for Research on Cancer). 1993. Cadmium and certain cadmium compounds. Pp.119-146, 210-236 in IARC Monographs on the Evaluation of the Carcinogenic Risk of Chemicals to Humans. Beryllium, Cadmium, Mercury and Exposures in the Glass Manufacturing Industry. Vol. 58. Lyon, France: World Health Organization, International Agency for Research on Cancer.

IAWG (The International Ash Working Group). 1995. An International Perspective on Characterisation and Management of Residues from Municipal Solid Waste Incineration, Final Document, Volume 1. The International Ash Working Group.

Ide, C.W., and G.R. Bullough. 1988. Arsenic and old glass. J. Soc. Occup. Med. 38(3):85-88.

IOM (Institute of Medicine). 1999. Toward Environmental Justice: Research, Education and Health Policy Needs. Washington, D.C.: National Academy Press.

Ito, K., P.L. Kinney, and G.D. Thurston. 1995. Variations in PM-10 concentrations within two metropolitan areas and their implications for health effects analyses. Inhalation Toxicol. 7(5):735-745.

IWSA (Integrated Waste Services Association). 1999. Fast Facts about Waste-to-Energy. [Online]. Available: HtmlResAnchor http://www.wte.org/facts.html [July 13, 1999].

Jaffe, K.M., D.B. Shurtleff, and W.O. Robertson. 1983. Survival after acute mercury vapor poisoning. Am. J. Dis. Child. 137(8):749-751.

JAMA (Journal of the American Medical Association). 1994. Blood lead levels–United States, 1988-1991. 272(13):999.

Jarup, L., C.G. Elinder, and G. Spang. 1988. Cumulative blood-cadmium and tubular proteinuria: A dose-response relationship. Int. Arch. Occup. Environ. Health 60(3):223-229.

Johnson, B.B. 1993. Advancing understanding of knowledge's role in lay risk perception. Risk Issues Health Safety 4(3):189.

Kagamimori, S., M. Watanabe, H. Nakagawa, Y. Okumura, and S. Kawano. 1986. Case-control study on cardiovascular function in females with a history of heavy exposure to cadmium. Bull. Environ. Contam. Toxicol. 36(4):484-490.

Kahneman, D., and A. Tversky. 1982. The psychology of preferences. Sci. Am. 246(1):160-173.

Kahneman, P., P. Slovic, and A. Tversky. 1982. Judgment Under Uncertainty: Heuristics and Biases. New York: Cambridge University Press.

Kang-Yum, E., and S.H. Oransky. 1992. Chinese patent medicine as a potential source of mercury poisoning. Vet. Hum. Toxicol. 34:235-238.

Kaplan, H.L. 1987. Effects of irritant gases on avoidance/escape performance and respiratory response of the baboon. Toxicology 47(1-2):165-179.

Kaplan, H.L., A. Anzueto, W.G. Switzer, and R.K. Hinderer. 1988. Effects of hydrogen chloride on respiratory response and pulmonary function of the baboon. J. Toxicol. Environ. Health 23(4):473-493.

Karickhoff, S.W. 1981. Semiempirical estimation of sorption of hydrophobic pollutants on natural sediments and soils. Chemosphere 10(8):833-846.

Kasperson, R.E., O. Renn, P. Slovic, H.S. Brown, J. Emel, R. Goble, J.X. Kasperson, and S. Ratick. 1988. The social amplification of risk—A conceptual-framework. Risk Anal. 8(2):177-187.

Kasperson, R.E., and I. Palmlund. 1989. Evaluating Risk Communication. Pp. 143-158 in Effective Risk Communication: The Role and Responsibility of Government and Nongovernment Organizations, V.T. Covello, D.B. McCallum, and M.T. Pavlova, eds. New York: Plenum.

Kasperson, R., D. Golding, and S. Tuler. 1992. Social distrust as a factor in siting hazardous facilities and communicating risks. J. Soc. Issues 48(4):161-187.

Kazantzis, G., T.H. Lam, and K.R. Sullivan. 1988. Mortality of cadmium-exposed workers. A five-year update. Scand. J. Work Environ. Health 14:220-223.

Kehoe, R.A. 1961. The Harben lectures, 1960: The metabolism of lead in man in health and disease. J. R. Inst. Public Health Hyg. 24(4):81-97;129-143;177-203.

Keskinen, H., P.L. Kalliomaki, and K. Alanko. 1980. Occupational asthma due to stainless steel welding fumes. Clin. Allergy 10(2):151-159.

Kimmel, G.L. 1988. Reproductive and Developmental Toxicity of 2,3,7,8-TCDD, Appendix C in A Cancer Risk-Specific Dose Estimate for 2,3,7,8-TCDD, Appendices A Through F. U.S. EPA, External Review Draft.

Kimmel, C.A., and J. Buelke-Sam. 1994. Target Organ Toxicology Series: Developmental Toxicology, Second Ed. New York: Raven Press.

Kinney, P.L., K. Ito, and G.D. Thurston. 1995. A sensitivity analysis of mortality/PM-10 associations in Los Angeles. Inhalation Toxicol. 7(1):59-69.

Klaassen, C.D., M.O. Amdur, and J. Doull, eds. 1995. Casarett and Doull's Toxicology: The Basic Science of Poisons, 5th Ed. New York: McGraw-Hill.

Kleinman, M.T., D.M. Davidson, R.B. Vandagriff, V.J. Caiozzo, and J.L. Whittenberger. 1989. Effects of short-term exposure to carbon monoxide in subjects with coronary artery disease. Arch. Environ. Health 44(6):361-369.

Kociba, R.J., P.A. Keeler, C.N. Park, and P.J. Gehring. 1976. 2,3,7,8-Tetrachlorodibenzo-*p*-dioxin (TCDD): Results of a 13-week oral toxicity study in rats. Toxicol. Appl. Pharmacol. 35:553-574.

Kociba, R.J., D.G. Keyes, J.E. Beyer, R.M. Carreon, C.E. Wade, D.A. Dittenber, R.P. Kalnins, L.E. Frauson, C.N. Park, S.D. Barnard, R.A. Hummel, and C.G. Humiston. 1978. Results of a two-year chronic toxicity and oncogenicity study of 2,3,7,8- tetrachlorodibenzo-*p*-dioxin in rats. Toxicol. Appl. Pharmacol. 46:279-303.

Kociba, R.J., D.G. Keyes, J.E. Beyer, R.M. Carreon, and P.J. Gehring. 1979. Long-term toxicologic studies of 2,3,7,8-tetrachlorodibenzo-*p*-dioxin (TCDD) in laboratory animals. Ann. N.Y. Acad. Sci. 320:397-404.

Koenig, J.Q., W.E. Pierson, and M. Horike. 1983. The effects of inhaled sulfuric acid on pulmonary function in adolescent asthmatics. Am. Rev. Respir. Dis. 128(2):221-225.

Koenig, J.Q., D.S. Covert, and W.E. Pierson. 1989. Effects of inhalation of acidic compounds on pulmonary function in allergic adolescent subjects. Environ. Health Perspect. 79:173-178.

Koshland, C.P. 1997. Combustion Processes–Emissions Monitoring and Intervention. CRISP/98/ES04705-110022. CRISP Data Base, National Institutes of Health.

Koutrakis, P., J.M. Wolfson, and J.D. Spengler. 1988. An improved method for measuring aerosol strong acidity: Results from a nine-month study in St. Louis, Missouri and Kingston, Tennessee. Atmos. Environ. 22(1):157-162.

Koutrakis, P., S.L.K. Briggs, and B.P. Leaderer. 1992. Source apportionment of indoor aerosols in Suffolk and Onondaga Counties, New York. Environ. Sci. Technol. 26(3):521-527.

Kriebel, D., J.D. Brain, N.L. Sprince, and H. Kazemi. 1988a. The pulmonary toxicity of beryllium. Am. Rev. Respir. Dis. 137(2):464-473.

Kriebel, D., N.L. Sprince, E.A. Eisen, and I.A. Greaves. 1988b. Pulmonary function in beryllium workers: Assessment of exposure. Br. J. Ind. Med. 45(2):83-92.

Kuehn, R.R. 1994. Remedying the unequal enforcement of environmental laws. Environmental justice: The merging of civil rights and environmental activism. St. John's J. Leg. Commentary 9(2):625-668.

Kunreuther, H.C., and J. Linnerooth. 1983. Risk Analysis and Decision Processes: The Siting of Liquefied Energy Gas Facilities in Four Countries. Berlin: Springer-Verlag.

Kunreuther, H., D. Easterling, W. Desvousges, and P. Slovic. 1990. Public attitudes toward siting a high-level nuclear waste repository in Nevada. Risk Anal. 10(4):469-484.

Kunreuther, H., K. Fitzgerald, and T.D. Aarts. 1993. Siting noxious facilities: A test of the facility siting credo. Risk Anal. 13(3):301-318.

Kurttio, P., J. Pekkanen, G. Alfthan, M. Paunio, J.J. Jaakkola, and O.P. Heinonen. 1998. Increased mercury exposure in inhabitants living in the vicinity of a hazardous waste incinerator: A 10-year follow-up. Arch. Environ. Health 53(2):129-137.

Kyle, R.A., and G.L. Pease. 1965. Hematologic aspects of arsenic intoxication. New Engl. J. Med. 273(1):18-23.

Ladd, J. 1959. The concept of community: A logical analysis. Pp. 269-293 in Community, NOMOS II, C. J. Friedrich, ed. New York: Liberal Arts Press.

Lahl, U., M. Wilken, B. Zeschmar-Lahl, and J. Jager. 1990. PCDD/PCDF Balance of Different Municipal Waste Management Methods. Tenth International Symposium on Chlorinated Dioxins and Related Compounds. Bayreuth, Germany, September 10-14, 1990.

Laird, F.N. 1989. The decline of deference: The political context of risk communication. Risk Anal. 9(4):543-550.

Lansdown, R., W. Yule, M.-A. Urbanowicz, and J. Hunter. 1986. The relationship between blood-lead concentrations, intelligence, attainment and behaviour in a school population: The second London study. Int. Arch. Occup. Environ. Health 57:225-235.

Last, J.A., D.M. Hyde, D.J. Guth, and D.L. Warren. 1986. Synergistic interaction of ozone and respirable aerosols on rat lungs I. Importance of aerosol acidity. Toxicology 39(3):247-258.

Lavelle, M., and M. Coyle. 1992. Unequal protection—The racial divide in environmental law. Natl. Law J. (Sept. 21), S1.

Lee-Feldstein, A. 1986. Cumulative exposure to arsenic and its relationship to respiratory cancer among copper smelter employees. J. Occup. Med. 28(4):296-302.

Lerner, B.J. 1993. Mercury Emission Control in Medical Waste Incineration. 93-MP-21.05. Presentation at the 86th Annual Meeting & Exhibition, Air & Waste Management Association, Denver, CO., June 13-18, 1993.

Lindberg, E., and G. Hedenstierna. 1983. Chrome plating: Symptoms, findings in the upper airways, and effects on lung function. Arch. Environ. Health 38:367-374.

Lindberg, S.E., and R.C. Harriss. 1981. The role of atmospheric deposition in an eastern U.S.A. deciduous forest. Water Air Soil Pollut. 16:13-31.

Lindquist, O. 1991. Mercury in the Swedish Environment. Water Air Soil Pollut. 55(1-2):1-261.

Lioy, P.J., J.M. Waldman, T. Buckley, J. Butler, and C. Pietarinen. 1990. The personal, indoor and outdoor concentrations of PM-10 measured in an industrial community during the winter. Atmos. Environ. B Urban Atmos. 24(1):57-66.

Lloyd, O.L., M.M. Lloyd, F.L. Williams, and A. Lawson. 1988. Twinning in human populations and in cattle exposed to air pollution from incinerators. Br. J. Ind. Med. 45(8):556-560.

Lorber, M., D. Cleverly, J. Schaum, L. Phillips, G. Schweer, and T. Leighton. 1994. Development and validation of an air-to-beef food chain model for dioxin-like compounds. Sci. Total Environ. 156:39-65.

Lorber, M., P. Pinsky, P. Gehring, C. Braverman, D. Winters, and W. Sovocool. 1998. Relationships between dioxins in soil, air, ash, and emissions from a municipal solid waste incinerator emitting large amounts of dioxins. Chemosphere 37(9-12):2173-2197.

Lucas, J.B., and R.S. Kramkowski. 1975. Health hazard evaluation determination. Report no. 74-87-221. Cincinnati, OH: U.S. Department of Health, Education, and Welfare, Center for Disease Control, National Institute for Occupational Safety and Health.

Lyman, W.J., W.F. Reehl, and D.H. Rosenblatt. 1982. Handbook of Chemical Property Estimation Methods. New York: McGraw-Hill.

Lynn, F.M. 1987. Citizen involvement in hazardous waste sites: Two North Carolina success stories. Environ. Impact Assess. Rev. 7:347-361.

Lynn, F., and G. Busenberg. 1995. Citizen advisory committees and environmental policy: What we know, what's left to discover. Risk Anal. 15(2):147-162.

Ma, X.F., J.G. Babish, J.M. Scarlett, W.H. Gutenmann, and D.J. Lisk. 1992. Mutagens in urine sampled repetitively from municipal refuse incinerator workers and water treatment workers. J. Toxicol. Environ. Health 37:483-494.

Machle, W., K.V. Kitzmiller, E.W. Scott, and J.F. Treon. 1942. The effect of the inhalation of hydrogen chloride. J. Ind. Hyg. Toxicol. 24:222-225.

Mackay, D. 1991. Multimedia Environmental Models: The Fugacity Approach. Boca Raton, Fla.: Lewis.

Maclaren, V.W. 1987. The use of social surveys in environmental impact assessment. Environ. Impact Assess. Rev. 7:363-375.

Malkin, R., P. Brandt-Rauf, J. Graziano, and M. Parides. 1992. Blood lead levels in incinerator workers. Environ. Res. 59(1):265-270.

Mancuso, T.F. 1951. Occupational cancer and other health hazards in a chromate plant: A medical appraisal. II. Clinical and toxicologic aspects. Ind. Med. Surg. 20(9):393-407.

Mancuso, T.F., and W.C. Hueper. 1951. Occupational cancer and other health hazards in a chromate plant: A medical appraisal. I. Lung cancers in chromate workers. Ind. Med. Surg. 20(8):358-363.

Manz, A., J. Berger, J.H. Dwyer, D. Flesch-Janys, S. Nagel, and H. Waltsgott. 1991. Cancer mortality among workers in a chemical plant contaminated with dioxin. Lancet 338(8773):959-964.

Maranelli, G., and P. Apostoli. 1987. Assessment of renal function in lead poisoned workers. Occup. Environ. Chem. Hazards 344-348.

Marcus, W.L., and A,S. Rispin. 1988. Threshold carcinogenicity using arsenic as an example. Pp. 133-158 in Risk Assessment and Risk Management of Industrial and Environmental Chemicals. Vol. XV. Cothern C.R., M.A. Mehlman and W.L. Marcus, eds. Princeton, N.J.: Princeton Scientific Publishing.

Marlowe, M,. A. Cossairt, C. Moon, J. Errera, A. MacNeel, R. Peak, J. Ray, and C. Schroeder. 1985. Main and interaction effects of metallic toxins on classroom behavior. J. Abnorm. Child Psychol. 13(2):185-198.

Marsh, D.O., G.J. Myers, T.W. Clarkson, L. Amin-Zaki, S. Tikriti, and M.A. Majeed. 1980. Fetal methylmercury poisoning: Clinical and toxicological data on 29 cases. Ann. Neurol. 7(4):348-353.

Marsh, D.O., G.J. Myers, T.W. Clarkson, L. Amin-Zaki, S. Tikriti, M.A. Majeed, and A.R. Dabbagh. 1981. Dose-response relationship for human fetal exposure to methylmercury. Clin. Toxicol. 18(11):1311-1318.

Marsh, D.O., T.W. Clarkson, C. Cox, G.J. Myers, L. Amin-Zaki, and S. Al-Tikriti. 1987. Fetal methylmercury poisoning. Relationship between concentration in single strands of maternal hair and child effects. Arch. Neurol. 44(10):1017-1022.

Marsh, D.O., M.D. Turner, J.C. Smith, P. Allen, and N. Richdale. 1995. Fetal methylmercury study in a Peruvian fish-eating population. Neurotoxicology 16(4):717-726.

Matte, T.D., J.P. Figueroa, G. Burr, J.P. Flesch, R.A. Keenlyside, and E.L. Baker. 1989. Lead exposure among lead-acid battery workers in Jamaica. Am. J. Ind. Med. 16(2):167-177.

Max, S.R., and E.K. Silbergeld. 1987. Skeletal muscle glucocorticoid receptor and glutamine synthetase activity in the wasting syndrome of rats treated with 2,3,7,8- tetrachlorodibenzo-p-dioxin. Toxicol. Appl. Pharmacol. 87:523-527.

McCrady, J.K. 1994. Vapor-phase 2,3,7,8-TCDD sorption to plant foliage—a species comparison. Chemosphere 28(1):207-216.

McCrady, J.K., and S.P. Maggard. 1993. Uptake and photodegradation of 2,3,7,8-tetrachlorodibenzo-p-dioxin sorbed to grass foliage. Environ. Sci. Technol. 27(2):343-350.

McKone, T.E. 1993. The precision of QSAR methods for estimating intermedia transfer factors in exposure assessments. SAR QSAR Environ. Res. 1:41-51.

McKone, T.E., and J.I. Daniels. 1991. Estimating human exposure through multiple pathways from air, water, and soil. Regul. Toxicol. Pharmacol. 13(1):36-61.

McKone, T.E., and P.B. Ryan. 1989. Human exposures to chemicals through food chains: An uncertainty analysis. Environ. Sci. Technol. 23:1154-1163.

McLaughlin, D.L., R.G. Pearson, and R.E. Clement. 1989. Concentrations of chlorinated dibenzo-p-dioxins (CDD) and dibenzofurans (CDF) in soil from the vicinity of a large refuse incinerator in Hamilton, Ontario. Chemosphere 18:851-854.

McMichael, A.J., G.V. Vimpani, E.F. Robertson, P.A. Baghurst, and P.D. Clark. 1986. The Port Pirie cohort study: Maternal blood lead and pregnancy outcome. J. Epidemiol. Commun. Health 40(1):18-25.

Menaghan, E.G. 1989. Role changes and psychological well-being: Variations in effects by gender and role repertoire. Social Forces 67(3):693-714.

Michelozzi, P., D. Fusco, F. Forastiere, C. Ancona, V. Dell'Orco, and C.A. Perucci. 1998. Small area study of mortality among people living near multiple sources of air pollution. Occup. Environ. Med. 55(9):611-615.

Milne, J., A. Christophers, and P. De Silva. 1970. Acute mercurial pneumonitis. Br. J. Ind. Med. 27:334-338.

Minar, D.W., and S. Greer. 1969. The Concept of Community: Readings with Interpretations. Chicago: Aldine Publishing Company.

Mishonova, V.N., P.A. Stepanova, and V.V. Zarudin. 1980. Characteristics of the course of pregnancy and births in women with occupational contact with small concentrations of metallic mercury vapors in industrial facilities. Gig. Truda. Prof. Zabol. 24(2):21-23.

MOEE (Ministry of the Environment). 1989. Upper Limits of Normal Contaminant Guidelines for Phytotoxicity Studies. ARB-138-88-Phyto. Toronto, Ontario.

Montgomery, L.E., and O. Carter-Pokras. 1993. Health status by social class and/or minority status: Implications for environmental equity research. Toxicol. Ind. Health. 9(5):729-773.

Moore, R.E., C.F. Baes, III, L.M. McDowell-Boyer, A.P. Watson, F.O. Hoffman, J.C. Pleasant, and C.W. Miller. 1979. AIRDOS-EPA: A Computerized Methodology for Estimating Environmental Concentrations and Dose to Man from Airborne Releases of Radionuclides. ORNL-5532. Oak Ridge, Tenn.: Oak Ridge National Laboratory.

Morgan, M.G., and M. Henrion. 1990. Uncertainty: A Guide to Dealing with Uncertainty in Quantitative Risk and Policy Analysis. New York : Cambridge University Press.

Morgan, M.G., B. Fischhoff, A. Bostrom, L. Lave, and C. J. Atman. 1992. Communicating risk to the public. Environ. Sci. Technol. 26(11):2048-2056.

Morrell, D. 1987. Siting and the politics of equity. Pp. 117-136 in Resolving Locational Conflict, R.W. Lake, ed. New Brunswick, N.J.: Center for Urban Policy Research. Reprinted from Hazardous Waste, 1984, 1:555-571.

Morrell, D., and C. Magorian. 1982. Siting Hazardous Waste Facilities: Local Opposition and the Myth of Preemption. Cambridge, Mass.: Ballinger Publishing.

Morris, R.D., E.N. Naumova, and R.L. Munasinghe. 1995. Ambient air pollution and hospitalization for congestive heart failure among elderly people in seven large U.S. cities. Am. J. Public Health 85(10):1361-1365.

Morton, W.E., and G.A. Caron. 1989. Encephalopathy: An uncommon manifestation of workplace arsenic poisoning? Am. J. Ind. Med. 15:1-5.

Murphy, M.J., E.J. Culliford, and V. Parsons. 1979. A case of poisoning with mercuric chloride. Resuscitation 7(1):35-44.

Murray, C.J.L., and A.D. Lopez. 1996. The Global Burden of Disease: A Comprehensive Assessment of Mortality and Disability from Diseases, Injuries, and Risk Factors in 1990 and Projected to 2020. Cambridge, Mass.: Harvard University Press.

Murray, F.J., F.A. Smith, K.D. Nitschke, C.G. Humiston, R.J. Kociba, and B.A. Schwetz. 1979. Three-generation reproduction study of rats given 2,3,7,8-Tetrachlorodibenzo-*p*-dioxin (TCDD) in the diet. Toxicol. Appl. Pharmacol. 50:241-252.

Nakachi, K., K. Imai, S-I. Hayashi, J. Watanabe, and K. Kawajiri. 1991. Genetic susceptibility to squamous cell carcinoma of the lung in relation to cigarette smoking dose. Cancer Res. 51(19):5177-5180.

Nakajima, D., Y. Yoshida, J. Suzuki, and S. Suzuki. 1995. Seasonal changes in the concentration of polycyclic aromatic hydrocarbons in azalea leaves and relationship to atmospheric concentration. Chemosphere 30(3):409-418.

National Conference of State Legislatures Environmental Justice Group. 1995. Environmental justice: A matter of perspective. Denver, Colo.: National Conference of State Legislatures.

Neas, L.M., D.W. Dockery, P. Koutrakis, D.J. Tollerud, and F.E. Speizer. 1995. The association of ambient air pollution with twice daily peak expiratory flow rate measurements in children. Am. J. Epidemiol. 141(2):111-122.

Needleman, H.L. 1979. Lead levels and children's psychologic performance. N. Engl. J. Med. 1301(3):163.

Needleman, H.L., and D.C. Bellinger. 1989. Type II fallacies in the study of childhood exposure to lead at low dose: A critical and quantitative review. In Lead Exposure and Child Development: An International Assessment. M. Smith, L.D. Grant, and A. Sors, eds. Lancaster, U.K.: Kluwer Academic Publishers.

Needleman, H.L., M. Rabinowitz, A. Leviton, S. Linn, and S. Schoenbaum. 1984. The relationship between prenatal exposure to lead and congenital anomalies. JAMA 251:2956-2959.

Neutra, R., J. Lipscomb, K. Satin, and D. Shusterman. 1991. Hypotheses to explain the higher symptom rates observed around hazardous waste sites. Environ. Health Perspect. 94:31-38.

Ng, Y.C., C.S. Colsher, and S.E. Thompson. 1982. Transfer Coefficients for Assessing the Dose from Radionuclides in Meat and Eggs. U.S. Nuclear Regulatory Commission. NUREG/CR-2976.

NIOSH (National Institute for Occupational Safety and Health). 1995. NIOSH Health Hazard Evaluation Report. HETA 90-0329-2482. New York City Department of Sanitation, New York, New York. U.S. Department of Health and Human Services, Public Health Service, Centers for Disease Control and Prevention, National Institute for Occupational Safety and Health.

Nordberg, G., S. Slorach, and T. Steinstrom. 1973. [Cadmium poisoning caused by a cooled-soft-drink machine.] Lakartidingen 70:601-604.

Nordberg, G.F., T. Kjellstrom, and M. Nordberg. 1985. Kinetics and metabolism. Pp. 103-178 in Cadmium and Health: A Toxicological and Epidemiological Appraisal. Vol. I. Exposure, Dose, and Metabolism. L. Friberg, C.G. Elinder, T. Kjellstrom, et al., eds. Boca Raton, Fla.: CRC Press.

Novey, H.S., M. Habib, and I.D. Wells. 1983. Asthma and IgE antibodies induced by chromium and nickel salts. J. Allergy Clin. Immunol. 72(4):407-412.

NRC (National Research Council). 1977. Medical and Biological Effects of Environmental Pollutants: Nitrogen Oxides. Washington, D.C.: National Academy Press. 333pp.

NRC (National Research Council). 1982. Risk and Decision Making: Perspectives and Research. Washington, D.C.: National Academy Press.

NRC (National Research Council). 1983. Risk Assessment in the Federal Government: Managing the Process. Washington, D.C.: National Academy Press.

NRC (National Research Council). 1988a. Complex Mixtures: Methods for In Vivo Toxicity Testing. Washington, D.C.: National Academy Press.

NRC (National Research Council). 1988b. Acceptable Levels of Dioxin Contamination in an Office Building Following a Transformer Fire. Washington, D.C.: National Academy Press.

NRC (National Research Council). 1989a. Improving Risk Communication. Washington, D.C: National Academy Press.

NRC (National Research Council). 1989b. Diet and Health: Implications for Reducing Chronic Disease Risk. Washington, D.C.: National Academy Press.

NRC (National Research Council). 1991a. Environmental Epidemiology: Vol. 1, Public Health and Hazardous Wastes. Washington, D.C.: National Academy Press.

NRC (National Research Council). 1991b. Human Exposure Assessment for Airborne Pollutants: Advances and Opportunities. Washington, D.C.: National Academy Press.

NRC (National Research Council). 1991c. Permissible Exposure Levels and Emergency Exposure Guidance Levels for Selected Airborne Contaminants. Washington, D.C.: National Academy Press.

NRC (National Research Council). 1992a. Biologic Markers in Immunotoxicology. Washington, D.C.: National Academy Press.

NRC (National Research Council). 1992b. Environmental Neurotoxicology. Washington, D.C.: National Academy Press.

NRC (National Research Council). 1993. Measuring Lead Exposure in Infants, Children, and Other Sensitive Populations. Washington, D.C.: National Academy Press.

NRC (National Research Council). 1994. Science and Judgment in Risk Assessment. Washington, D.C.: National Academy Press.

NRC (National Research Council). 1995. Biologic Markers in Urinary Toxicology. Washington, D.C.: National Academy Press.

NRC (National Research Council). 1996. Understanding Risk: Informing Decisions in a Democratic Society. Washington, D.C.: National Academy Press.

NRC (National Research Council). 1997. Environmental Epidemiology, Volume 2: Use of the Gray Literature and Other Data in Environmental Epidemiology. Washington, D.C.: National Academy Press.

NRC (National Research Council). 1998. Research Priorities for Airborne Particulate Matter. I. Immediate Priorities and a Long-Range Research Portfolio. Washington, D.C.: National Academy Press. .

NRC (National Research Council). 1999a. Carbon Filtration for Reducing Emissions from Chemical Agent Incineration. Washington, D.C.: National Academy Press.

NRC (National Research Council). 1999b. Arsenic in Drinking Water. Washington, D.C.: National Academy Press.

NRC (National Research Council). 1999c. Research Priorities for Airborne Particulate Matter. II. Evaluating Research Progress and Updating the Portfolio. Washington, D.C.: National Academy Press.

Nriagu, J.O. 1980. Cadmium in the Environment. Part I: Ecological Cycling. New York: John Wiley & Sons.

Nriagu, J.O. 1990. Global metal pollution: Poisoning the biosphere? Environment 32(7):7-11; 28-33.

NTP (National Toxicology Program). 1982a. Carcinogenesis Bioassay of 2,3,7,8-tetrachlorodibenzo-p-dioxin (CAS No 1746-01-6) in Swiss-Webster Mice (Dermal Study), Tech. Report No. 201. Bethesda, MD: Carcinogenesis Testing Program, National Cancer Institute, National Institute of Health. Reserch Triangle Park, NC: National Toxicology Program (NIH) DHHS 82-1757.

NTP (National Toxicology Program). 1982b. Carcinogenesis Bioassay of 2,3,7,8-tetrachlorodibenzo-p-dioxin (CAS No 1746-01-6) in Osborne-Mendel Rats and B6C3F1(Gavage Study), Tech. Report No. 209. Bethesda, MD: Carcinogenesis Testing Program, National Cancer Institute, National Institute of Health. Reserch Triangle Park, NC: National Toxicology Program (NIH) DHHS 82-1765.

NYSERDA (New York State Energy Research and Development Authority). 1987. Results of the Combustion and Emissions Research Project at the Vicon Incinerator Facility in Pittsfield, Massachusetts. Final Report, #87-16. Prepared by Midwest Research Institute.

NYSERDA (New York State Energy Research and Development Authority). 1989. Combustion and Emissions Testing at the Westchester County Solid Waste Incinerator. Volume 1. Report No. NYSERDA-89-4-VOL-1. Prepared for NYSERDA by Radian Corp., Research Triangle Park, NC.

NYSERDA (New York State Energy Research and Development Authority). 1990. Combustion and Emissions Testing at the Oswego County Municipal Solid Waste Incinerator. Report No. NYSERDA-90-10. Prepared for NYSERDA by Radian Corp., Research Triangle Park, NC.

O'Hare, M., L. S. Bacow, and D. Sanderson. 1983. Facility Siting and Public Opposition. New York: Van Nostrand Reinhold.

OECD (Organisation for Economic Co-operation and Development). 1996. Environmental Performance Reviews: United States. Paris: Organisation for Economic Co-operation and Development.

Oldiges, H., D. Hochrainer, and U. Glaser. 1989. Long-term inhalation study with Wistar rats and four cadmium compounds. Toxicol. Environ. Chem. 19(3/4):217-222.

Ong, C.N., G. Endo, K.S. Chia, et al. 1987. Evaluation of renal function in workers with low blood lead levels. Pp. 327-333 in Occupational and Environmental Chemical Hazards. V. Fao, E.A. Emmett, M. Maroni, et al., eds. Chinchester: Ellis Horwood Limited, 327-333.

Ontario Ministry of Agriculture and Food, Ontario Ministry of the Environment, Toxics in Food Steering Committee. 1988. Polychlorinated Dibenz-p-dioxins and Polychlorinated Dibenzofurans and other Organochlorine Contaminants in Food. Toronto.

Oreskes, N., K. Shrader-Frechette, and K. Belitz. 1994. Verification, validation, and confirmation of numerical models in the earth sciences. Science 263(5147):641-646.

Organ, J.M. 1995. Limitations of state agency authority to adopt environmental standards more stringent than federal standards: Policy considerations and interpretive problems. Md. L. Rev. 54(4):1373-1434.

Ostro, B. 1993. The association of air pollution and mortality: Examining the case for inference. Arch. Environ. Health 48(5):336-342.

OTA (Office of Technology Assessment). 1990. Finding the Rx for Managing Medical Wastes. OTA-O-459. Washington, D.C.: U.S. Government Printing Office.

Otway, H. 1987. Experts, risk communication, and democracy. Risk Anal. 7(2):125-129.

Otway, H., and B. Wynne. 1989. Risk communication: Paradigm and paradox. Risk Anal. 9(2):141-145.

Ozkaynak, H., and G.D. Thurston. 1987. Associations between 1980 U.S. mortality rates and alternative measures of airborne particle concentration. Risk Anal. 7(4):449-461.

Pani, B., U. Laureni, N. Babudri, A. Collareta, S. Venturini, R. Ferri, M. Carozzi, F. Burlini, and C. Monti-Bragadin. 1983. Mutagenicity test of extracts of airborne dust from the municipal incinerator of Trieste. Environ. Mutagen. 5:23-32.

Pasquill, F. 1961. The estimation of the dispersion of windborne material. Meteorol. Mag. 90(1063):33-49.

Pastides, H., R. Austin, S. Lemeshow, et al. 1991. An Epidemiological Study of Occidental Chemical Corporations Castle Hayne Chromate Production Facility. Occupational Epidemiology Unit, Univ. Mass. School of Public Health.

Paterson, S., and D. Mackay. 1991. Correlation of the equilibrium and kinetics of leaf-air exchange of hydrophobic organic chemicals. Environ. Sci. Technol. 25(5):866-871.

Paterson, S., W.Y. Shiu, D. Mackay, and J.D. Phyper. 1990. Dioxins from combustion processes: Environmental fate and transport. Pp. 405-423 in Emissions from Combustion Processes: Origin, Measurement, Control, R. Clement, and R. Kagel, eds. Boca Raton, Fla.: Lewis.

Paterson, S., D. Mackay, and C. McFarlane. 1994. A model of organic chemical uptake by plants from soil and the atmosphere. Environ. Sci. Technol. 28(13):2259-2266.

Pavlova, T.E. 1976. Disorders in the development of offspring following exposure of rats to hydrogen chloride. Biull Eksp Biol Med. 82(7):866-8.

Payne, B.A., S.J. Olshansky, and T.E. Segel. 1987. The effects on property-values of proximity to a site contaminated with radioactive-waste. Nat. Resour. J. 27(3):579-590.

Petruzzelli, S., A.M. Camus, L. Carrozzi, L. Ghelarducci, M. Rindi, G. Menconi, C.A. Angeletti, M. Ahotupa, E. Hietanen, A. Aitio, R. Saracci, H. Bartsch, and C. Giuntini. 1988. Long-lasting effects of tobacco smoking on pulmonary drug-metabolizing enzymes: A case-control study on lung cancer patients. Cancer Res. 48(16):4695-4700.

Piikivi, L. 1989. Cardiovascular reflexes and low long-term exposure to mercury vapor. Int. Arch. Occup. Environ. Health 61:391-395.

Piikivi, L., and A. Ruokonen. 1989. Renal function and long-term low mercury vapor exposure. Arch. Environ. Health 44(3):146-149.

Pirkle, J.L., J. Schwartz, J.R. Landis, and W.R. Harlan. 1985. The relationship between blood lead levels and blood pressure and its cardiovascular risk implications. Am. J. Epidemiol. 121(2):246-258.

Pirkle, J.L., L.L. Needham, and K. Sexton. 1995. Improving exposure assessment by monitoring human tissues for toxic chemicals. J. Expo. Anal. Environ. Epidemiol. 5(3):405-424.

Pocock, S.J., D. Ashby, and M.A. Smith. 1989. Lead exposure and children's intellectual performance: The Institute of Child Health/Southampton Study. Pp. 149-165 in Lead Exposure and Child Development: An International Assessment. M.A. Smith, L.D. Grant, and A.I. Sors, eds. Lancaster, U.K.: Kluwer Academic.

Pollock, C.A., and L.S. Ibels. 1986. Lead intoxication in paint removal workers on the Sidney Harbour Bridge. Med. J. Aust. 145:635-639.

Pope, C.A., III, D.W. Dockery, J.D. Spengler, and M.E. Raizenne. 1991. Respiratory health and PM10 pollution: A daily time series analysis. Am. Rev. Respir. Dis. 144(3):668-674.

Pope, C.A., III, J. Schwartz, and M.R. Ransom. 1992. Daily mortality and PM10 pollution in Utah Valley. Environ. Res. 47(3):211-217.

Pope, C.A., III, M.J. Thun, M.M. Namboodiri, D.W. Dockery, J.S. Evans, F.E. Speizer, and C.W. Heath, Jr. 1995. Particulate air pollution as a predictor of mortality in a prospective study of U.S. adults. Am. J. Respir. Crit. Care Med. 151(3 Pt. 1):669-674.

Poplin, D.E. 1972. Communities: A Survey of Theories and Methods of Research. New York: Macmillan.

Porcella, D.B. 1990. Mercury in the environment. EPRI J. April/May:46-49.

Porcella, D.B. 1994. Mercury in the environment: Biogeochemistry. Pp. 3-19 in Mercury Pollution Integration and Synthesis. Boca Raton, Fla: Lewis Publishers.

Portney, K.E. 1991. Siting Hazardous Waste Treatment Facilities: The Nimby Syndrome. New York: Auburn House.

Presidential/Congressional Commission on Risk Assessment and Risk Management. 1997. Framework for Environmental Health Risk Management. Final Report.

Pueschel, S.M., L. Kopito, and H. Schwachman. 1972. Children with an increased lead burden: A screening and follow-up study. JAMA 222:462-466.

Raiffa, H., W. Schwartz, and M. Weinstein. 1977. On evaluating health effects of societal programs in Volume IIb, Decision Making in the Environmental Protection Agency: Selected Working Papers. Commission on Natural Resources of the National Academy of Sciences. Washington, D.C.: National Academy Press.

Raizenne, M., L.M. Neas, A.I. Damokosh, D.W. Dockery, J.D. Spengler, P. Koutrakis, J.H. Ware, and F.E. Speizer. 1996. Health effects of acid aerosols on North American children: Pulmonary function. Environ. Health Perspect. 104(5):506-514.

Ramos, L., E. Eljarrat, L.M. Hernandez, L. Alonso, J. Rivera, and M.J. Gonzalez. 1997. Levels of PCDDs and PCDFs in farm cow's milk located near potential contaminant sources in Asturias (Spain). Comparison with levels found in control, rural farms and commercial pasteurized cow's milks. Chemosphere 35(10):2167-2179.

Rao, M.S., V. Subbarao, J.D. Prasad, and D.G. Scarpelli. 1988. Carcinogenicity of 2,3,7,8-tetrachlorodibenzo-p-dioxin in the Syrian golden hamster. Carcinogenesis 9(9):1677-1679.

Rapiti, E., A. Sperati, V. Fano, V. Dell'Orco, and F. Forastiere. 1997. Mortality among workers at municipal waste incinerators in Rome: A retrospective cohort study. Am. J. Ind. Med. 31(5):659-661.

Rau, J.G., and D.C. Wooten. 1980. Socioeconomic impact analysis. In Environmental Impact Analysis Handbook, J.G. Rau, and D.C. Wooten, eds. New York: McGraw Hill.

Rawls, J. 1971. A Theory of Justice. Cambridge, Mass.: Belknap Press of Harvard University Press.

Reitze, A.W., Jr., and A.N. Davis. 1993. Regulating municipal solid waste incinerators under the clean air act: History, technology and risks. 21 B. C. Environ. Aff. L. Rev. 1.

Renn, O., and D. Levine. 1991. Credibility and trust in risk communication. Pp. 175-218 in Communicating Risks to the Public: International Perspectives, R.E. Kasperson, and P.J.M. Stallen, eds. Dordrecht, The Netherlands: Kluwer Academic.

Reuber, B., D. Mackay, S. Paterson, and P. Stokes. 1987. A discussion of chemical equilibria and transport at the sediment-water interface. Environ. Toxicol. Chem. 6(10):731-740.

Richey, G. W. 1995. Compliance Strategies for Employee Exposures to Lead and Cadmium at Modular Mass Burn Municipal Incinerators. Presented at the 1995 American Industrial Hygiene Conference and Exposition, May, 1995, Kansas City, MO.

Rickson, R.E., R.J. Burdge, T. Hundloe, and G.T. MacDonald. 1990. Institutional constraints to adoption of social impact assessment as a decision-making and planning tool. Environ. Impact Assess. Rev. 10(1-2):233-243.

Riederer, M. 1990. Estimating partitioning and transport of organic chemicals in the foliage/atmosphere system: Discussion of a fugacity-based model. Environ. Sci. Technol. 24(6):829-837.

Robinson, G., S. Baumann, D. Kleinbaum, C. Barton, S. Schroeder, P. Mushak, and D. Otto. 1985. Effects of low to moderate lead exposure on brainstem auditory evoked potentials in children. Pp. 177-182 in Neurobehavioural Methods in Occupational and Environmental Health: Extended abstracts from the Second International Symposium, Copenhagen, 6-9 August 1985. Copenhagen: World Health Organization Regional Office for Europe.

Rodamilans, M., M.J. Osaa, J. To-Figueras, F. Rivera Fillat, J.M. Marques, P. Perez, and J. Corbella. 1988. Lead toxicity on endocrine testicular function in an occupationally exposed population. Hum. Toxicol. 7(2):125-128.

Roels, H.A., and R. Lauwerys. 1987. Evaluation of dose-effect and dose-response relationships for lead exposure in different Belgian population groups (fetus, child, adult men and women). Trace Elem. Med. 4:80-87.

Roels, H.A., R. Lauwerys, and A.N. Dardenne. 1983. The critical level of cadmium in human renal cortex: A re-evaluation. Toxicol. Lett. 15:357-360.

Roemer, W., G. Hoek, and B. Brunekreef. 1993. Effect of ambient winter air pollution on respiratory health of children with chronic respiratory symptoms. Am. Rev. Respir. Dis. 147(1):118-124.

Rom, W.N. 1992. Pp. 745-746 in Environmental and Occupational Medicine, 2nd Ed., W.N. Rom, ed. Boston: Little, Brown, and Co.

Ross, R.K., J.M. Yuan, M.C. Yu, G.N. Wogan, G.S. Qian, J.T. Tu, J.D. Groopman, Y.T. Gao, and B.E. Henderson. 1992. Urinary aflatoxin biomarkers and risk of hepatocellular carcinoma. Lancet 339(8799):943-946.

Rossman, M.D., J.A. Kern, J.A. Elias, M.R. Cullen, P.E. Epstein, O.P. Preuss, T.N. Markham, and R.P. Daniele. 1988. Proliferative response of bronchoalveolar lymphocytes to beryllium. A test for chronic beryllium disease. Ann. Intern. Med. 108(5):687-693. (Published erratum appears in Ann. Intern. Med. 1989, Apr. 15; 110(8)672.)

Ruckelshaus, W.D. 1985. Risk, science, and democracy. Issues Sci. Technol. 1(3):19-38.

Rummo, J.H. 1974. Intellectual and behavioral effects of lead poisoning in children. Chapel Hill, NC: University of North Carolina. University Microfilms, Ann Arbor MI, Publication No. 74-26-930.

Rummo, J.H., D.K. Routh, N.J. Rummo, and J.F. Brown. 1979. Behavioral and neurological effects of symptomatic and asymptomatic lead exposure in children. Arch. Environ. Health 34(2):120-124.

Rusch, G.M., J.S. O'Grodnick, and W.E. Rinehart. 1986. Acute inhalation study in the rat of comparative uptake, distribution and excretion for different cadmium containing materials. Am. Ind. Hyg. Assoc. J. 47(12):754-763.

Russell, C.S. 1990. Monitoring and enforcement. Pp. 243-274 in Public Policies for Environmental Protection, P.R. Portney, R.C. Dower, A.M. Freeman III, C.S. Russell, and M. Shapiro, eds. Washington, D.C.: Resources for the Future.

Sabljić, J., H. Güsten, J. Schönherr, and M. Riederer. 1990. Modeling plant uptake of airborne organic chemicals. 1. Plant cuticle/water partitioning and molecular connectivity. Environ. Sci. Technol. 24(9):1321-1326.

Saenger, P., M.E. Markowitz, and J.F. Rosen. 1984. Depressed excretion of 6 beta-hydroxycortisol in lead-toxic children. J. Clin. Endocrinol. Metab. 58(2):363-367.

Saldiva, P.H., A.J. Lichtenfels, P.S. Paiva, I.A. Barone, M.A. Martins, E. Massad, J.C. Pereira, V.P. Xavier, J.M. Singer, and G.M. Bohm. 1994. Association between air pollution and mortality due to respiratory diseases in children in Sao Paulo, Brazil: A preliminary report. Environ. Res. 65(2):218-225.

Samet, J.M., and M.J. Utell. 1990. The risk of nitrogen dioxide: What have we learned from epidemiological and clinical studies? Toxicol. Ind. Health 6(2):247-262.

Samuels, E.R., H.M. Heick, P.N. McLaine, and J.P. Farant. 1982. A case of accidental inorganic mercury poisoning. J. Anal. Toxicol. 6(3):120-122.

Sandman, P.M. 1985. Getting to maybe: Some communication aspects of siting hazardous waste facilities. Seton Hall Legislative J. 9:437-465.

Sassi, C. 1956. Occupational pathology in a chromate plant. Med. Lav. 47(5):314-327.

Scarlett, J.M., J.G. Babish, J.T. Blue, S.E. Voekler, and D.J. Lisk. 1990. Urinary mutagens in municipal refuse incinerator workers and water treatment workers. J. Toxicol. Environ. Health 31:11-27.

Schaum, J., D. Cleverly, M. Lorber et al. 1994. Updated analysis of U.S. sources of dioxin-like compounds and background exposure levels. Organohalogen Compounds 20:178-184.

Schecter, A., R. Malkin, O. Papke, M. Ball, and P.W. Brandt-Rauf. 1991. Dioxin levels in blood of municipal incinerator workers. Med. Sci. Res. 19(11):331-332.

Schecter, A., J.J. Ryan, Y. Masuda, P. Brandt-Rauf, J. Constable, H.D. Cau, L.C. Dai, H.T. Quynh, N.T.N. Phuong, and P.H. Phiet. 1994. Chlorinated and brominated dioxins and dibenzofurans in human tissue following exposure. Environ. Health Perspect. 102(Suppl. 1):135-147.

Scheringer, M. 1996. Persistence and spatial range as endpoints of an exposure-based assessment of organic chemicals. Environ. Sci. Technol. 30:1652-1659.

Schnoor, J.J., and D.C. McAvoy. 1981. Pesticide Transport and Bioconcentration Model. J. Environ. Eng. Div. ASCE 107(EE6):1229-1246.

Schreiber, L., and J. Schönherr. 1992. Uptake of organic chemicals in conifer needles: Surface adsorption and permeability of cuticles. Environ. Sci. Technol. 26:153-159.

Schroeder, S.R., and B. Hawk. 1987. Psycho-social factors, lead exposure, and IQ. Monogr. Am. Assoc. Ment. Defic. 8:97-137.

Schroeder, W.H., and F. J. Fanaki. 1988. Field measurement of water-air exchange of mercury in freshwater systems. Environ. Technol. Lett. 9(5):369-374.

Schroeder, W.H., M. Dobson, D.M. Kane, and N.D. Johnson. 1987. Toxic trace elements associated with airborne particulate matter: A review. J. Air Pollut. Control Assoc. 37(11):1267-1285.

Schuhmacher, M., S. Granero, A. Xifro, J.L. Domingo, J. Rivera, and E. Eljarrat. 1998. Levels of PCDD/Fs in soil samples in the vicinity of a municipal solid waste incinerator. Chemosphere 37(9-12):2127-2137.

Schwartz, J. 1988. The relationship between blood lead and blood pressure in the NHANES II survey. Environ. Health Perspect. 78:15-22.

Schwartz, J. 1991. Particulate air pollution and daily mortality: A synthesis. Public Health Rev. 19(1-4):39-60.

Schwartz, J. 1993. Air pollution and daily mortality in Birmingham, Alabama. Am. J. Epidemiol. 137(10):1136-1147.

Schwartz, J. 1994a. Air pollution and daily mortality: A review and meta analysis. Environ. Res. 64(1):36-52.

Schwartz, J. 1994b. Air pollution and hospital admissions for the elderly in Birmingham, Alabama. Am. J. Epidemiol. 139(6):589-598.

Schwartz, J. 1994c. Air pollution and hospital admissions for the elderly in Detroit, Michigan. Am. J. Respir. Crit. Care Med. 150(3):648-655.

Schwartz, J. 1994d. PM10, ozone and hospital admissions for the elderly in Minneapolis-St. Paul, Minnesota. Arch. Environ. Health 49(5):366-374.

Schwartz, J., and R. Levin. 1992. Lead: Example of the job ahead. EPA J. 18(1):42-44.

Schwartz, J., and D.A. Otto. 1987. Blood lead, hearing thresholds, and neurobehavioral development in children and youth. Arch. Environ. Health 42:153-160.

Schwartz, J., D. Wypij, D. Dockery, J. Ware, S. Zeger, J. Spengler, and B. Ferris, Jr. 1991. Daily diaries of respiratory symptoms and air pollution: Methodological issues and results. Environ. Health Perspect. 90:181-187.

Schwartz, J., D.W. Dockery, L.M. Neas, D. Wypij, J.H. Ware, J.D. Spengler, P. Koutrakis, F.E. Speizer, and B.G. Ferris, Jr. 1994. Acute effects of summer air pollution on respiratory symptom reporting in children. Am. J. Respir. Crit. Care Med. 150(5 Pt. 1):1234-1242.

Schwarz, M., and M. Thompson. 1990. Divided We Stand: Redefining Politics, Technology, and Social Choice. Philadelphia: University of Pennsylvania Press.

Sellakumar, A.R., C.A.Snyder, J.J. Solomon, and R.E. Albert. 1985. Carcinogenicity of formaldehyde and hydrogen chloride in rats. Toxicol. Appl. Pharmacol. 81(3 Pt. 1):401-6.

Seppäläinen, A.M., S. Hernberg, R. Vesanto, and B. Kock. 1983. Early neurotoxic effects of occupational lead exposure: A prospective study. NeuroToxicology 4(2):181-192.

Sexton, D.J., J.C. Smith, K.E. Powell, T.W. Clarkson, J. Liddle, and A. Smrek. 1978. A nonoccupational outbreak of inorganic mercury vapor poisoning. Arch. Environ. Health 33(4):186-191.

Sexton, K., D.K. Wagener, S.G. Selevan, T.O. Miller, and J.A. Lybarger. 1994. An inventory of human exposure-related data bases. J. Expo. Anal. Environ. Epidemiol. 4(1):95-109.

Seyfrit, C. L. 1988. A need for post-impact and policy studies: The case of the "Shetland experience". Sociol. Inquiry 58(2):206-215.

Shigematsu, I. 1984. The epidemeological approach to cadmium pollution in Japan. Ann. Acad. Med. Singapore 13:231-236.

Shy, C.M., D. Degnan, D.L. Fox, S. Mukerjee, M.J. Hazucha, B.A. Boehlecke, D. Rothenbacher, P.M. Briggs, R.B. Devlin, D.D. Wallace, R.K. Stevens, and P.A. Bromberg. 1995. Do waste incinerators induce adverse respiratory effects? An air quality and epidemiological study of six communities. Environ. Health Perspect. 103(7-8):714-724.

Sieber, S.D. 1974. Toward a theory of role accumulation. Am. Sociol. Rev. 39(4):567-578.

Siebert, P.C., D.R. Alston, J.F. Walsh, and K.H. Jones. 1988. Effect of Control Equipment and Operating Parameters on Municipal Solid Waste (MSW) Incinerator Trace Emissions. Paper No. 88-98.3 presented at the 81st Annual Meeting of APCA, June 19-24, 1988, Dallas, TX.

Simonich, S.L., and R.A. Hites. 1994a. Importance of vegetation in removing polycyclic aromatic hydrocarbons from the atmosphere. Nature 370(6484):49-51.

Simonich, S.L., and R.A. Hites. 1994b. Vegetation—atmosphere partitioning of polycyclic aromatic hydrocarbons. Environ. Sci. Technol. 28(5):939-943.

Simonich, S.L., and R.A. Hites. 1995. Global distribution of persistent organochlorine compounds. Science 269(5232):1851-1854.

Slob, W., O. Klepper, and J.A. van Jaarsveld. 1993. A Chain Model for Dioxins: From Emission to Cow's Milk. Report No. 730501039. National Institute of Public Health and Environmental Protection (RIVM), Bilthoven, The Netherlands.

Slovic, P. 1987. Perception of risk. Science 236:280-285.

Slovic, P. 1993. Perceived risk, trust, and democracy. Risk Anal. 13(6):675-681.

Slovic, P., B. Fischoff, and S. Lichtenstein. 1982. Rating the risks: The structure of expert and lay perceptions. Pp. 141-166 in Risk in the Technological Society, C. Hohenemser and J.X. Kasperson, eds. Boulder, Colo.: Westview.

Slovic, P., M. Layman, and J.H. Flynn. 1991. Risk perception, trust, and nuclear waste: Lessons from Yucca Mountain. Environment 33(3):6-11; 28-30.

Slovic, P., J. Flynn, and R. Gregory. 1994. Stigma happens: Social problems in the siting of nuclear waste facilities. Risk Anal. 14(5):773 -777.

Smolen, G.E., G. Moore, and L.V. Conway. 1992. Hazardous waste landfill impacts on local property values. Real Estate Appraiser 58(1):4-11.

Sobel, W., G.G. Bond, C.L. Baldwin, and D.J. Ducommun. 1988. An update of respiratory cancer and occupational exposure to arsenicals. Am. J. Ind. Med. 13:263-270.

Soni, J.P., R.U. Singhania, A. Bansa, et al. 1992. The uptake of methylmercury (203 Hg) in different tissues related to its neurotoxic effects. J. Pharmacol. Exp. Ther. 187:602-611.

Sovocool, G.W., J.R. Donnelly, W.D. Munslow, T.L. Vonnahme, N.J. Nunn, Y. Tondeur, and R.K. Mitchum. 1989. Analysis of municipal incinerator fly ash for bromo- and bromochloro-dioxins, dibenzofurans, and related compounds. Chemosphere 18(1-6):193-200.

Speizer, F.E. 1989. Studies of acid aerosols in six cities and in a new multi-city investigation: Design issues. Environ. Health Perspect. 79:61-67.

Speizer, F.E. 1999. Acid sulfate aerosols and health. Pp. 603-616 in Air Pollution and Health, S.T. Holgate, H.S. Koren, J.M. Samet, and R.L. Maynard, eds. San Diego: Academic Press.

Spengler, J.D., D.W. Dockery, W.A. Turner, J.M. Wolfson, and B.G. Ferris, Jr. 1981. Long-term measurements of respirable sulphates and particles inside and outside homes. Atmos. Environ. 15:23-30.

Stanley, J.S., K.E. Boggess, J. Onstot, T.M. Sack, J.C. Remmers, J. Breen, F.W. Kurz, J. Carra, P. Robinson, and G.A. Mack. 1986. PCDDS and PCDFS in human adipose tissue from the EPA FY82 NHATS repository. Chemosphere 15(9-12):1605-1612.

Steenland, K., and L. Stayner. 1997. Silica, asbestos, man-made mineral fibers, and cancer. Cancer Causes Control 8(3):491-503.

Steenland, K., and E. Ward. 1991. Lung cancer incidence among patients with beryllium disease: A cohort mortality study. J. Natl. Cancer Inst. 83(19):1380-1385.

Stern, F.B., W.E. Halperin, R.W. Hornung, V.L. Ringenburg, and C.S. McCammon. 1988. Heart disease mortality among bridge and tunnel officers exposed to carbon monoxide. Am. J. Epidemiol. 128(6):1276-1288.

Steverson, E.M. 1994. The U.S. approach to incinerator regulation. Pp. 113-135 in Waste Inciner-ation and the Environment, R.E. Hester and R.M. Harrison, eds. Cambridge, U.K.: Royal Society of Chemistry.

Stieglitz, L., and H. Vogg. 1987. On formation conditions of PCDD/PCDF in fly ash from munici-pal waste incinerators. Chemosphere 16(8-9):1917-1922.

Stokinger, H.E. 1981. The halogens and nonmetals boron and silicon. Pp. 2937-3043 in Patty's Industrial Hygiene and Toxicology, Vol. IIB., 3rd. Ed., G.D. Clayton, and F.E. Clayton, eds. New York: John Wiley & Son.

Stone, R. 1995. Dioxin receptor knocked out. Science 268:638-639.

Struempler, R.E., G.E. Larson, and B. Rimland. 1985. Hair mineral analysis and disruptive behav-ior in clinically normal young men. J. Learn. Disabil. 18(10):609-612.

Stubbs, K.P. 1993. The Impact of the GVRD Municipal Refuse Incinerator Emissions on Ambient Air Quality. Paper 93-RA-120.03 in the Proceedings of the 86th Annual Meeting of the Air & Waste Management Association. Pittsburgh, PA.: Air & Waste Management Association.

Stubbs, K.P., and O. Knizek. 1993. The Impact of the GVRD Municipal Refuse Incinerator Emis-sions on Regional Soils and Vegetation. Paper 93-RA-120.04 in the Proceedings of the 86th Annual Meeting of the Air & Waste Management Association. Pittsburgh, PA.: Air & Waste Management Association.

Studnicka, M.J., T. Frischer, R. Meinert, A. Studnicka-Benke, K. Hajek, J.D. Spengler, and M.G. Neumann. 1995. Acidic particles and lung function in children. A summer camp study in the Austrian Alps. Am. J. Respir. Crit. Care Med. 151(2 Pt. 1):423-430.

Sudol, F. 1994. The Newark Experience: A Case Study in Newark, New Jersey's Recycling Program. Presented at the Manhattan Citizens' Solid Waste Advisory Board Meeting, New York , N.Y., December 7, 1994.

Suskind, R.R., and V.S. Hertzberg. 1984. Human health effects of 2,4,5-T and its toxic contaminants. JAMA 251:2372-2380.

Susskind, L., and J.L. Cruikshank. 1987. Breaking the Impasse: Consensual Approaches to Resolving Public Disputes. New York: Basic Books.

Takenaka, S., H. Oldiges, H. Konig, D. Hochrainer, and G. Oberdorster. 1983. Carcinogenicity of cadmium chloride aerosols in W rats. J. Natl. Cancer Inst. 70(2):367-373.

Tarlock, A.D. 1987. State versus local control of hazardous waste facility siting: Who decides in whose backyard? Pp. 137-158 in Resolving Locational Conflict, R.W. Lake, ed. New Brunswick, N.J.: Center for Urban Policy Research. Reprinted from Zoning and Planning Law Report, 1984. New York: Clark Boardman.

Taueg, C., D.J. Sanfilippo, B. Rowens, J.Szejda, and J.L. Hesse. 1992. Acute and chronic poisoning from residential exposures to elemental mercury—Michigan, 1989-1990. J. Toxicol. Clin. Toxicol. 30(1):63-67.

Taylor, F.H. 1966. The relationship of mortality and duration of employment as reflected by a cohort of chromate workers. Am. J. Public Health 56:218-229.

ten Bruggen Cate, H.J. 1968. Dental erosion in industry. Br. J. Ind. Med. 25(4):249-266.

Tennant, R., H.J. Johnston, and J.B. Wells. 1961. Acute bilateral pneumonitis associated with the inhalation of mercury vapor: Report of five cases. Conn. Med. 25(2):106-109.

Theelen, R.M.C. 1991. Modeling of human exposure to TCDD and I-TEQ in the Netherlands: Background and occupational. Pp. 277-288 in Banbury Report, 35. Biological Basis for Risk Assessment of Dioxin and Related Compounds, M.A. Gallo, R.J. Scheuplein, and K.A. Van Der Heijden, eds. Plainview, N.Y.: Cold Spring Harbor Laboratory Press.

Thoits, P.A. 1983. Multiple identities and psychological well-being: A reformulation and test of the social isolation hypothesis. Am. Sociol. Rev. 48(2):174-187.

Thomas, K.W., E.D. Pellizzari, C.A. Clayton, D.A. Whitaker, R.C. Shores, J. Spengler, H. Ozkaynak, S.E. Froehlich, and L.A. Wallace. 1993. Particle Total Exposure Assessment Methodology (PTEAM) 1990 study: Method performance and data quality for personal, indoor, and outdoor monitoring. J. Expo. Anal. Environ. Epidemiol. 3(2):203-226.

Thun, M.J., T.M. Schnorr, A.B. Smith, W.E. Halperin, and R.A. Lemen. 1985. Mortality among a cohort of U.S. cadmium production workers—an update. J. Natl. Cancer Inst. 74(2):325-333.

Thurston, G. 1996. A critical review of PM_{10}-mortality time-series studies. J. Expos. Anal. Environ. Epidemiol. 6(1):3-21.

Thurston, G.D., K. Ito, P.L. Kinney, and M. Lippmann. 1992. A multi-year study of air pollution and respiratory hospital admissions in three New York State metropolitan areas: Results for 1988 and 1989 summers. J. Expos. Anal. Environ. Epidemiol. 2(4):429-450.

Thurston, G.D., K. Ito, C.G. Hayes, D.V. Bates, and M. Lippmann. 1994. Respiratory hospital admissions and summertime haze air pollution in Toronto, Ontario: Consideration of the role of acid aerosols. Environ. Res. 65(2):271-290.

TNRCC (Texas Natural Resource Conservation Commission). 1995. Critical Evaluation of the Potential Impact of Emissions from Midlothian Industries. Toxicology Risk Assessment Section, Office of Air Quality, Austin.

Tolls, J., and M.S. McLachlan. 1994. Partitioning of semivolatile organic compounds between air and Lolium multiflorum (Welsh ray grass). Environ. Sci. Technol. 28(10):159-166.

Topp, E., I. Scheunert, A. Attar, and F. Korte. 1986. Factors affecting the uptake of carbon-14 labeled organic chemicals by plants from soil. Ecotoxicol. Environ. Saf. 11(2):219-228.

Toth, K., S. Somfai-Relle, J. Sugar, and J. Bence. 1979. Carcinogenicity testing of herbicide 2,4,5-trichlorophenoxy ethanol containing dioxin and of pure dioxin in Swiss mice. Nature 278(5704):548-549.

Townshend, R.H. 1982. Acute cadmium pneumonitits: A 17-year follow-up. Br. J. Ind. Med. 39:411-412.

Trapp, S. and M. Matthies. 1997. Modeling volatilization of PCDD/F from soil and uptake into vegetation. Environ. Sci. Technol. 31(1):71-74.

Trapp, S., M. Matthies, I. Scheunert, and E.M. Topp. 1990. Modeling the bioconcentration of organic chemicals in plants. Environ. Sci. Technol. 24:1246-1252.

Travis, C.C., and A.D. Arms. 1988. Bioconcentration of organics in beef, milk, and vegetation. Environ. Sci. Technol. 22(3):271-274.

Travis, C.C, and B.P. Blaylock. 1992. Validation of terrestrial food chain models. J. Expo. Anal. Environ. Epidemiol. 2(2):221-239.

Travis, C.C., and S.C. Cook. 1989. Hazardous Waste Incineration and Human Health. Boca Raton, Fla.: CRC.

Travis, C.C., and H.A. Hattemer-Frey. 1987. Human exposure to 2,3,7,8-TCDD. Chemosphere 16(10-12):2331-2343.

Travis, C.C., and H.A. Hattemer-Frey. 1988. Uptake of organics by aerial plant parts: A call for research. Chemosphere 17(2):277-284.

Travis, C.C., and S.T. Hester. 1991. Global chemical pollution. Environ. Sci. Technol. 25(5):814-819.

Trenholm, A.R., and R. Thurnau. 1987. Total mass emissions from a hazardous waste incinerator. Pp. 304-317 in Land Disposal, Remedial Action, Incineration and Treatment of Hazardous Waste: Proceedings of the Thirteenth Annual Research Symposium, Cincinnati, Ohio, May 6-8, 1987. Report No. EPA/600/9-87/015. U.S. EPA, Office of Research and Development, Hazardous Waste Engineering Research Laboratory, Cincinnati, OH.

Trenholm, A., P. Gjorman, and G. Jungclaus. 1984. Performing Evaluation of Full-Scale Hazardous Waste Incinerators, Vols. 1-5. EPA/600/2-84/181a-3. U.S. Environmental Protection Agency, Industrial Environmental Research Laboratory, Cincinnati,Ohio.

Troen, P., S.A. Kaufman, and K.H. Katz. 1951. Mercuric bichloride poisoning. N. Engl. J. Med. 244:459-463.

Tsubaki, T., and H. Takahashi. 1986. Recent Advances in Minamata Disease Studies: Methylmercury poisoning in Minamata and Niigata, Japan. Tokyo: Kodansha.

Turner, D.B. 1970. Workbook of Atmospheric Dispersion Estimates. U.S. Environmental Protection Agency, Office of Air Programs, Research Triangle Park, N.C.

Tversky, A., and D. Kahneman. 1991. Loss aversion and riskless choice: A reference-dependent model. Q. J. Econ. 106(4):1039-1061.

U.K. Ministry of Health. 1954. Mortality and morbidity during the London fog of December 1952. Report by a committee of departmental officers and expert advisers appointed by the Minister of Health. Report on public health and medical subjects; v95. London: Her Majesty's Stationery Office.

U.S. Congress, House of Representatives. 1994. Public Health Impacts of Incineration, Hearings Before the Human Resources and Intergovernmental Relations Subcommittee of the House Committee on Governmental Operations, 103d Cong., 2d Sess.

United Church of Christ Commission for Racial Justice. 1987. Toxic Wastes and Race in the United States: A National Report on the Racial and Socio-economic Characteristics of Communities with Hazardous Waste Sites. New York: Public Data Access: Inquiries to the Commission.

Urabe, H., H. Koda, and M. Asahi. 1979. Present state of yusho patients. Ann. N.Y. Acad. Sci. 320:273-276.

USNRC (U.S. Nuclear Regulatory Commission). 1975. Reactor Safety Study: an Assessment of Accident Risks in U.S. Commercial Nuclear Power Plants. WASH-1400 (NUREG-75-014). Washington, D.C.

USNRC (U.S. Nuclear Regulatory Commission). 1977. Regulatory Guide 1.109, Calculation of Annual Doses to Man from Routine Releases of Reactor Effluents for the Purpose of Evaluating Compliance with 10 CFR Part 50 appendix I (Revision 1). Office of Standards Development.

Utell, M.J., P.E. Morrow, D.M. Speers, J. Darling, and R.W. Hyde. 1983. Airway responses to sulfate and sulfuric acid aerosols in asthmatics. An exposure-response relationship. Am. Rev. Respir. Dis. 128(3):444-450.

Van Miller, J.P., J.J. Lalich, and J.R. Allen. 1977. Increased incidence of neoplasms in rats exposed to low levels of 2,3,7,8-tetrachlorodibenzo-p-dioxin. Chemosphere 9:537-544.

Van Ordstrand, H.S., R. Hughes, J.M. DeNardi, and M.G. Carmody. 1945. Beryllium poisoning. JAMA 129(16):1084-1090.

Verschoor, M., A. Wibowo, R. Herber, J. van Hemmen, and R. Zielhuis. 1987. Influence of occupational low-level lead exposure on renal parameters. Am. J. Ind. Med. 12(4):341-351.

Verschueren, K. 1983. Handbook of Environmental Data on Organic Chemicals. New York: Van Nostrand Reinhold.

Visalli, J.R. 1987. A comparison of dioxin, furan and combustion gas data from test programs at three MSW incinerators. JAPCA 37(12):1451-63.

Vogg, H., H. Hunsinger, A. Merz, and L. Stieglitz. 1992. Influencing the production of dioxin-furan in solid waste incineration plants by measures affecting the combustion as well as the flue gas cleaning systems. Chemosphere 25(1-2):149-152.

Von Winterfeldt, D., and W. Edwards. 1984. Patterns of conflict about risky technologies. Risk Anal. 4(1):55-68.

Vroom, F.Q., and M. Greer. 1972. Mercury vapour intoxication. Brain 95(2):305-318.

Walsh, E., R. Warland, and D.C. Smith. 1993. Backyards, NIMBYs, and incinerator sitings: Implications for social movement theory. Soc. Probl. 40(1):25-38.

Wang, J.Y., T.R. Hsiue, and H.I. Chen. 1992. Bronchial responsiveness in an area of air pollution resulting from wire reclamation. Arch. Dis. Child. 67(4):488-490.

Wania, F., and D. Mackay. 1993. Global fractionation and cold condensation of low volatility organochlorine compounds in polar regions. Ambio 22(1):10-18.

Warkany, J., and D.M. Hubbard. 1953. Acrodynia and mercury. J. Pediatr. 42:365-386.

Waste-Tech (Waste Energy Technologies). 1991. The New York City Medical Waste Management Study, Task 4 Final Report: The New York City Medical Waste Management Plan. Prepared by Waste Energies Technologies, Houston, TX - New York, New York for The City of New York, New York City Health and Hospitals Corporation.

Watson, J.G., C.F. Rogers, and J.C. Chow. 1995. PM10 and PM2.5 Variations in Time and Space. Report No. 4204.1F. Prepared for TRC Environmental Corporation, Chapel Hill, N.C., by Desert Research Institute, Reno, NV.

Weast, R.C., M.J. Astle, and W.H. Beyer, eds. 1986. CRC Handbook of Chemistry and Physics: A Ready Reference Book of Chemical and Physical Data, 67th ed. Boca Raton, Fla.: CRC Press.

Webster, T., and P. Connett. 1998. Dioxin emission inventories and trends: The importance of large point sources. Chemosphere 37(9-12):2105-2118.

Weden, R.P., D.K. Mallik, and V. Batuman. 1979. Detection and treatment of occupational lead nephropathy. Arch. Intern. Med. 139:53-57.

Weinstein, N. 1988. Attitudes of the Public and the Department of Environmental Protection Toward Environmental Hazards. New Brunswick, N.J.: Department of Human Ecology and Psychology, Rutgers University.

Wernette, D.R., and L.A. Nieves. 1992. Breathing Polluted Air. EPA J. 18(1):16-17.

Whicker, F.W., and T.B. Kirchner. 1987. Pathway: A dynamic food-chain model to predict radionuclide ingestion after fallout deposition. Health Phys. 52(6):717-737.

Whitby, K.T. 1978. The physical characteristics of sulfur aerosols. Atmos. Environ 12:135-159.

WHO (World Health Organization). 1995. Update and Revision of the Air Quality Guidelines for Europe. Meeting of the Working Group "Classical" Air Pollutants, Bilthoven, The Netherlands, 11-14 October 1994. Report No. EUR/ICP/EHAZ 94 05/PB01. Copenhagen: World Health Organization Regional Office for Europe.

Williams, F.L.R., A.B. Lawson, and O.L. Lloyd. 1992. Low sex ratios of births in areas at risk from air pollution from incinerators, as shown by geographical analysis and 3-dimensional mapping. Int. J. Epidemiol. 21(2):311-319.

Wilson, W.E., and H.H. Suh. 1997. Fine particles and coarse particles: Concentration relationships relevant to epidemiologic studies. J. Air Waste Manage. Assoc. 47(12):1238-1249.

Winger, P.V., D.P. Schultz, and W.W. Johnson. 1990. Environmental contaminant concentrations in biota from the lower Savannah River, Georgia and South Carolina. Arch. Environ. Contam. Toxicol. 19(1):101-117.

Wohlslagel, J., L.C. DiPasquale, and E.H. Vernot. 1976. Toxicity of solid rocket motor exhaust: Effects of HCl, HF, and alumina on rodents. J. Combust. Toxicol. 3:61-70.

Wolf, C.P. 1980. Getting social impact assessment into the policy arena. Environ. Impact Assess. Rev. 1(1):27-36.

Woodruff, T.J., J. Grillo, and K.C. Schoendorf. 1997. The relationship between selected causes of postneonatal infant mortality and particulate air pollution in the United States. Environ. Health Perspect. 105(6):608-612.

Wu, M.-M., T.-L. Kuo, Y.-H. Hwang, and C.-J. Chen. 1989. Dose-response relation between arsenic concentration in well water and mortality from cancers and vascular diseases. Am. J. Epidemiol. 130(6):1123-1132.

Wynne, B. 1992. Risk and social learning: Reification to engagement. Chapter 12 in Social Theories of Risk, S. Krimsky, and D. Golding, eds. Westport, Conn.: Praeger.

Yanders, A.F., C.E. Orazio, R.K. Puri, and S. Kapila. 1989. On translocation of 2,3,7,8 tetrachlorodibenzo-p-dioxin time dependent analysis at the Times Beach experimental site. Chemosphere 19(1-6):429-432.

Yasuhara, A., H. Ito, and M. Morita. 1987. Isomer-specific determination of polychlorinated dibenzo-p-dioxins and dibenzofurans in incinerator-related environmental samples. Environ. Sci. Technol. 21:971-979.

Zaldivar, R. 1974. Arsenic contamination of drinking water and foodstuffs causing endemic chronic poisoning in man: Dose-response curve. Beitr. Pathol. 151:384-400.

Zeiss, C. 1991. Community decision-making and impact management priorities for siting waste facilities. Environ. Impact Assess. Rev. 11(3):231-255.

Zeiss, C., and J. Atwater. 1989. Waste facility impacts on residential property-values. J. Urban Plan. Dev. ASCE 115(2):64-80.

Zmirou, D., B. Parent, and J.-L. Potelon. 1984. Etude épidémiologique des effets sur la santé des rejets atmosphériques d'une usine d'incinération de déchets industriels et ménagers. Rev. Epidém. Santé Publ. 32:391-397.

Zober, A., P. Messerer, and P. Huber. 1990. Thirty-four-year mortality follow-up of BASF employees exposed to 2,3,7,8-TCDD after the 1953 accident. Int. Arch. Occup. Environ. Health 62:139-157.

Appendixes

APPENDIX A

Biographical Information on the Committee on Health Effects of Waste Incineration

Donald R. Mattison (Chair) is medical director at the March of Dimes in White Plains, New York. Previously, he was Dean of the Graduate School of Public Health and Professor of Environmental and Occupational Health and Obstetrics and Gynecology at the University of Pittsburgh. Dr. Mattison received his undergraduate education at Augsburg College (Minneapolis, Minnesota) majoring in chemistry and mathematics, and an M.S. in chemistry from the Massachusetts Institute of Technology. His medical education was received at The College of Physicians and Surgeons, Columbia University and clinical training in obstetrics and gynecology at Sloane Hospital for Women, Columbia Presbyterian Medical Center in New York. Dr. Mattison obtained postgraduate research training at the National Institutes of Health. From 1978 to 1984, Dr. Mattison was director of the Reproductive Toxicology Program in the Pregnancy Research Branch, National Institute of Child Health and Human Development, National Institutes of Health. From 1984 to 1990, Dr. Mattison was on the faculty of the University of Arkansas for medical sciences and advanced to professor of obstetrics and gynecology and professor of interdisciplinary toxicology. During this period, he was also acting director of the Human Risk Assessment Program at the National Center for Toxicological Research, a component of the Food and Drug Administration. Dr. Mattison moved to the University of Pittsburgh in 1990. Dr. Mattison is a diplomat of the American Board of Toxicology. He is a member of the Commission on Life Sciences and the Board on Environmental Studies and Toxicology (National Research Council), chair of the Board on Health Promotion and Disease Prevention (Institute of Medicine) and many other local and national boards. Dr. Mattison has chaired or co-chaired National Research Council committees on biomarkers in reproductive and developmental toxicology, pesticides in the diets of infants and children, and risk assessment methodology. He has published more than 150 papers, chapters, and reviews in the areas of public health, reproductive and developmental toxicology, risk assessment, and clinical obstetrics and gynecology.

Regina Austin is the William A. Schnader Professor of Law at the University of Pennsylvania. She received a B.A. from the University of Rochester in 1970 and a J.D. from Pennsylvania in 1973. She is a member of the Order of the Coif, the legal honorary society. Before joining the Penn faculty in 1977, Professor Austin was an associate with the firm of Schnader, Harrison, Segal & Lewis. She has also been a visiting professor at Harvard and Stanford Law Schools. Professor Austin has written on various topics including the working conditions of low status minority and/or female workers; law, culture and black women's lack of wealth; governmental restraints on black leisure; and the minority grassroots environmental movement.

Paul C. Chrostowski is principal of CPF Associates, Inc. He received a B.S. from the University of California, Berkeley and an M.S. and Ph.D. from Drexel University. Formerly, he was director of risk management programs for the Weinberg Group, on the faculty at Vassar College, a consultant in private practice, and vice president and director of research and development at ICF/Clement. He is a registered Qualified Environmental Professional. Dr. Chrostowski has over 25 years experience in industry, academia, and consulting in the fields of risk analysis, environmental chemistry, and environmental engineering. He has conducted environmental impact studies on risk assessments for over 100 waste management facilities on behalf of regulatory agencies or regulated parties.

Marjorie J. Clarke is an instructor in the Department of Geography at Rutgers University. She has been an environmental consultant specializing in waste incineration emissions control and waste prevention techniques and a fellow at the Center for Applied Studies of the Environment at the City University of New York. She received a B.A. in geology from Smith College, an M.A. in geography and environmental engineering from Johns Hopkins University, an M.S. in energy technologies from New York University, and a Ph.D. in earth and environmental sciences at the City University of New York. She has been chair of the Integrated Waste Management Technical Committee of the Air and Waste Management Association since 1994, was past chair of AWMA's Solid Waste Thermal Treatment technical committee, and past chair of the Manhattan Citizens' Solid Waste Advisory Board.

Edmund A. Crouch is a senior scientist for Cambridge Environmental, Inc., and an associate of the Department of Physics at Harvard University. Dr. Crouch holds a B.A. in natural sciences (Theoretical Physics) and a Ph.D. in high energy physics, both from Cambridge University, United Kingdom. Dr. Crouch has published widely in the areas of environmental quality, risk assessment, and presentation and analysis of uncertainties. He has co-authored a major text in risk assessment, *Risk/Benefit Analysis*. Dr. Crouch has served as an advisor to various local and national agencies concerned with public health and the environment. He has written computer programs for the sophisticated analysis of results from carcinogenesis bioassays; has developed algorithms (on the levels of both theory and computer implementation) for the objective quantification of waste site contamination; and has designed Monte Carlo simulations for purposes of fully characterizing uncertainties and variabilities inherent in health risk assessment.

Mary R. English is a research leader at the Energy, Environment, and Resources Center (EERC) at the University of Tennessee, Knoxville, as well as a member of the UT Waste Management Research and Education Institute. She also co-directs EERC's Program for Environmental Issues Analysis and Dialogue (Pro-Dialogue). She currently serves on the National Research Council's Board on Radioactive Waste Management and completed a 3-year term on the U.S. Environmental Protection Agency's National Environmental Justice Advisory Council. Her research since the 1970s has focused on environmental policy, particularly in the areas of waste management, land use, and energy. She has a Ph.D. in sociology and an M.S. in regional planning.

Dominic Golding is a research assistant professor at the George Perkins Marsh Institute at Clark University. He received his Ph.D. in geography from Clark University in 1988, where his research focused on occupational hazards and the social issues of risk assessment and risk management, especially with regard to nuclear power. His current research interests include the history and development of risk research, environmental equity, risk communication, and social trust. He is the author of "The Differential Susceptibility of Workers to Occupational Hazards: A Comparison of Policies in Sweden, Britain, and the United States," co-author of "Managing Nuclear Accidents: A Model Emergency Response Plan for Power Plants and Communities," and co-editor of "Social Theories of Risk," "Worst Things First: The Debate over Risk-Based National Environmental Policies," and "Preparing for Nuclear Power Plant Accidents."

Ian A. Greaves is associate professor in the Division of Environmental and Occupational Health and associate dean in the University of Minnesota School of Public Health. He received an undergraduate degree in biomedical science and a medical degree from Monash University, Australia. He was a National Health and Medical Research (Australia) traveling fellow and subsequently a faculty member in the Harvard School of Public Health. He is a fellow of the

Royal Australasian College of Physicians and of the American Association for the Advancement of Science, and a member of the American College of Occupational and Environmental Medicine. He directs the Midwest Center for Occupational Health and Safety at the University of Minnesota.

S. Katherine Hammond is an associate professor of environmental health sciences at the University of California, Berkeley, School of Public Health. She received her B.A. from Oberlin College, her Ph.D. in chemistry from Brandeis University, and her M.S. in environmental health sciences from Harvard School of Public Health, where she holds an appointment as Visiting Lecturer in Industrial Hygiene. Her research has focused on assessing exposure to complex mixtures for epidemiologic studies; among the exposures she has evaluated are those associated with work in the semiconductor industry, diesel exhaust, and environmental tobacco smoke. She served as a consultant to the U.S. Environmental Protection Agency Scientific Advisory Board in its review of the environmental tobacco smoke documents that culminated in the publication of *Respiratory Health Effects of Passive Smoking: Lung Cancer and Other Disorders,* and she is currently on the Acrylonitrile Advisory Panel for the National Cancer Institute.

Allen Hershkowitz is a senior scientist in NRDC's Urban program, specializing in solid waste management, recycling, medical wastes, and sludge. He is the originator, principal coordinator, and co-chairman of the Board of the Bronx Community Paper Company, a half-billion dollar paper recycling and community development project in the South Bronx area of New York City, which will convert waste paper from the metropolitan area (and beyond) into newsprint. Dr. Hershkowitz has served as the chairman of the New York State Department of Environmental Conservation Commissioner's Advisory Board on Operating Requirements for Municipal Solid Waste Incinerators. In the past he served on the EPA's Science Advisory Board Subcommittee on Sludge Incineration, as well as the Agency for Toxic Substances and Disease Registry's Peer Review Panel for its Report to Congress on the Health Implications of Medical Waste. Dr. Hershkowitz was the principal contractor for the United States Congress' Office of Technology Assessment's Report to Congress on Municipal Solid Waste Management. He was a member of the U.S. EPA's Regulatory Negotiations on Fugitive Emissions from Equipment Leaks at Synthetic and Organic Chemical Manufacturing Industries. Prior to going to NRDC he was the director of solid waste research at INFORM, an environmental research group. His training into advanced solid waste management strategies includes more than two dozen research visits to Japan and Europe, as well as extensive on-site research at solid waste management facilities throughout the United States.

Robert J. McCormick is founder and president of Franklin Engineering Group, Inc., an engineering firm specializing in the fields of waste combustion and air pollution control. He received a B.S. degree in chemical engineering from Washington University. He has over 20 years experience with industrial pollution control, and is a co-author of EPA's *Engineering Handbook for Hazardous Waste Incineration*.

Thomas E. McKone is group leader for exposure and risk analysis at the Lawrence Berkeley National Laboratory and an adjunct professor and researcher with the School of Public Health at the University of California, Berkeley. He has a Ph.D. and M.S. in engineering from UCLA. In addition to his membership on the Committee on Health Effects of Waste Incinerators, he is a member of the NAS Committee on Toxicology. He is also a member of the EPA Science Advisory Board, president of the International Society of Exposure Analysis (ISEA), and on the Council of the Society for Risk Analysis (SRA). He has served as a consultant to the International Atomic Energy Agency, the World Health Organization, and the Food and Agriculture Organization. He is an associate editor of the *Journal of Exposure Analysis and Environmental Epidemiology* and on the editorial board of the journal *Risk Analysis*.

Adel F. Sarofim is Presidential Professor of Chemical Engineering, University of Utah. He was affiliated with MIT from 1958 to 1996 where he held the position of Lammot du Pont Professor of Chemical Engineering from 1989 to 1996, emeritus from October 1, 1996. He has been a visiting professor at Sheffield University, England, the University of Naples, Italy; and at the California Institute of Technology. His awards include the Sir Alfred Egerton Gold Medal from the Combustion Institute; the Kuwait Prize for Petrochemical Engineering; the Walter Ahlström Environmental Prize of the Finnish Academies of Technology; the University of Pittsburgh's 1995 Award for Innovation in Coal Conversion; the DOE's 1996 Homer H. Lowry Award in Fossil Energy and the AIME/ASME 1996 Percy Nicholls Award.

Carl M. Shy received an M.D. degree from Marquette University School of Medicine (1962) and a doctor of public health degree in epidemiology from the University of Michigan School of Public Health (1967). Formerly, he was director of the Human Studies Laboratory at the U.S. EPA, and since 1974 has been on the faculty of the Department of Epidemiology, School of Public Health, University of North Carolina at Chapel Hill, where he is now a full professor. His career research and teaching interests have been focused on environmental and occupational health, with particular emphasis on air pollution and environmentally-related respiratory disease.

George D. Thurston is an associate professor on the faculty of the Department of Environmental Medicine at the New York University School of Medicine. His research primarily involves the measurement and health effects assessment of air pollutants. He has recently been called upon several times by the U.S. Congress during Clean Air Act hearings to testify regarding the known human health effects of air pollution. His past research has included field studies of the effects of ozone and acid aerosols on the incidence of asthma attacks in children, as well as time-series studies of the effects of air pollution on the incidence of daily hospital admissions and human mortality. Dr. Thurston has served as an associate editor of the *Journal of Exposure Analysis and Environmental Epidemiology* since 1993. Prior to joining the NYU faculty in 1984, he was a research fellow at Harvard University's Kennedy School of Government. Dr. Thurston received a B.S. in engineering from Brown University in 1974, and his doctorate from the Harvard School of Public Health in 1983.

APPENDIX B

Off-Normal Operations of Six Facilities

Box B-1
Prince Edward Island (Concord Scientific Corporation 1985)

The Prince Edward Island facility consists of three, two-stage incinerators, each rated at about 36 tons of waste per day. The incinerator design uses controlled or "starved"-air combustion (as contrasted with the excess-air operations used at Pittsfield and Westchester discussed later). Municipal solid waste is burned in the primary chamber, where a fraction of the total air needed for complete combustion is provided. The combustible gases enter the secondary chamber, where pre-heated air is added to complete combustion.

During testing of the primary chamber temperature was maintained at a relatively constant 1292°F — within $\pm 104^\circ$F, except for the low-temperature test, where it was maintained at 1250°F. The secondary-chamber temperatures were kept at 840° (1550°F) for two tests, at 1900°F for the high-temperature test, and at 1350°F for the low-temperature test. The percent excess air differed by about 40% between the tests involving normal and low secondary-chamber temperatures. The test data showed a tendency for dioxin concentrations to increase with increasing excess-oxygen concentrations, which occurs in conjunction with lower furnace temperature. This relationship was also observed in the Pittsfield data. See Figure B-1.

Conclusions from a comparative study of dioxin emissions vs. operating conditions at Westchester, Pittsfield, and Prince Edward Island were that "test results indicate that levels of dioxins and furans in the flue gas entering a pollution-control device are affected by different plant operating conditions if the conditions deviate sufficiently from normal operations." This study also indicated that furnace temperature might be a gross indicator of total dioxin and furan formation, and that operating an incinerator at excess oxygen levels below about 5% may cause an increase in dioxin and furan emissions (Visalli 1987).

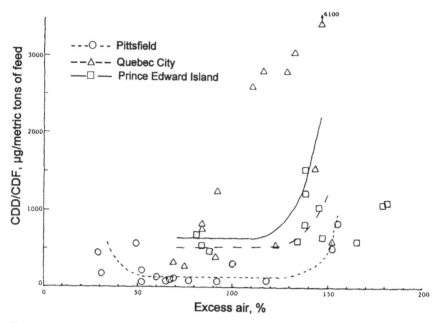

FIGURE B-1   Excess air and CDD/CDF emissions. Source: Visalli 1987.

Box B-2
Pittsfield (Midwest Research Institute 1987)

The Pittsfield, MA facility consists of three 120 ton of waste/day, two-stage, refractory-lined incinerators with two waste-heat boilers. Municipal solid waste is burned under excess-air conditions in the primary chamber; hot effluent gases pass into a secondary combustion chamber where any remaining uncombusted gases are burned.

Though no data were collected for startup or shutdown, the Pittsfield study included runs at different temperatures and oxygen levels to show how emissions varied when operating conditions were not optimized (i.e., upset conditions). The data showed a tendency for dioxin and furan emissions to increase with excess oxygen below and above certain levels. In other words, dioxin concentration in the flue gases was at a minimum when excess oxygen was between 9 and 11% in the hot zone (see Figure B-1). Total dioxin rose to over 50 ng/dscm corrected to 7% O_2 when excess oxygen was below 5% or above 12%. In fact, when the excess oxygen rose to over 11% the dioxins escalated quickly to over 100 ng/dscm and beyond.

In addition, a clear pattern was found with respect to temperature impacts on dioxin. The optimal temperature range for the Pittsfield facility, measured at the tertiary duct some distance from primary combustion, was roughly between 1500 and 1650°F. At temperature below and above that window dioxins increased. Below 1500°F, dioxins increased dramatically to over 120 ng/dscm at just below 1300°F. The corresponding dioxin concentrations for the two low-temperature runs were much higher than those for all other runs by a factor averaging more than four, and they were statistically different than those emitted under normal operating conditions. The two low furnace temperature runs (1300°F and 1350°F) also produced CO levels that were more than a factor of 10 higher than the rest of the test runs, showing CO as a useful indicator in this.

As important as the level of CO emissions in a medical-waste combustor is, an equally important issue is the averaging time over which these emissions are evaluated. It is important to note, in this regard, that the Pittsfield combustion tests showed that CO levels above 100 ppm were associated with a greater certainty of higher dioxin levels. If new and existing incinerators exceed this 100 ppm level routinely, by virtue of a 4- or 24-hour averaging time, the effect of the MACT regulation would not be to minimize dioxin emissions in these incinerators. The Pittsfield research demonstrates the importance of minimizing the number, intensity, and duration of CO spikes, and thus, of limiting the length of the averaging period for CO. Thus, to minimize the opportunity for formation of products of incomplete combustion, an average limit for CO is needed that would result in a strict limitation on the frequency, intensity, and duration of excursions. For example, New York state requires a one-hour averaging time for evaluating CO from medical-waste incinerators and permits typically specify a 100 ppm limitation.

A study managed by the American Society of Mechanical Engineers (ASME 1995) on the relationship between chlorine in waste streams and dioxin emissions, indicates that because combustion control is limited in most batch-mode medical-waste incinerators they "can be expected to emit relatively high PCDD/F levels associated with incomplete combustion." That finding points to a need to ensure that batch-fed incinerators provide good combustion.

Box B-3
Westchester (New York State Energy Research and Development Authority 1989)

The Peekskill incinerator in Westchester County, NY, consists of three mass-burn waterwall incinerators, each rated at 750 tons/per day. Each has a transverse reciprocating grate made up of modular sections: the drying zone, two burning zones, and two finishing or burnout zones. Rates of underfire air and grate speed can be set for each zone. Overfire air is supplied through nozzles on the front and rear of each furnace.

A study conducted in the 1980s includes two test runs in which dioxin emissions were recorded during cold starts, as well as several under more-normal operating conditions. The study was not intended to examine cold starts in great detail. In fact, the report excluded the cold-start runs from most of its analyses because analysis of variance (ANOVA) results showed that dioxin emissions during cold starts were statistically different (higher) than those emitted under normal operating conditions at a significance level of 0.0001 for both CDD and CDF. Run 4 was a normal cold-start condition where the auxiliary gas burner was used to get the furnace up to normal-operating temperature, the garbage was ignited, and the gas was turned off. This test sample was taken over a 65 minute period once the furnace was at "elevated temperature". Run 14 was research-oriented, in an attempt to determine if adding more natural gas than usual would lower emissions during cold starts. The report stated that the purpose of the cold-start tests was to observe the effect on CDD and CDF emissions of feeding refuse to a furnace that was not at thermal equilibrium. These tests reflect continually changing operating conditions (non-steady state).

The quantities of dioxins and furans generated during startup are striking. "The testing results show that the average CDD and CDF concentrations measured at the superheater exit during the first cold starts are two to three times higher than the average of the other 12 runs that were at steady operating conditions. For the electrostatic precipitator (ESP) inlet, the increase in dioxins from normal operations to normal cold start is between 18 and 51 times, and these increase further to between 40 and 96 times at the ESP outlet." The temperature at the superheater exit was considerably less during the normal cold start ($892^\circ F$) than the other normal runs ($995^\circ F$ – $1150^\circ F$, with most over $1100^\circ F$). Temperature was not recorded at the superheater for run 14.

Specifically, the normal cold start generated 124 ng/dscm of CDD @7% O_2 at the superheater outlet as compared with a range of 11.4 to 84.3 ng/dscm for the other runs. However, at the ESP inlet, the dioxins measured were 7226 ng/dscm for the first cold start vs. a range of 43.8 to 209 ng/dscm for the other runs. This precipitous rise shows that secondary dioxin formation was taking place near the ESP inlet due to temperatures conducive to such information and the presence of dioxin precursors and catalysts. The secondary dioxin generation for the cold start run occurred at a rate between one and two orders of magnitude higher than for the other runs.

Why should the dioxin generation rate increase faster under cold-start conditions than under steady-state conditions? There are two variables that might account for this. The temperature at the ESP inlet for the normal cold start run was $383^{\circ}F$ vs. 434 to $472^{\circ}F$ for the others. Because research by Stieglitz and Vogg (see Chapter 3) indicated the optimal temperature for secondary dioxin formation to be between $430^{\circ}F$ and $750^{\circ}F$, peaking at $570^{\circ}F$, it would seem that the normal runs would have greater secondary dioxin formation if temperature were the only variable. The variable that likely distinguishes the normal cold start run from the others, and explains the result observed, is the furnace temperature, which is decidedly lower for the cold start. Because dioxin precursors are created in the furnace at the highest rate in the few hundred degrees below optimal furnace temperatures, the lower furnace temperature during cold start is likely to have caused a higher generation rate for dioxin precursors and a lower rate of destruction. A greater generation rate during the cold start run of dioxin precursors (e.g., chlorobenzenes and chlorophenols), some of which require higher destruction temperatures (e.g., $1800^{\circ}F$, vs. $1300^{\circ}F$ for dioxins), would seem to be the cause of the tremendously higher amount of dioxins in the flue gas further downstream.

Carbon Monoxide (CO) was also measured for the various runs at Westchester. The mean CO concentration during the normal cold start was 180 ppmv at the superheater exit, not an astoundingly high figure considering the quantity of dioxins formed. The second, modified cold start had a mean of 114 ppm CO, and was characterized by dioxin emissions nearly as high as the first, normal cold start. By comparison, the CO levels for the other 12 runs ranged from 6 ppm to 57 ppm at the superheater exit. (While the NYSERDA report pointed out that the mean CO level does not adequately characterize the range of CO experienced during the one hour test, it is nonetheless of great interest because EPA standards for existing municipal solid-waste incinerators specify averaging times of *4 hours* for compliance with the CO emissions standard for four types of municipal solid-waste incinerators and *24 hours* for compliance by four other types of municipal solid-waste incinerators. The CO standard for municipal solid-waste incinerators for all plant types is 40 ppm, but over a *12-hour* rolling average. Further, only one of the eight municipal solid-waste incinerator plant types is required to meet a 50 ppmv CO standard for existing plants. Averaging times are similar for new plants, and the range of CO emissions permitted is 50 to 150 ppmv, with only one of eight plant types being required to meet 50 ppmv.

Box B-4
Quebec City (Stieglitz and Vogg 1987)

The Quebec City mass burn incinerator includes four incinerators/boilers rated at 227 tonnes of waste/per day each. Each of those has a vibrating feeder-hopper, drying/burning/burnout grates, refractory-lined lower burning zone, and water-walled upper burning zone.

The second of two studies of this incinerator conducted by Environment Canada compared the combustor performance at a variety of operating conditions: low-, medium- and high-load; percent excess air; furnace temperature; and primary/secondary air ratio. Some of these conditions were characterized as "very poor" (where primary/secondary air ratio was 90/10 and excess air was considered "high" (115%)), and "poor" (where furnace temperature was 1562°F, excess air was very high at 130%, and primary/secondary air ratio was 60/40). Three other combinations, under low-, design-, and high-load, were considered to be "good" operations.

Dioxin and furan emissions were measured for each of these test combinations, and statistical analysis of the data showed a fairly strong correlation between high excess air levels and dioxin/furan. See Figure B-1.

In addition, the best single parameter correlation ($r^2 = 0.876$) was a comparison of uncontrolled particulate matter entering the ESP versus dioxin/furan in the stack. Two other variables with extremely good single parameter fits were flue gas flow concentrations rate ($r^2 = 0.771$) and primary air flow rate ($r^2 = 0.723$). Notice that the rate of increase in dioxin and furan becomes exponential at around 123% excess air, indicating a move towards upset conditions. Data for Pittsfield and PEI test are also shown. It was also found that load has an effect on dioxin. This effect is shown in Figure B-2.

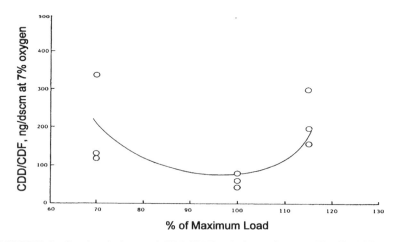

FIGURE B-2   Load variations and CDD/CDF emissions. Source: Visalli 1987.

Box B-5
Oswego (Radian Corporation 1990)

The Oswego facility consists of four, two-stage mass-burn units each of which are rated at 50 tons of waste/day. Batch loads are fed to the primary chamber and moved through the chamber along the stepped bottom by air-cooled transfer rams on a cycle of approximately seven to eight minutes. Combustible gases and entrained particles exit the primary chamber to the secondary chamber where the flue gas is mixed with preheated secondary air to complete combustion of unburned gases and particulate matter.

The Oswego study compared dioxin and furan generation in groups of three runs under each of the following four conditions:

(1) Clean combustor, right after startup (start of campaign), at which time the secondary chamber (SC) was 1837 to 1875°F.
(2) Dirty combustor, ready for maintenance shutdown (end of campaign), at which the SC was 1817 to 1834°F.
(3) Mid-range secondary chamber temperature ranged from 1738 to 1752°F.
(4) Low secondary chamber temperature runs ranged from 1617 to 1634°F. (The temperatures at the secondary chamber *exit* were lower from two to four hundred degrees (the low temperature secondary chamber exit was 1336°F).)

The low furnace temperature condition had a negative effect on dioxin emissions, increasing dioxin emissions over the normal temperature condition by about a factor of six at the secondary chamber exit and also at the ESP Inlet. At the secondary chamber exit total dioxins ranged from 67.7 to 110.1 ng/dscm @ 7%O_2 (averaging 84.5) at the secondary chamber exit. At the ESP inlet the range was 255.1 to 349.8 ng/dscm @7%O_2 (averaging 289.9). As compared with the normal operating condition (mid-range), averaging 13.6 ng/dscm @ 7%O_2 at the secondary chamber exit, and 53.4 ng/dscm @ 7%O_2 at the ESP inlet, the aforementioned means for the dioxin emissions for the low temperature runs were found to be statistically different and significantly correlated. Using one-way ANOVA, the chance that the dioxin means for the four conditions measured at the secondary chamber exit are not really different is 0.0049 or half of one percent. The chance that the dioxin means for the conditions measured at the ESP inlet are not really different is even lower at 0.0001. Thus, it can be stated definitively that the lower furnace temperature tested here is associated with a six-fold increase in dioxins.

Dioxin emissions were also correlated with CO emissions. Because CO measurements were taken continuously, different assumptions could be made about representation of CO as single value representations (SVRs). The correlation between total dioxins and CO at the secondary chamber exit that was most pronounced was at the 90th percentile CO value, where r = 0.921, a nearly perfect positive correlation. In general, the SVRs representing extreme values of CO (i.e., 90th, 95th and 99th percentile) correlated most frequently with the dioxin and furan levels measured, indicating that it is frequency and duration of the highest values of CO that best predict changes in dioxin and furan concentrations in the flue gas. All relationships were significant at the 0.05 level. These data highlight the importance of avoiding both CO spikes and poor combustion efficiency, and accurately recording the amount of time during which elevated CO occurs—not simply averaging CO over long periods of time to mask such excursions.

Box B-6
Hartford (EPA 1994)

The Mid-Connecticut facility is a refuse-derived fuel facility (RDF) and consists of three RDF-fired, spreader-stoker boilers that process a total of 2,000 tons/day of municipal solid waste. Four pneumatic distributors spread the RDF across the width of the combustion grate. Ten underfire air zones allow the operator to optimize combustion and to respond quickly to "piling" situations by manual adjustment of underfire air dampers. The overfire air system is equipped with four tangential assemblies located in the furnace corners; each assembly includes three levels that are separately controlled. Preheated combustion air enters the furnace forming a vortex, providing longer residence times for the combustion gases.

A major goal of the project was to determine generation of trace organics and metals in the furnace under different process operating conditions, not under "upset" conditions, per se. Steam flow rate (an indicator of load) and combustion air flow rates were the primary independent variables defining operating conditions as "good," "poor," "very poor." However, as compared with the other studies discussed in this appendix, the variation in combustion conditions was smaller, because this study was focused on a more-efficient range of operation than these other studies. Dioxins, furans, CO, total hydrocarbons, PCB, cholorobenzenes, chlorophenols, and PAHs were measured.

Multiple-regression analysis was used to study the effect of various continuously monitored emission and process parameters on dioxin emissions (prediction models) and the effect of various combustion control measures on dioxin emissions (control models). The best prediction model showed that CO, NO_x, moisture in the flue gas at the spray dryer inlet, and furnace temperature explained 93% of the variation in uncontrolled dioxin emissions, with CO explaining 79% by itself. The best control model showed that RDF moisture, rear-wall overfire air, underfire air flow, and total air explained 67% of the variation in uncontrolled dioxin emissions.

Because CO was found to be such a strong predictor of dioxin emissions, the relationship was explored further. It was found that the percent of time the CO level was over 400 ppm was quite strongly correlated with the amount of uncontrolled dioxins generated, particularly when examining only those runs where there was poor combustion. The authors of the report indicated that "Poor combustion implies that greater amounts of organic material escape the combustor unburned. In the correlation between CO and PCDD/PCDF, use of only the poor combustion tests would improve R^2 from 0.70 to 0.95." The correlation between total hydrocarbons and dioxin/furan improved from a R^2 value of 0.68 to 0.97 for the poor combustion tests. This indicates that for the poor combustion tests, 95% of the change in dioxin/furan values are explained by the change in CO emissions. This is consistent with the theory that, during periods of incomplete combustion, the amount of organic matter leaving the furnace strongly influences the formation of PCDD/PCDF. Thus, these data indicate that CO is an important surrogate for dioxin, and that allowing longer averaging times for CO levels for compliance with standards will more likely result in higher dioxin/furan emissions because, under these conditions, more CO spikes can occur without exceeding standards.

The furnace temperature was at 1,789°F for "poor" conditions and at 1,920°F for "good" combustion conditions under high load. The resulting dioxin emissions at the spray dryer inlet were 317 ng/Sm^3 for the poor combustion conditions vs. 67 ng/Sm^3 @ 12%CO_2 for the good combustion conditions, a factor of almost five. The same relationship was true for furans, total hydrocarbons, PAHs, chloroben-zenes and chlorophenols. CO was 397 ppm for "poor" combustion and 116 ppm for "good" combustion under high load. Under intermediate load, the underfire/overfire ratio was .923 under "good" conditions and 1.632 under "very poor" condi-tions. This resulted in a ten-fold increase in CO from 93 to 903 ppm. At this load total hydrocarbons increased from 2.5 to 52.4, PAH from 7,330 to 112,000, chlo-rophenols from 14,300 to 114,000, chlorobenzenes from 6,050 to 15,800, dioxins from 228 to 580, and furans from 579 to 1,280, all in units of ng/Sm^3 @CO_2.

REFERENCES

ASME (The American Society of Mechanical Engineers). 1995. The Relationship Between Chlo-rine in Waste Streams and Dioxin Emissions from Waste Combustor Stacks, CRTD-Vol. 36. Fairfield, N.J.: ASME Press.

Concord Scientific Corporation. 1985. National Incinerator Testing and Evaluation Program: Two-Stage Combustion (Prince Edward Island). Environment Canada Report EPS 3/UP/1, Volumes I-IV.

EPA (U.S. Environmental Protection Agency). 1994. National Incinerator Testing and Evaluation Program: The Environmental Characterization of Refuse-derived Fuel (RDF) Combustion Tech-nology: Mid-Connecticut Facility, Hartford, Connecticut, Summary Report. Environment Canada Report EPS 3/UP/7 and U.S. Environmental Protection Agency Report EPA-600/R-94-140, December.

Midwest Research Institute. 1987. Results of the Combustion and Emissions Research Project at the Vicon Incinerator Facility in Pittsfield, Massachusetts. Final Report 87-16. Prepared for New York State Energy Research and Development Authority by Midwest Research Institute.

New York State Energy Research and Development Authority. 1989. Combustion and Emissions Testing at the Westchester County Solid Waste Incinerator, Volume I, Final Report. Report 89-4. New York State Energy Research and Development Authority.

Radian Corporation. 1990. Results from the Analysis of MSW Incinerator Testing at Oswego County, New York, Volume I, Final Report. Prepared for the New York State Energy Re-search and Development Authority and the New York State Department of Environmental Conservation by Radian Corporation. Energy Authority Report 90-10.

Stieglitz, L. and H. Vogg. 1987. New Aspects of PCDD/PCDF Formation in Incineration Processes. Preliminary Proceedings, Municipal Waste Incineration, October 1-2, 1987, Montreal, Quebec.

Visalli, J.R. 1987. A comparison of dioxin, furan and combustion gas data from test programs at three MSW incinerators. JAPCA 37(12):1451-1463.

List of Abbreviations

ACS	American Chemical Society
APCD	air pollution control device
As	arsenic
ATSDR	Agency for Toxic Substances and Disease Registry of the U.S. Public Health Service
BaP	benzo(a)pyrene
BIF	boiler and industrial furnace
CAA	Clean Air Act
CARB	California Air Resources Board
Cd	cadmium
CDD	chlorinated dibenzo-p-dioxin
CDF	chlorinated dibenzofuran
CFR	Code of Federal Regulations
CO	carbon monoxide
COHb	carboxyhemoglobin
COPD	chronic obstructive pulmonary disease
Dioxins	polychlorinated dibenzo-p-dioxins
DOE	U.S. Department of Energy
DRE	destruction and removal efficiency
dscm	dry standard cubic meters (at 14.7 pounds per square inch, 68°F)
EPA	U.S. Environmental Protection Agency
ESP	electrostatic precipitator

FEV$_1$	forced expiratory volume in 1 sec—amount of air an individual can exhale in 1 sec
Furans	polychlorinated dibenzofurans
FVC	forced vital capacity
g	grams
GGT	gamma-glutamyltransferase
HCB	hexachlorobenzene
HCl	hydrogen chloride
Hg	mercury
IARC	International Agency for Research on Cancer
IRIS	Integrated Risk Information System
kg	kilogram
L	liter
LCA	life cycle assessment
m	meters
MACT	maximum achievable control technology
mg	milligram
Mg	megagram
MRL	minimal risk level
MSW	municipal solid waste
MWC	municipal waste combustor
MWI	medical waste incinerator
NIOSH	National Institute for Occupational Safety and Health
NOAEL	no-observed-adverse-effect level
NO$_x$	oxides of nitrogen
NO$_2$	nitrogen dioxide
OEL	occupational exposure limit
OSHA	U.S. Occupational Safety and Health Administration
PAH	polycyclic aromatic hydrocarbon
Pb	lead
PCB	polychlorinated biphenyl
PCDD	polychlorinated dibenzo(p)dioxin
PCDF	polychlorinated dibenzofuran
PEL	permissible exposure limit
PIC	product of incomplete combustion
PM	particulate matter
PM$_{2.5}$	particulate matter less than 2.5 μm in aerodynamic diameter
PM$_{10}$	particulate matter less than 10 μm in aerodynamic diameter
ppb	parts per billion by weight
ppm	parts per million by weight
ppmv	parts per million by volume
ppt	parts per trillion by weight
RCRA	Resource Conservation and Recovery Act

RfD	reference dose
SMSA	standard metropolitan statistical area
SO_2	sulfur dioxide
TCDD	2,3,7,8-tetrachlorodibenzo-*p*-dioxin
TEQ	2,3,7,8-tetrachlorinated dibenzo-*p*-dioxin toxic equivalent based on the 1989 international toxic equivalency factors
TLV	threshold limit value
TSP	total suspended particulate matter
VOC	volatile organic compound

Index

A

Acidic gases and aerosols, 2, 23, 50, 51-52, 104, 110
    air pollution control devices, 42, 43, 45-46, 49, 51, 103
    fly ash, 64-65
    Maximum Achievable Control Technology (MACT), 166, 178
    risk assessments, 113, 134-139, 166, 178
    standards, 113, 188, 193
    urban areas, 138
    see also Hydrogen chloride; Nitrogen and nitrogen oxides; Sulfates
Accidents, *see* Upset conditions, accidents and malfunctions
Adolescents, 139
Advocacy, 13, 21, 193, 209-210, 211-212, 218-219, 221-223, 227, 229
African-Americans, 162, 231
Age factors, human, 118, 123, 126, 127, 128, 175
    adolescents, 139
    elderly persons, 132, 140, 162, 193
    food contamination effects, 90
    particulates, 132, 133
    see also Children; Elderly persons
Age of incinerators, 8, 38, 42, 126, 130, 184, 188, 197-198, 214, 215, 234

Agency for Toxic Substances and Disease Registry (ATSDR)
    arsenic, 88
    cadmium, 86, 141-142
    dioxins and furans, 93
    lead, 91
    local population studies, 120
    mercury, 90
    risk assessments, 116, 141-142
    Superfund sites, 200
Agriculture, *see* Rural areas
AIDS, 24
Air dispersion coefficients, 6, 70, 76-77, 99, 132
Air-injection systems, 2-3, 40, 46, 48, 66-69 (passim)
Air pollution, general, 4-5, 14, 71, 109, 248
    air-dispersion coefficients, 6, 70, 76-77, 99, 132
    cadmium, 86, 87, 106-107
    dioxins, 93-94, 95, 96
    hazardous waste, 22, 23
    lead, 91, 92, 106
    multimedia transport models, 79-80, 108-109, 255
    plant contamination and, 77, 107
    process emissions, 1, 2-4, 50-56, 211
    see also Air pollution control devices; Ambient pollutant concentrations; *specific pollutants*

Y

Z